Spring+MyBatis 企业应用实战

（第2版）

疯狂软件　编著

電子工業出版社
Publishing House of Electronics Industry
北京•BEIJING

内容简介

本书介绍了 Java EE（现已改名为 Jakarta EE）领域的两个开源框架：Spring 的 MVC 和 MyBatis。其中 Spring 的版本是 5.0.1，MyBatis 的版本是 3.4.5。本书的示例建议在 Tomcat 8 上运行。

本书重点介绍如何整合 Spring MVC 5+MyBatis 3 进行 Java EE 开发，主要包括三个部分。第一部分详细介绍了 Spring MVC 框架的用法。第二部分详细介绍了 MyBatis 框架的用法。第三部分重点介绍了 Spring MVC 5＋MyBatis 3 的整合，并示范开发了一个包含 6 个表、表之间具有复杂的关联映射关系，且业务功能也相对完善的 HRM 人事管理系统案例，希望读者理论联系实际，将这两个框架真正运用到实际开发当中去。该案例采用目前最流行、最规范的 Java EE 架构，整个应用分为 DAO 持久层、领域对象层、业务逻辑层、控制器层和视图层，各层之间分层清晰，层与层之间以松耦合的方法组织在一起。所有代码完全基于 Eclipse IDE 来完成，一步步带领读者深入两个框架的核心。

对于没有 Java 基础的初学者，建议先阅读疯狂软件教育的《疯狂 Java 讲义》一书。而本书适合有较好的 Java 编程基础，JSP、Servlet、JDBC 基础，Spring 框架基础的读者，尤其适合于对 Spring MVC 和 MyBatis 了解不够深入，或对 Spring MVC+MyBatis 整合开发不太熟悉的开发人员阅读。

图书在版编目（CIP）数据

Spring+MyBatis 企业应用实战 / 疯狂软件编著. —2 版. —北京：电子工业出版社，2018.4
ISBN 978-7-121-33780-2

Ⅰ. ①S… Ⅱ. ①疯… Ⅲ. ①JAVA 语言－程序设计 Ⅳ. ①TP312.8

中国版本图书馆 CIP 数据核字（2018）第 038902 号

策划编辑：张月萍
责任编辑：牛 勇
印　　刷：三河市良远印务有限公司
装　　订：三河市良远印务有限公司
出版发行：电子工业出版社
北京市海淀区万寿路 173 信箱　　邮编：100036
开　　本：787×1092　1/16　　印张：24　　字数：630 千字
版　　次：2017 年 1 月第 1 版
2018 年 4 月第 2 版
印　　次：2020 年 6 月第 7 次印刷
印　　数：9701~10700 册　　定价：69.00 元

凡所购买电子工业出版社图书有缺损问题，请向购买书店调换。若书店售缺，请与本社发行部联系，联系及邮购电话：（010）88254888，88258888。

质量投诉请发邮件至 zlts@phei.com.cn，盗版侵权举报请发邮件至 dbqq@phei.com.cn。

本书咨询联系方式：010-51260888-819　faq@phei.com.cn。

前　　言

时至今日，以 Spring 为核心的轻量级 Java EE 企业开发平台在企业开发中占有绝对的优势，Java EE 应用以其稳定的性能、良好的开放性以及严格的安全性，深受企业应用开发者的青睐，应用的性能、稳定性都有很好的保证。

轻量级 Java EE 开发大致可分为两种方式：以 Struts+Spring+Hibernate 三大框架为核心的轻量级 Java EE 和以 Spring MVC+MyBatis 为核心的轻量级 Java EE。这两种组合都在保留经典 Java EE 应用架构、高度可扩展性、高度可维护性的基础上，降低了 Java EE 应用的开发、部署成本，对于大部分中小型企业应用来说是首选。

本书重点介绍以 Spring MVC+MyBatis 为核心的轻量级 Java EE。本书采用 Tomcat 8 作为 Web 服务器，Eclipse IDE 作为开发工具，详细介绍了 Spring MVC 和 MyBatis 框架。Spring MVC 的配置全部使用注解方式，不再采用旧版本中传统的 XML 配置方式；MyBatis 则基于企业开发实际情况，首先介绍了传统的 XML 持久化映射，之后又介绍了升级的注解持久化映射。

随着 Spring 的不断发展，Spring MVC 已有取代 Struts 的能力，而 MyBatis 相对 Hibernate 而言则更为轻便、简单，越来越多的公司选择使用 Spring MVC+MyBatis 的轻量级框架组合来开发 Java EE 项目，因此掌握 Spring MVC+MyBatis 技术将为 Java 开发者带来更多的就业机会与竞争力。

本书有什么特点

本书是一本介绍 Spring MVC+MyBatis 技术以及两者整合的实用图书，全面介绍了最新的 Spring MVC 和 MyBatis 各方面的知识。

本书针对每一个知识点都通过相应的程序给出了示范，第 13 章的实战项目“HRM 人事管理系统”采用目前最流行、最规范的 Java EE 架构，整个应用分为 DAO 持久层、领域对象层、业务逻辑层、控制器层和视图层，各层之间分层清晰，层与层之间以松耦合的方式组织在一起。笔者既担任过软件开发的技术经理，也担任过软件公司的培训导师，现如今从事专业、高端的职业技术培训。本书所有应用范例都密切契合企业开发实际场景，例如用户权限验证、文件上传下载等都是企业开发中要实现的实际功能，同时采用目前企业中颇为流行、规范的开发架构，严格遵守 Java EE 开发规范。读者参考本书的架构，完全可以身临其境地感受企业实际项目开发。

本书并不是一本关于所谓“思想”的书，也没有一堆“深奥”的新名词和“高深”的理论，只是让读者学会实际的 Spring MVC 和 MyBatis 技术。本书的特点是操作步骤详细，编程思路清晰，语言平实易懂。只要读者认真阅读本书，并掌握书中知识，那么就完全可以胜任企业中的 Spring MVC+MyBatis 项目开发。

阅读本书需要具备一定的计算机知识以及编程功底。熟练掌握 Java 语言和 Spring 框架的 IOC、AOP 和持久层的 ORM 设计模式等知识对于学习本书是很有必要的。

可访问 www.crazyit.org 或 www.broadview.com.cn/33780 下载本书配套资源。

本书写给谁看

如果你已经掌握了 Java SE 的内容，或已经学完了疯狂软件教育的《疯狂 Java 讲义》一书，那么你非常适合阅读此书。除此之外，如果你已有初步的 JSP、Servlet、JDBC 基础，甚至对 Spring、MyBatis 等框架有所了解，但希望掌握它们在实际开发中的应用，本书也非常适合你。如果你对 Java 的掌握还不熟练，则建议遵从学习规律，循序渐进，暂时不要购买、阅读此书，而是按照“疯狂 Java 学习路线图”中的建议顺序学习。

衷心感谢

衷心感谢李刚老师，他是笔者非常好的朋友，在本书的创作过程中，他提供了大量切实、有用的帮助。同时衷心感谢疯狂软件教育中心所有同事提供的帮助。

感谢我的妻子王丹娜女士，她给了我生活上、精神上无微不至的关怀和照顾，并承担几乎所有的日常家务，使我能够心无旁骛地专注于程序的研究和书籍的撰写。

感谢所有参加疯狂软件实训的学生，他们在实际工作场景的应用中证明了本书的价值，他们的反馈让本书更加实用。

编者
2018 年 2 月

目录 CONTENTS

CHAPTER 1

第 1 章 Java EE 应用

本章要点

- Java EE 应用的基础知识
- Java EE 应用的模型和相关组件
- Java EE 应用的结构和优势
- 轻量级 Java EE 应用的相关技术

时至今日，轻量级 Java EE 平台在企业开发中占有绝对的优势，Java EE 应用以其稳定的性能、良好的开放性以及严格的安全性，深受企业应用开发者的青睐。实际上，对于信息化要求较高的行业，如银行、电信、证券以及电子商务等，都不约而同地选择了 Java EE 作为开发平台。

对于一个企业而言，选择 Java EE 构建信息化平台，更体现了一种长远的规划：企业的信息化是不断整合的过程，在未来的日子里，经常会有不同平台、不同的异构系统需要整合。Java EE 应用提供的跨平台、开放性以及各种远程访问技术，为异构系统的良好整合提供了保证。

一些有高并发、高稳定要求的电商网站（如淘宝、京东等），公司创立之初并没有采用 Java EE 技术架构（淘宝早期用 PHP，京东早期用.NET），但当公司的业务一旦真正开始，他们马上就发现 PHP、.NET 无法支撑公司业务运营，立即全部改为使用 Java EE 技术架构。就目前的局面来看，Java EE 已经成为真正企业级应用的不二之选。

1.1 Java EE 应用概述

今天所说的 Java EE 应用，超出了 Sun 所提出的经典 Java EE 应用规范，而是一种更广泛的开发规范。经典 Java EE 应用往往以 EJB（企业级 JavaBean）为核心，以应用服务器为运行环境，所以开发、运行成本较高。本书所介绍的 Spring MVC + MyBatis 作为轻量级 Java EE 应用不仅具备 Java EE 规范的种种特征，例如面向对象建模的思维方式、优秀的应用分层及良好的可扩展性、可维护性，而且保留了经典 Java EE 应用的架构，但其开发、运行成本更低。

1.1.1 Java EE 应用的分层模型

不管是经典的 Java EE 架构，还是本书介绍的轻量级 Java EE 架构，大致上都可分为如下几层：

- **Domain Object（领域对象）层**。此层由一系列的 POJO（Plain Old Java Object，普通的、传统的 Java 对象）组成，这些对象是该系统的 Domain Object（领域对象），往往包含了各自所需实现的业务逻辑方法。
- **DAO（Data Access Object，数据访问对象）层**。此层由一系列的 DAO 组件组成，这些 DAO 实现了对数据库的创建、查询、更新和删除（CRUD）等原子操作。

在经典 Java EE 应用中，DAO 层也被称为 EAO 层，EAO 层组件的作用与 DAO 层组件的作用基本相似。只是 EAO 层主要完成对实体（Entity）的 CRUD 操作，因此简称为 EAO 层。

DAO 层在 MyBatis 中也被称为 Mapper 层，其通过 SQL 语句的映射完成 CRUD 操作。

- **Service（业务逻辑）层**。此层由一系列的业务逻辑对象组成，这些业务逻辑对象实现了系统所需要的业务逻辑方法。这些业务逻辑方法可能仅仅用于暴露 Domain Object 对象所实现的业务逻辑方法，也可能是依赖 DAO 组件实现的业务逻辑方法。
- **Controller（控制器）层**。此层由一系列控制器组成，这些控制器用于拦截用户请求，并调用业务逻辑组件的业务逻辑方法，处理用户请求，并根据处理结果向不同的表现层组件转发。

➢ **View（表现）层**。此层由一系列的 JSP 页面、Velocity 页面、PDF 文档视图组件组成，负责收集用户请求，并显示处理结果。

大致上，Java EE 应用的架构如图 1.1 所示。

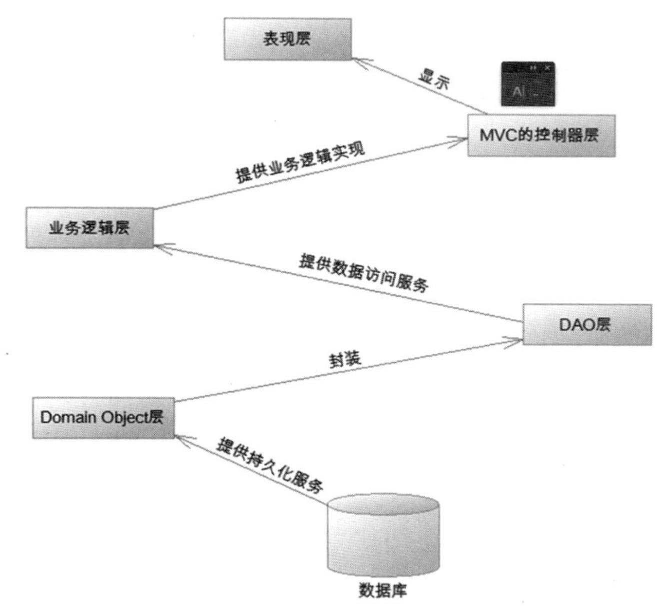

图 1.1　Java EE 应用的架构

各层的 Java EE 组件之间以松耦合的方式组织在一起，各组件并不以硬编码方式耦合，这种方式是为了应用以后的扩展性。从上向下，上面组件的实现依赖于下面组件的功能；从下向上，下面组件支持上面组件的实现。

1.1.2　Java EE 应用的组件

通过上一节的介绍，我们可以看到 Java EE 应用实现了系统架构上的飞跃。Java EE 架构实现了良好的分离，隔离了各组件之间的代码依赖。

总体而言，Java EE 应用大致包括如下几类组件：

➢ **表现层组件**。主要负责收集用户输入数据，或者向客户显示系统状态。最常用的表现层技术是 JSP，但 JSP 并不是唯一的表现层技术。表现层还可由 Velocity、FreeMarker 和 Tapestry 等技术完成，或者使用普通的应用程序充当表现层组件，甚至可以是小型智能设备。

➢ **控制器组件**。关于 Java EE 的 MVC 框架，其提供一个前端核心控制器，核心控制器负责拦截用户请求，并将请求转发给用户实现的控制器组件。这些用户实现的控制器组件则负责调用业务逻辑方法，处理用户请求。

➢ **业务逻辑组件**。这是系统的核心组件，实现系统的业务逻辑。通常，一个业务逻辑方法对应一次用户操作。一个业务逻辑方法应该是一个整体，因此要求对业务逻辑方法增加事务性。业务逻辑方法仅仅负责实现业务逻辑，不应该进行数据库访问。因此，业务逻辑组件中不应该出现原始的 MyBatis、Hibernate 和 JDBC 等 API。

提示

保证业务逻辑组件之中不出现 MyBatis、Hibernate 和 JDBC 等 API，有一个更重要的原因：保证业务逻辑方法的实现与具体的持久层访问技术分离。当系统需要

> 在不同持久层技术之间切换时，系统的业务逻辑组件无须任何改变。有时会见到一些所谓的 Java EE 应用，居然在 JSP 页面里面调用 SqlSessionFactory、SqlSession 等 API，这无疑是非常荒唐的，这种应用仅仅是使用 MyBatis，完全没有脱离 Model 1 的 JSP 开发模式，这是相当失败的结构。实际上，不仅 JSP，Servlet 中也不应出现持久层 API，包括 JDBC、MyBatis、Hibernate API。最理想的情况是，业务逻辑组件中都不应出现持久层 API。

- **DAO 组件**。这个类型的对象比较缺乏变化，每个 DAO 组件都提供 Domain Object 对象基本的创建、查询、更新和删除等操作，这些操作对应于数据库的 CRUD（创建、查询、更新和删除）等原子操作。当然，如果采用不同的持久层访问技术，DAO 组件的实现会完全不同。为了业务逻辑组件的实现与 DAO 组件的实现分离，程序应该为每个 DAO 组件都提供接口，业务逻辑组件面向 DAO 接口编程，这样才能提供更好的解耦。
- **领域对象组件**。领域对象（Domain Object）抽象了系统的对象模型。通常而言，这些领域对象的状态都必须保存在数据库里。因此，每个领域对象通常对应一个或多个数据表，领域对象通常需要提供对数据记录的访问方式。

1.1.3　Java EE 应用的结构和优势

作为 Java EE 的初学者，常常有一个问题：明明可以使用 JSP 完成这个系统，为什么还要使用 MyBatis 和 Hibernate 等技术？难道仅仅是为了听起来高深一些？明明可以使用纯粹的 JSP 完成整个系统，为什么还要将系统分层？

要回答这些问题，就不能仅仅考虑系统开发过程，还需要考虑系统后期的维护、扩展；而且不能仅仅考虑小型系统，还要考虑大型系统的协同开发。如果是用于个人学习、娱乐的个人站点，的确没有必要使用复杂的 Java EE 应用架构，采用纯粹的 JSP 就可以实现整个系统。

但对于大型的信息化系统，采用 Java EE 应用架构则有很大的优势。

对于信息化系统，前期开发工作对整个系统工作量而言，仅仅是小部分，而后期的维护、升级往往占更大的比重。更极端的情况是，可能在前期开发期间，企业需求已经发生变化，而这种改变是客观的，软件系统必须适应这种改变，这要求软件系统具有很好的扩展性。

这种框架结构其目的是让应用的各组件以松耦合的方式组织在一起，让应用之间的耦合停留在接口层次，而不是代码层次。

1.2　轻量级 Java EE 应用相关技术

轻量级 Java EE 应用以传统的 JSP 作为表现层技术，以一系列开源框架作为 MVC 层、中间层、持久层解决方案，并将这些开源框架有机地组合在一起，使得 Java EE 应用具有高度的可扩展性、可维护性。

1.2.1　JSP、Servlet 和 JavaBean 及替代技术

JSP 是最早的 Java EE 规范之一，也是最经典的 Java EE 技术之一。直到今天，JSP 依然广泛地应用于各种 Java EE 应用中，充当 Java EE 应用的表现层角色。

JSP 具有简单、易用的特点，JSP 的学习路线平坦，而且国内有大量 JSP 学习资料，所以大部分 Java 学习者学习 Java EE 开发都会选择从 JSP 开始。

Servlet 和 JSP 其实是完全统一的，二者底层的运行原理是完全一样的。实际上，JSP 必须被 Web 服务器编译成 Servlet，真正在 Web 服务器内运行的是 Servlet。从这个意义上来看，JSP 相当于一个“草稿”文件，Web 服务器根据该“草稿”文件生成 Servlet，真正提供 HTTP 服务的是 Servlet，因此广义的 Servlet 包含了 JSP 和 Servlet。

从目前的 Java EE 应用来看，纯粹的 Servlet 已经很少使用了，毕竟 Servlet 的开发成本太高，而且使用 Servlet 充当表现层将导致表现层页面难以维护，不利于美工人员参与 Servlet 开发，所以在实际开发中大都使用 JSP 充当表现层技术。

Servlet 3.x 规范的出现，再次为 Java Web 开发带来了巨大的便捷。Servlet 3.x 提供了异步请求、注解、增强的 Servlet API、非阻塞 I/O 功能，这些功能都极大地简化了 Java Web 开发。

由于 JSP 只负责简单的显示逻辑，因此 JSP 无法直接访问应用的底层状态，Java EE 应用会选择使用 JavaBean 来传输数据。在严格的 Java EE 应用中，中间层的组件会将应用底层的状态信息封装成 JavaBean 集，这些 JavaBean 也被称为 DTO（Data Transfer Object，数据传输对象），并将这些 DTO 集传到 JSP 页面，从而让 JSP 可以显示应用的底层状态。

在目前阶段，Java EE 应用除了可以使用 JSP 作为表现层技术之外，还可以使用 FreeMarker 或 Velocity 作为表现层技术，这些表现层技术更加纯粹，使用起来更加便捷，完全可作为 JSP 的替代。

1.2.2 MyBatis 3 及替代技术

传统的 Java 应用都是采用 JDBC 来访问数据库的，但传统的 JDBC 采用的是一种基于 SQL 的操作方式，这种操作方式与 Java 语言的面向对象特性不太一致，所以 Java EE 应用需要一种技术，通过这种技术能让 Java 以面向对象的方式操作关系数据库。

这种特殊的技术就是 ORM（Object Relation Mapping），最早的 ORM 是 Entity EJB（Enterprise JavaBean），EJB 就是经典 Java EE 应用的核心，从 EJB1.0 到 EJB 2.x，许多人会觉得 EJB 非常烦琐，所以导致 EJB 备受诟病。

在这种背景下，Hibernate 框架应运而生。Hibernate 框架是一种开源的、轻量级的 ORM 框架，它允许将普通的、传统的 Java 对象（POJO）映射成持久化类，允许应用程序以面向对象的方式来操作 POJO，而 Hibernate 框架则负责将这种操作转换成底层的 SQL 操作。

大多数情况下（特别是对新项目、新系统的开发而言），Hibernate 这样的机制无往不利，大有一统天下的势头。但是，在一些特定的环境下，Hibernate 这种一站式的解决方案却未必适合。如：

- 系统的部分或全部数据来自现有数据库，出于安全考虑，只对开发团队提供几条 Select SQL（或存储过程）以获取所需数据，具体的表结构不予公开。
- 开发规范中要求，所有牵涉到业务逻辑部分的数据库操作，必须在数据库层由存储过程实现（就金融行业而言，工商银行、中国银行、交通银行等商业银行都曾在开发规范中严格指定）。
- 系统数据处理量巨大，性能要求极为苛刻，这往往意味着我们必须通过经过高度优化的 SQL 语句（或存储过程）才能达到系统性能设计指标。

面对这样的需求，Hibernate 不再适合，甚至无法使用。此时，直接使用 JDBC 进行数据库操作实际上也是不错的选择，只是拖沓的数据库访问代码、乏味的字段读取操作令人厌烦，而“半自动化”的 MyBatis，却正好解决了这个问题。

这里的“半自动化”，是相对 Hibernate 等提供了全面的数据库封装机制的“全自动化”ORM 实现而言的，“全自动”ORM 实现了 POJO 和数据库表之间的映射，以及 SQL 的自动生成和执行。而 MyBatis 的着力点，则在于 POJO 与 SQL 之间的映射关系。也就是说，使用 MyBatis 提供的 ORM 机制，对业务逻辑实现人员而言，面对的是纯粹的 Java 对象，这与通过 Hibernate 实现 ORM 基本一致，而对于具体的数据操作，Hibernate 会自动生成 SQL 语句，但 MyBatis 则并不会自动生成 SQL 语句。具体的 SQL 需要程序员编写，然后通过映射配置文件，将 SQL 所需的参数以及返回的结果字段映射到指定的 POJO。

相对 Hibernate 等“全自动”ORM 机制而言，MyBatis 以 SQL 开发的工作量和数据库移植性上的让步，为系统设计提供了更大的自由空间。作为对“全自动”ORM 实现的一种有益补充，MyBatis 的存在具有特别的意义。

MyBatis 是 Apache 组织提供的一个轻量级持久层框架，是一个支持普通 SQL 查询、存储过程和高级映射的优秀持久层框架。MyBatis 消除了几乎所有的 JDBC 代码和参数的手工设置过程以及对结果集的检索封装。MyBatis 可以使用简单的 XML 或注解来进行配置和原始映射，将接口和 Java 的 POJO 映射成数据库中的记录。

MyBatis 作为持久层框架，其主要思想是将程序中的大量 SQL 语句剥离出来，配置在配置文件中，实现 SQL 的灵活配置。这样做的好处是将 SQL 与程序代码分离，可以在不修改程序代码的情况下，直接在配置文件中修改 SQL。

MyBatis 最新版本是 3.4.5，这也是本书所使用的 MyBatis 版本。

除此之外，Oracle 的 TopLink、Apache 的 OJB 都可作为替代方案。但由于种种原因，它们并未得到广泛的市场支持，所以这两个框架的资料、文档相对比较少，选择它们需要一定的勇气和技术功底。

1.2.3 Spring 5 及替代技术

如果你有 5 年以上的 Java EE 开发经验，并主持过一些大型项目的设计，你会发现 Spring 框架似曾相识。Spring 甚至没有太多的新东西，它只是抽象了大量 Java EE 应用中的常用代码，将它们抽象成一个框架。通过使用 Spring 可以大幅度地提高开发效率，并可以保证整个应用具有良好的设计。

Spring 框架里充满了各种设计模式的应用，如单例模式、工厂模式、抽象工厂模式、命令模式、职责链模式、代理模式等，Spring 框架的用法、源码则更是一道丰盛的 Java 大餐。

Spring 框架号称 Java EE 应用的一站式解决方案，Spring 本身提供了一个设计优良的 MVC 框架：Spring MVC。使用 Spring 框架可以直接使用该 MVC 框架。由于 Spring 框架拥有极高的市场占有率，因此越来越多的 Spring 框架的使用者使用 Spring MVC 替代曾经的 MVC 框架的王者 Struts 2。当然，Spring 也可以无缝地整合 Struts 2、JSF 等优秀的 MVC 框架。

Spring 框架并未提供完整的持久层框架，可以将其理解成一种“空”，但这种“空”正是 Spring 框架的魅力所在。Spring 能与大部分持久层框架无缝整合：MyBatis、Hibernate、JPA、TopLink，更甚至直接使用 JDBC，随便你喜欢，无论选择哪种持久层框架，Sping 都会为你提供无缝的整合和极好的简化。

从这个意义上来看，Spring 更像一种中间层容器，Spring 向上可以与 MVC 框架无缝整合，向下可以与各种持久层框架无缝整合，其的确具有强大的生命力。由于 Spring 框架的特殊地位，轻量级 Java EE 应用通常都不会拒绝使用 Spring。实际上，轻量级 Java EE 这个概念也是由 Spring 框架衍生出来的。Spring 框架暂时没有较好的替代框架。

Spring 的最新版本是 5.0.1，本书所介绍的 Spring 也基于该版本。

1.2.4 使用开源框架的好处

以上提到的 Struts 2、MyBatis 3、Hibernate 5、Spring 5 等都是 Java 领域最常见的框架，这些框架得到开发者广泛的支持，它们能极大地提高 Java EE 应用的开发效率，并能保证应用具有稳定的性能。

越来越多的企业开始选择 Spring MVC+MyBatis 来构建系统架构，在电商热门的今天，Spring MVC+MyBatis 已成为电商项目架构的最佳搭配。本书将重点讲解 Spring MVC+ MyBatis 如何无缝整合开发 Java EE 项目。

常常有些初学者，甚至包括一些所谓的企业开发人士提出：为什么需要使用框架？用 JSP 和 Servlet 已经足够了。

提出这些疑问的人通常还未真正进入企业开发，或者从未开发一个真正的项目。因为真实的企业应用开发有两个重要的关注点：可维护性和复用。

先从软件的可维护性来考虑这种说法。全部采用 JSP 和 Servlet 的应用，因为分层不够清晰，业务逻辑的实现没有单独分离出来，从而造成系统后期维护困难。

从软件复用角度来考虑。这是一个企业开发的生命，企业以追求利润为最大目标，企业希望以最快的速度，开发出最稳定、最实用的软件。因为系统没有使用任何框架，每次开发系统都需要重新开发，重新开发的代码具有更多的漏洞，这就增加了系统出错的风险；另外，每次开发新代码都需要投入更多的人力和物力。

以笔者多年的实际开发经验来看，每个公司都会有自己的基础类库，这就是软件的复用，这些基础类库将在后续开发中多次被重复使用。例如，信息化系统，其中总有一些开发过程是重复的，为什么不将这些重复开发工作抽象成基础类库呢？这种抽象既提高了开发效率，而且因为重复使用，也降低了引入错误的风险。

因此只要是一个有实际开发经验的软件公司，就一定有自己的一套基础类库，这就是需要使用框架的原因。从某个角度来看，框架也是一套基础类库，它抽象了软件开发的通用步骤，让实际开发人员可以直接利用这部分实现。当然，即使使用 JSP 和 Servlet 开发的公司，也可以抽象出自己的一套基础类库，那么这也是框架！一个从事实际开发的软件公司，不管它是否意识到，它已经在使用框架。区别只有：使用的框架到底是别人提供的，还是自己抽象出来的。

到底是使用第三方提供的框架更好，还是使用自己抽象的框架更好？这个问题就见仁见智了。通常而言，第三方提供的框架更稳定，更有保证，因为第三方提供的框架往往经过了更多人的测试。而使用自己抽象的框架，则更加熟悉底层运行原理，在处理问题上更有方向性。如果不是有非常特殊的理由，还是推荐使用第三方框架，特别是那些流行的、广泛使用的、开源的框架。

1.3 本章小结

本章主要介绍了 Java EE 应用的相关基础知识，其中，简要介绍了 Java EE 应用应该遵循怎样的架构模型，通常应该具有哪些组件，以及这些组件通常使用什么样的技术来实现。本章还简单归纳了 Java EE 应用所具有的优势和吸引力。

本书使用的是 Apache Tomcat Web 服务器，使用的开发工具是 Eclipse。关于 Tomcat 的安装和 Eclipse 工具的具体用法，请参考“疯狂 Java 系列”之《轻量级 Java EE 企业应用实战》，这里不做讨论。

第 2 章将重点介绍 Spring MVC。

CHAPTER 2

第2章 Spring MVC 简介

本章要点

- Model1 和 Model2
- MVC 思想及其优点
- Spring MVC 的优势
- Spring MVC 的前端控制器 DispatcherServlet
- 开发第一个 Spring MVC 应用
- 基于 Controller 接口的控制器
- 基于注解的控制器
- Spring MVC 的工作流程

2.1 MVC思想概述

2.1.1 传统Model1和Model2

Java Web应用的结构经历了Model1和Model2两个时代，从Model1发展到Model2既是技术发展的必然，也是无数程序员的心血结晶。

在Model1模式下，整个Web应用几乎全部由JSP页面组成，JSP页面接收处理客户端请求，对请求处理后直接做出响应。用少量的JavaBean来处理数据库连接、数据库访问等操作。

Model1模式的实现比较简单，适用于快速开发小规模项目。但从工程化的角度看，它的局限性非常明显：JSP页面身兼View和Controller两种角色，将控制逻辑和表现逻辑混杂在一起，从而导致代码的重用性非常低，增加了应用的扩展和维护的难度。

早期由大量JSP页面所开发出来的Web应用，大都采用了Model1架构。实际上，早期绝大部分ASP应用也属于这种Model1架构。

Model2是基于MVC架构的设计模式。在Model2架构中，Servlet作为前端控制器，负责接收客户端发送的请求。在Servlet中只包含控制逻辑和简单的前端处理；然后，调用后端JavaBean来完成实际的逻辑处理；最后，将其转发到相应的JSP页面来处理显示逻辑。其具体的实现方式如图2.1所示。

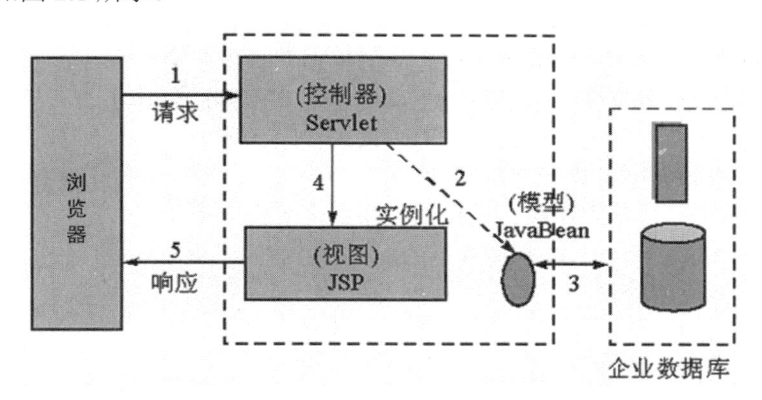

图2.1 Model2的处理流程

正如在图2.1中看到的，Model2下的JSP不再承担控制器的责任，它仅仅是表现层角色，仅仅用于将结果呈现给用户。JSP页面的请求与Servlet（控制器）交互，而Servlet负责与后台的JavaBean通信。在Model2模式下，模型（Model）由JavaBean充当，视图（View）由JSP页面充当，而控制器（Controller）则由Servlet充当。

由于引入了MVC模式，使得Model2具有组件化的特点，从而更适用于大规模应用的开发，但也增加了应用开发的复杂程度。原本需要一个简单的JSP页面就能实现的应用，在Model2中被分解成多个协同工作的部分，程序员需要花更多时间才能真正掌握其设计和实现过程。

Model2是MVC设计思想下的架构。下面简要介绍MVC设计思想的优势。

提示

对于非常小型的Web站点，如果后期的更新、维护工作不是特别多，则可以使用Model1模式来开发应用，而不是Model2模式。虽然Model2提供了更好的可扩展性及可维护性，但其增加了前期开发成本。从某种程度上讲，Model2为了降低系统后期维护的复杂度，而导致前期开发的高复杂度。

2.1.2 MVC 思想及其优势

MVC 并不是 Java 语言所特有的设计思想，也并不是 Web 应用所特有的思想，它是所有面向对象程序设计语言都应该遵守的规范。

MVC 思想将一个应用分成三个基本部分：Model（模型）、View（视图）和 Controller（控制器），这三个部分以最少的耦合协同工作，从而提高应用的可扩展性及可维护性。

在经典的 MVC 模式中，事件由控制器处理，控制器根据事件的类型改变模型或视图，反之亦然。具体地说，每个模型对应一系列的视图列表，这种对应关系通常采用注册来完成，即把多个视图注册到同一个模型，当模型发生改变时，模型向所有注册过的视图发送通知，接下来，视图从对应的模型中获得信息，然后完成视图显示的更新。

从设计模式的角度来看，MVC 思想非常类似于观察者模式，但其与观察者模式存在少许差别：在观察者模式下，观察者和被观察者可以是两个互相对等的对象；但在 MVC 中，被观察者往往只是单纯的数据体，而观察者则是单纯的视图页面。

概括起来，MVC 有如下特点：

- 多个视图可以对应一个模型。按 MVC 设计模式，一个模型对应多个视图，可以减少代码的复制及代码的维护量，这样，一旦模型发生改变，也易于维护。
- 模型返回的数据与显示逻辑分离。模型数据可以应用任何的显示技术，例如，使用 JSP 页面、Velocity 模板或者直接产生 Excel 文档等。
- 应用被分隔为三层，这降低了各层之间的耦合，提供了应用的可扩展性。
- 控制层的概念也很有效，由于它把不同的模型和不同的视图组合在一起，完成不同的请求。因此，控制层可以说包含了用户请求权限的概念。
- MVC 更符合软件工程化管理的精神。不同的层各司其职，同一层的组件具有相同的特征，这有利于通过工程化和工具化的方法产生管理程序代码。

相对于早期的 MVC 思想，Web 模式下的 MVC 思想则又存在一些变化。对于一个普通应用程序，可以将视图注册给模型，当模型数据发生改变时，即时通知视图页面发生了改变；而对于 Web 应用，即使将多个 JSP 页面注册给一个模型，但当模型发生变化时，模型也无法主动给 JSP 页面发送消息（因为 Web 应用都是基于请求/响应模式的），只有当用户请求浏览该页面时，控制器才负责调用模型数据来更新 JSP 页面。图 2.2 显示了遵循 MVC 模式的 Java Web 的运行流程。

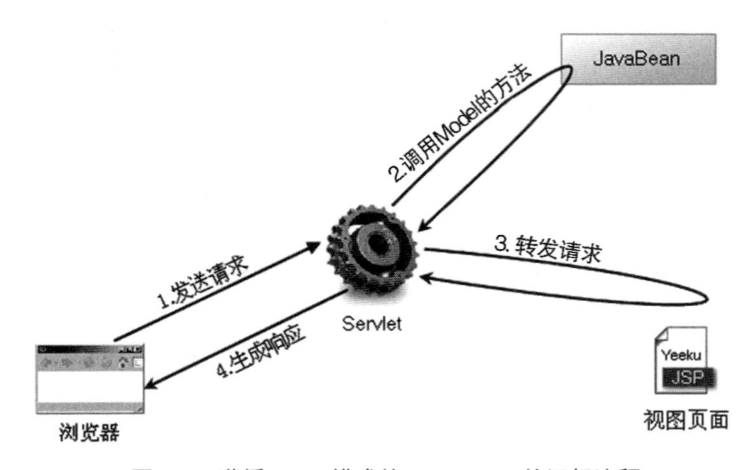

图 2.2 遵循 MVC 模式的 Java Web 的运行流程

提示

MVC思想与观察者模式有一定的相似之处，但并不完全相同。经典的MVC思想与Web应用的MVC思想也存在一定的差别，引起差别的主要原因是因为Web应用是一种请求/响应模式下的应用，对于请求/响应应用，如果用户不对应用发出请求，视图将无法主动更新自己。

2.2 Struts 2和Spring MVC

2.2.1 Spring MVC的优势

Spring框架提供了构建Web应用程序的全功能MVC模块——Spring MVC。Spring MVC框架提供了一个DispatcherServlet作为前端控制器来分派请求，同时提供灵活的配置处理程序映射、视图解析、语言环境和主题解析功能，并支持文件上传。Spring MVC还包含多种视图技术，例如Java Server Pages（JSP）、Velocity、Tiles、iText和POI等。Spring MVC分离了控制器、模型对象、分派器以及处理程序对象的角色，这种分离让它们更容易定制。

Spring MVC具有如下特点：

- Spring MVC拥有强大的灵活性、非侵入性和可配置性。
- Spring MVC提供了一个前端控制器DispatcherServlet，开发者无须额外开发控制器对象。
- Spring MVC分工明确，包括控制器、验证器、命令对象、模型对象、处理程序映射、视图解析器，等等，每一个功能实现由一个专门的对象负责完成。
- Spring MVC可以自动绑定用户输入，并正确地转换数据类型。例如，Spring MVC能自动解析字符串，并将其设置为模型的int或float类型的属性。
- Spring MVC使用一个名称/值的Map对象实现更加灵活的模型数据传输。
- Spring MVC内置了常见的校验器，可以校验用户输入，如果校验不通过，则重定向回输入表单。输入校验是可选的，并且支持编程方式及声明方式。
- Spring MVC支持国际化，支持根据用户区域显示多国语言，并且国际化的配置非常简单。
- Spring MVC支持多种视图技术，最常见的有JSP技术以及其他技术，包括Velocity和FreeMarker。
- Spring提供了一个简单而强大的JSP标签库，支持数据绑定功能，使得编写JSP页面更加容易。

2.2.2 Spring MVC和Struts 2的区别

- 从机制上来说，Spring MVC的入口是Servlet，而Struts 2的入口是filter，这样就导致了二者的机制不同。
- 从性能上来说，Struts 2是基于类的设计，每发一次请求都会创建一个Action实例，每个Action都会被注入属性；而Spring MVC是基于方法的设计，粒度更细，一个方法对应一个request上下文，而方法同时又跟一个url对应，从架构本身上Spring MVC就非常容易实现RESTful url，而Struts 2的架构实现起来相对麻烦，因为Struts 2中Action的一个方法可以对应一个url，但是类属性却被所有方法共享，这也就无法用注解或其他方式标识属性所属的方法。由于Struts 2需要针对每个request进行封装，把

request、session 等 Servlet 生命周期的变量封装成一个一个的 Map，提供给每个 Action 使用，并保证线程安全，所以在原则上，Struts 2 是比较耗费内存的，所以 Spring MVC 在性能上高于 Struts 2。

- 从参数上来说，Spring MVC 的方法之间基本上是独立的，独享 request 和 response 数据，请求数据通过参数获取，处理结果通过 Model 交回给框架，方法之间不共享变量；而 Struts 2 虽然方法之间也是独立的，但其所有 Action 变量是共享的，每次来了请求就创建一个 Action，一个 Action 对象对应一个 request 上下文。
- 从设计思想上来说，Struts 2 使用的是拦截器（Interceptor）机制，而 Spring MVC 使用的是独立的 AOP 方式，这样导致 Struts 2 的配置文件量还是比 Spring MVC 大，Spring MVC 的使用更加简洁。
- 从数据验证上来说，Spring MVC 的验证功能是一个亮点，支持 JSR 303，处理 Ajax 的请求更是方便，只需一个注解@ResponseBody，然后直接返回响应文本即可，而 Struts 2 的验证则比较烦琐。
- 从配置上来说，在实际项目开发中使用 Struts 2 时大多采用传统的配置文件的方式，Spring MVC 除了配置 spring mvc-servlet.xml 文件之外，已经是 100%的零配置开发，所以在开发效率上高于 Struts 2。
- 从项目管理上来说，Spring MVC 和 Spring 无缝结合，这个优势是 Struts 2 无法与之相提并论的。

提示

现在讨论 Spring MVC 和 Struts 2 的区别，对于还不了解 Spring MVC 的读者来说可能会感觉很难理解，建议初学者学习完 Spring MVC 的知识之后再回过头来看本节的内容。

2.3　开发第一个 Spring MVC 应用

本书成书之时，Spring 的最新稳定版本是 5.0.1，本书的代码都是基于该版本的。建议读者下载该版本或者更高版本的 Spring。

2.3.1　Spring 的下载和安装

Spring 是一个独立的框架，它不需要依赖于任何 Web 服务器或容器。它既可在独立的 Java SE 项目中使用，也可以在 Java Web 项目中使用。下面首先介绍如何为 Java 项目和 Java Web 项目添加 Spring 支持。

下载和安装 Spring 框架可按如下步骤进行：

① 登录 http://repo.springsource.org/libs-release-local/站点，该页面显示一个目录列表，读者沿着 org→springframework→spring 路径进入，即可看到 Spring 框架各版本的压缩包的下载链接。下载 Spring 的最新稳定版 5.0.1。

② 下载完成，得到一个 spring-framework-5.0.1.RELEASE-dist.zip 压缩文件，解压该压缩文件得到一个名为 spring-framework-5.0.1.RELEASE 的文件夹，该文件夹下有如下几个子文件夹：

- docs。该文件夹下存放 Spring 的相关文档，包含开发指南、API 参考文档。

- libs。该文件夹下的jar分为三类：Spring框架class文件的jar包；Spring框架源文件的压缩包，文件名以-source结尾；Spring框架API文档的压缩包，文件名以-javadoc结尾。
- schemas。该文件夹下包含了Spring各种配置文件的XML Schema文档。
- readme.txt、notice.txt、license.txt等说明性文档。

③ 将libs文件夹下所需模块的class文件的jar包复制添加到项目的类加载路径中，既可通过添加环境变量的方式来添加，也可使用Ant或IDE工具来管理应用程序的类加载路径。如果需要发布该应用，则将这些jar包一同发布即可。如果没有太多要求，建议将libs文件夹下所有模块的class文件的jar包添加进去。

④ 除此之外，Spring的核心容器必须依赖于common-logging的jar包，因此还应该登录http://commons.apache.org/站点，沿着Releases→Logging路径进入，下载最新的commons-logging工具，下载完成得到一个commons-logging-1.2-bin.zip压缩文件，将该文件解压路径下的commons-logging-1.2.jar也添加到项目的类加载路径中。

完成上面4个步骤后，接下来即可在Java Web应用中使用Spring MVC框架了。

2.3.2 Spring MVC 的 DispatcherServlet

在许多的MVC框架中，都包含一个用于调度控制的Servlet。Spring MVC也提供了一个名为org.springframework.web.servlet.DispatcherServlet的Servlet充当前端控制器，所有的请求驱动都围绕这个DispatcherServlet来分派请求。

DispatcherServlet是一个Servlet（它继承自 HttpServlet 基类），因此使用时需要把它配置在Web应用的部署描述符web.xml文件当中，配置信息如下：

```
<servlet>
    <!-- Servlet 的名称 -->
  <servlet-name>springmvc</servlet-name>
  <!-- Servlet 对应的 Java 类 -->
  <servlet-class>
    org.springframework.web.servlet.DispatcherServlet
  </servlet-class>
  <!-- 当前 Servlet 的参数信息 -->
  <init-param>
    <!-- contextConfigLocation 是参数名称，该参数的值包含 SpringMVC 的配置文件路径 -->
   <param-name>contextConfigLocation</param-name>
   <param-value>/WEB-INF/springmvc-config.xml</param-value>
  </init-param>
  <!-- 在 Web 应用启动时立即加载 Servlet -->
  <load-on-startup>1</load-on-startup>
</servlet>
<!-- Servlet 映射声明 -->
<servlet-mapping>
    <!-- 请求对应的 Servlet 的名称 -->
  <servlet-name>springmvc</servlet-name>
  <!-- 监听当前域的所有请求 -->
  <url-pattern>/</url-pattern>
</servlet-mapping>
```

以上是标准Java EE Servlet的配置。配置了一个DispatcherServlet，该Servlet在Web应用程序启动时立即加载，DispatcherServlet加载时会需要一个Spring MVC的配置文件，默认情况下，应用会去应用程序文件夹下的WEB-INF文件夹下查找对应的[servlet- name]-servlet.xml文件，例如本例的<servlet-name>是springmvc，默认查找的就是/WEB-INF/springmvc-servlet.xml。

也可以把 Spring MVC 的配置文件放到应用程序文件夹中的任何地方，用 servlet 元素的 init-param 子元素进行描述，本例的 param-name 元素的值 contextConfigLocation 表示参数名称，param-value 元素的值/WEB-INF/springmvc-config.xml 则表示 Spring MVC 的配置文件路径和名称。则 DispatcherServlet 会查找/WEB-INF/springmvc-config.xml 文件，作为 Spring MVC 的配置文件，解析该文件内容并根据文件配置信息创建一个 WebApplicationContext 容器对象，也称为上下文环境。WebApplicationContext 继承自 ApplicationContext 容器，它的初始化方式和 BeanFactory、ApplicationContext 有所区别，因为 WebApplicationContext 需要 ServletContext 实例，也就是说，它必须在拥有 Web 容器的前提下才能完成启动 Spring Web 应用上下文的工作。有了 WebApplicationContext 容器，开发者就可以很自然地使用 Spring 的 IOC、AOP 等其他功能了。

2.3.3 基于 Controller 接口的控制器

DispatcherServlet 在 Spring 当中充当一个前端控制器的角色，它的核心功能是分发请求。请求会被分发给对应处理的 Java 类，Spring MVC 中称为 Handle。在 Spring 2.5 以前，开发一个 Handle 的唯一方法是实现 org.springframework.web.servlet.mvc.Controller 接口。Controller 接口必须实现一个方法，该方法的签名如下：

```
ModelAndView handleRequest(HttpServletRequest request,
        HttpServletResponse response) throws Exception
```

Controller 接口的实现类可以通过 handleRequest 方法传递的参数访问对应请求的 HttpServletRequest 和 HttpServletResponse 对象，处理完业务请求之后，必须返回一个包含模型对象和视图路径的 ModelAndView 对象。

提示

Controller 接口的实现类只能处理一个单一请求动作，而 Spring 2.5 之后新增的基于注解的控制器可以支持同时处理多个请求动作，并且无须实现任何接口，其更加灵活。之后会详细介绍。

接下来我们演示一个基于 Controller 接口的 Spring MVC 控制器的 Web 应用，以便展示 Spring MVC 是如何工作的。

示例：第一个 Spring MVC 应用

1. 增加 Spring 的支持

首先，使用 Eclipse 新建一个 Dynamic Web Project，也就是新建一个动态 Web 项目，命名为 SpringMVCTest。

为了让 Web 应用具有 Spring 支持的功能，将 spring-framework-5.0.1.RELEASE 解压文件夹下的 libs 文件夹下所有 Spring 框架的 class 文件的 jar 包和 Spring 所依赖的 commons-logging-1.2.jar 复制到 Web 应用的 lib 文件夹下，也就是 SpringMVCTest\WebContent\WEB-INF\lib 路径下。

返回 Eclipse 主界面，此时在 Eclipse 主界面的左上角资源导航树中会看到 SpringMVCTest 节点，选中该节点，然后按 F5 键，将看到 Eclipse 主界面的左上角资源导航树中出现如图 2.3 所示的结构。

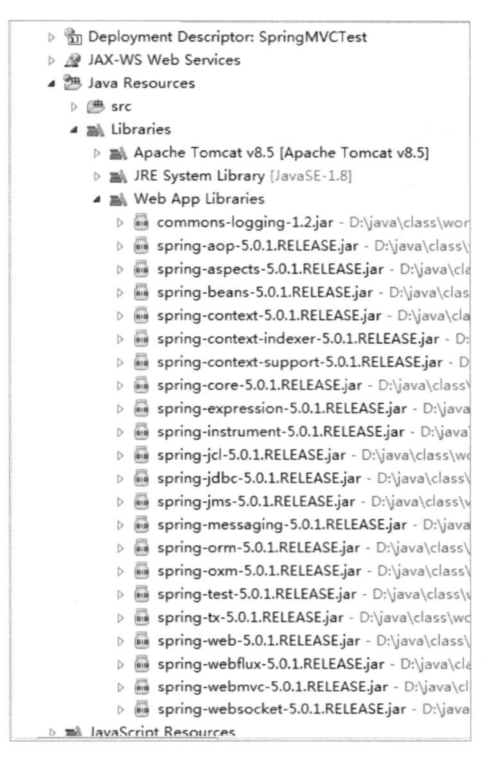

图 2.3　增加 Spring 5 支持

看到如图 2.3 所示的结构，即表明该 Web 应用已经加入了 Spring 的必需类库。但还需要修改 web.xml 文件，让该文件负责加载 Spring 框架。

2. 配置前端控制器 DispatcherServlet

在如图 2.3 所示的导航树中，单击 WebContent→WEB-INF 节点前的加号，展开该节点，会看到该节点下包含的 web.xml 文件子节点。

单击 web.xml 文件节点，编辑该文件，配置 Spring MVC 的前端控制器 DispatcherServlet。配置信息如下：

程序清单：codes/02/SpringMVCTest/WebContent/WEB-INF/ web.xml

```
<?xml version="1.0" encoding="UTF-8"?>
<web-app xmlns:xsi="http://www.w3.org/2001/XMLSchema-instance"
   xmlns="http://xmlns.jcp.org/xml/ns/javaee"
   xsi:schemaLocation="http://xmlns.jcp.org/xml/ns/javaee
   http://xmlns.jcp.org/xml/ns/javaee/web-app_3_1.xsd"
   id="WebApp_ID" version="3.1">
   <!-- 定义 Spring MVC 的前端控制器 -->
  <servlet>
   <servlet-name>springmvc</servlet-name>
   <servlet-class>
       org.springframework.web.servlet.DispatcherServlet
   </servlet-class>
   <init-param>
     <param-name>contextConfigLocation</param-name>
     <param-value>/WEB-INF/springmvc-config.xml</param-value>
   </init-param>
   <load-on-startup>1</load-on-startup>
  </servlet>
   <!-- 让 Spring MVC 的前端控制器拦截所有请求 -->
```

```
  <servlet-mapping>
    <servlet-name>springmvc</servlet-name>
    <url-pattern>/</url-pattern>
  </servlet-mapping>
</web-app>
```

web.xml 文件的内容告诉 Web 容器，将使用 Spring MVC 的 DispatcherServlet，并通过配置 url-pattern 元素的值为“/”，将所有的 URL 映射到该 Servlet。

3. 配置 Spring MVC 的 Controller

接下来是 Spring MVC 的配置文件，配置信息如下：

程序清单：codes/02/SpringMVCTest/WebContent/WEB-INF/springmvc-config.xml

```
<?xml version="1.0" encoding="UTF-8"?>
<beans xmlns="http://www.springframework.org/schema/beans"
    xmlns:xsi="http://www.w3.org/2001/XMLSchema-instance"
    xsi:schemaLocation="http://www.springframework.org/schema/beans
       http://www.springframework.org/schema/beans/spring-beans.xsd">
     <!-- 配置 Handle，映射"/hello"请求 -->
    <bean name="/hello" class="org.fkit.controller.HelloController"/>
    <!-- 处理器映射器将 bean 的 name 作为 url 进行查找，需要在配置 Handle 时指定 name（即 url）
-->
    <bean class="org.springframework.web.servlet.handler.BeanNameUrlHandlerMapping"/>
    <!-- SimpleControllerHandlerAdapter 是一个处理器适配器，所有处理器适配器都要实现
HandlerAdapter 接口-->
    <bean class="org.springframework.web.servlet.mvc.SimpleControllerHandlerAdapter"/>
    <!-- 视图解析器 -->
    <bean class="org.springframework.web.servlet.view.InternalResourceViewResolver"/>
</beans>
```

springmvc-config.xml 文件声明了 HelloController 业务控制器类，并将其映射到/hello 请求。

此处还配置了一个处理器映射器 BeanNameUrlHandlerMapping，这样可以 Bean 的名称为 url 进行查找；同时配置了处理器适配器 SimpleControllerHandlerAdapter，来完成对 HelloController 类的 handleRequest 方法的调用；最后配置了视图解析器 InternalResourceViewResolver 来解析视图，将 View 呈现给用户。需要注意的是，在 Spring 4.0 之后，如果不配置处理器映射器、处理器适配器和视图解析器，也会使用默认的完成 Spring 内部 MVC 工作，笔者在此处显示配置处理过程，是希望读者能够了解 Spring MVC 的每一个动作，之后可以更好地理解 Spring MVC 的工作流程。

4. Controller 类的实现

HelloController 类实现了 Controller 接口，用来处理/hello 请求。示例代码如下：

程序清单：codes/02/SpringMVCTest/src/org/fkit/controller/HelloController

```
package org.fkit.controller;
import javax.servlet.http.HttpServletRequest;
import javax.servlet.http.HttpServletResponse;
import org.springframework.web.servlet.ModelAndView;
import org.springframework.web.servlet.mvc.Controller;
/**
 *  HelloController 是一个实现 Controller 接口的控制器，
 *  可以处理一个单一的请求动作
 */
public class HelloController implements Controller{
```

```
    /**
     * handleRequest 是 Controller 接口必须实现的方法。
     * 该方法的参数是对应请求的 HttpServletRequest 和 HttpServletResponse。
     * 该方法必须返回一个包含视图名或视图名和模型的 ModelAndView 对象。
     * */
    @Override
    public ModelAndView handleRequest(HttpServletRequest request,
            HttpServletResponse response) throws Exception {
        System.out.println("handleRequest 被调用");
        // 创建准备返回的ModelAndView对象，该对象通常包含了返回视图名、模型的名称以及模型对象
        ModelAndView mv = new ModelAndView();
        //添加模型数据，可以是任意的 POJO 对象
        mv.addObject("message", "Hello World!");
        // 设置逻辑视图名，视图解析器会根据该名字解析到具体的视图页面
        mv.setViewName("/WEB-INF/content/welcome.jsp");
        // 返回 ModelAndView 对象。
        return mv;
    }
}
```

HelloController 是一个实现 Controller 接口的控制器，它可以处理一个单一的请求动作。handleRequest 是 Controller 接口必须实现的方法，Controller 调用该方法来处理请求。该方法的参数是对应请求的 HttpServletRequest 和 HttpServletResponse，该方法必须返回一个包含视图名或视图名和模型的 ModelAndView 对象。本例返回的模型中包含了一个名为 message 的字符串对象，返回的视图路径为/WEB-INF/content/welcome.jsp，因此，请求将被转发到 welcome.jsp 页面。

提示

Spring MVC 建议把所有的视图页面存放在 WEB-INF 文件夹下，这样可以保护视图页面，避免直接向视图页面发送请求。上面的 HelloController 类的 handleRequest 方法处理完请求后，Spring MVC 会调用/WEB-INF/content 文件夹下的 welcome.jsp 呈现视图。

5. View 页面

SpringMVCTest 包含一个视图页面 welcome.jsp，用来显示欢迎信息。

程序清单：codes/02/SpringMVCTest/WebContent/WEB-INF/content/welcome.jsp

```
<%@ page language="java" contentType="text/html; charset=UTF-8"
    pageEncoding="UTF-8"%>
<!DOCTYPE html PUBLIC "-//W3C//DTD HTML 4.01 Transitional//EN" "http://www.w3.org/TR/html4/loose.dtd">
<html>
<head>
<meta http-equiv="Content-Type" content="text/html; charset=UTF-8">
<title>welcome</title>
</head>
<body>
<!-- 页面可以访问 Controller 传递出来的 message -->
${requestScope.message}
</body>
</html>
```

此处的 JSP 页面使用了 EL 表达式来简化页面开发，关于 EL 表达式的使用可参考附录 A “EL 表达式和 JSTL 标签库”的内容。

6. 测试应用

使用 Eclipse 部署 SpringMVCTest 这个 Web 应用，在浏览器中输入如下 URL 来测试应用：

```
http://localhost:8080/SpringMVCTest/hello
```

将会看到如图 2.4 所示的界面，表示 Spring MVC 访问成功。

图 2.4 访问成功

提示

使用 MVC 框架就应该严格遵守 MVC 思想。MVC 框架不赞成浏览器直接访问 Web 应用的视图页面，用户的所有请求都只应向控制器发送，由控制器调用模型组件、视图组件向用户呈现数据。

示例：基于注解的控制器

从 Spring 2.5 开始新增了基于注解的控制器，也就是说控制器不用实现 Controller 接口，通过注释类型来描述。下面将 SpringMVCTest 这个 Web 应用进行修改，演示一个基于注解的控制器 Spring MVC 的 Web 应用。

新建一个 Dynamic Web Project，也就是新建一个动态 Web 项目，命名为 AnnotationTest。所有步骤和 2.3.3 节的“第一个 Spring MVC 应用”示例一样，只是修改两个地方。

1. Controller 类的实现

HelloController 类不需要 Controller 接口，改为使用注解类型来描述，处理/hello 请求。示例代码如下：

程序清单：codes/02/AnnotationTest/src/org/fkit/controller/HelloController

```
package org.fkit.controller;
import org.springframework.stereotype.Controller;
import org.springframework.web.bind.annotation.RequestMapping;
import org.springframework.web.servlet.ModelAndView;
/**
 * HelloController 是一个基于注解的控制器，
 * 可以同时处理多个请求动作，并且无须实现任何接口。
 * org.springframework.stereotype.Controller 注解用于指示该类是一个控制器
 */
@Controller
public class HelloController{

    /**
     * org.springframework.web.bind.annotation.RequestMapping 注解
     * 用来映射请求的 URL 和请求的方法等。本例用来映射"/hello"
     * hello 只是一个普通方法。
     * 该方法返回一个包含视图名或视图名和模型的 ModelAndView 对象。
     * */
    @RequestMapping(value="/hello")
    public ModelAndView hello(){
        System.out.println("hello 方法 被调用");
        // 创建准备返回的 ModelAndView 对象，该对象通常包含了返回视图名、模型的名称以及模型对象
```

```
        ModelAndView mv = new ModelAndView();
        // 添加模型数据，可以是任意的 POJO 对象
        mv.addObject("message", "Hello World!");
        // 设置逻辑视图名，视图解析器会根据该名字解析到具体的视图页面
        mv.setViewName("/WEB-INF/content/welcome.jsp");
       // 返回 ModelAndView 对象。
       return mv;
    }

}
```

HelloController 是一个基于注解的控制器，org.springframework.stereotype.Controller 注释类型用于指示 Spring 类的实例是一个控制器。org.springframework.web.bind.annotation.RequestMapping 注释类型用来映射一个请求和请求的方法，value="/hello"表示请求由 hello 方法进行处理。方法返回一个包含视图名或视图名和模型的 ModelAndView 对象，这和 2.3.4 节中的示例一样。

2. 修改 Spring MVC 的配置文件

```
<?xml version="1.0" encoding="UTF-8"?>
<beans xmlns="http://www.springframework.org/schema/beans"
    xmlns:xsi="http://www.w3.org/2001/XMLSchema-instance"
    xmlns:mvc="http://www.springframework.org/schema/mvc"
    xmlns:context="http://www.springframework.org/schema/context"
    xsi:schemaLocation="
        http://www.springframework.org/schema/beans
        http://www.springframework.org/schema/beans/spring-beans.xsd
        http://www.springframework.org/schema/mvc
        http://www.springframework.org/schema/mvc/spring-mvc.xsd
        http://www.springframework.org/schema/context
        http://www.springframework.org/schema/context/spring-context.xsd">

    <!-- spring 可以自动去扫描 base-pack 下面的包或者子包下面的 Java 文件,
        如果扫描到有 Spring 的相关注解的类，则把这些类注册为 Spring 的 bean -->
    <context:component-scan base-package="org.fkit.controller"/>
    <!-- 配置 annotation 类型的处理器映射器 -->
    <bean class="org.springframework.web.servlet.mvc.method.annotation.
RequestMappingHandlerMapping"/>
     <!-- 配置 annotation 类型的处理器适配器-->
    <bean class="org.springframework.web.servlet.mvc.method.annotation.
RequestMappingHandlerAdapter"/>
    <!-- 视图解析器 -->
    <bean class="org.springframework.web.servlet.view.InternalResourceViewResolver"/>
</beans>
```

由于使用了注解类型，因此不需要再在配置文件中使用 XML 描述 Bean。Spring 使用扫描机制查找应用程序中所有基于注解的控制器类。<context:component-scan base-package="org.fkit.controller"/>指定需要 Spring 扫描 org.fkit.controller 包及其子包下面的所有 Java 文件。

此处还配置了一个 annotation 类型的处理器映射器 RequestMappingHandlerMapping，它根据请求查找映射；同时配置了 annotation 类型的处理器适配器 RequestMappingHandlerAdapter，来完成对 HelloController 类的@RequestMapping 标注方法的调用；最后配置了视图解析器 InternalResourceViewResolver 来解析视图，将 View 呈现给用户。需要注意的是，在 Spring 4.0 之后，处理器映射器、处理器适配器的配置还可以使用更简便的方式，笔者在此处显示配置处理过程，是希望读者能够了解 Spring MVC 的每一个动作，之后可以更好地理解 Spring MVC 的工作流程。

3. 测试

使用 Eclipse 部署 AnnotationTest 这个 Web 应用，在浏览器中输入如下 URL 来测试应用：

```
http://localhost:8080/AnnotationTest/hello
```

会看到如图 2.4 所示的界面，表示 Spring MVC 访问成功。

2.4 详解 DispatcherServlet

2.3 节中的第一个 Spring MVC 应用已经成功运行了。那么，前端控制器 DispatcherServlet 截获请求后做了什么工作呢？DispatcherServlet 又是如何分派请求的呢？

分析 DispatcherServlet 的源代码如下：

程序清单：org/springframework/web/servlet/DispatcherServlet.java

```
protected void initStrategies(ApplicationContext context) {
    initMultipartResolver(context); // 初始化上传文件解析器
    initLocaleResolver(context); // 初始化本地化解析器
    initThemeResolver(context); // 初始化主题解析器
    initHandlerMappings(context); // 初始化处理器映射器，将请求映射到处理器
    initHandlerAdapters(context); // 初始化处理器适配器
    initHandlerExceptionResolvers(context);
    // 初始化处理器异常解析器，如果执行过程中遇到异常将交给 HandlerExceptionResolver 来解析
    initRequestToViewNameTranslator(context); // 初始化请求到视图名称解析器
    initViewResolvers(context);
    // 初始化视图解析器，通过 ViewResolver 解析逻辑视图名到具体视图实现
    initFlashMapManager(context); // 初始化 flash 映射管理器
}
```

提示

org/springframework/web/servlet/DispatcherServlet 是 Spring 框架的源代码，读者可在配套资源中找到 Spring 源代码或自行下载。

initStrategies 方法将在 WebApplicationContext 初始化后自动执行，自动扫描上下文的 Bean，根据名称或类型匹配的机制查找自定义的组件，如果没有找到则会装配一套 Spring 的默认组件。在 org.springframework.web.servlet 路径下有一个 DispatcherServlet.properties 配置文件，该文件指定了 DispatcherServlet 所使用的默认组件。

程序清单：org/springframework/web/servlet/DispatcherServlet.properties

```
    // 本地化解析器
    org.springframework.web.servlet.LocaleResolver=org.springframework.web.servlet.
i18n.AcceptHeaderLocaleResolver
    // 主题解析器
    org.springframework.web.servlet.ThemeResolver=org.springframework.web.servlet.t
heme.FixedThemeResolver
    // 处理器映射器（共 2 个）
    org.springframework.web.servlet.HandlerMapping=org.springframework.web.servlet.
handler.BeanNameUrlHandlerMapping,\
        org.springframework.web.servlet.mvc.annotation.DefaultAnnotationHandlerMapping
    // 处理器适配器（共 3 个）
    org.springframework.web.servlet.HandlerAdapter=org.springframework.web.servlet.
mvc.HttpRequestHandlerAdapter,\
        org.springframework.web.servlet.mvc.SimpleControllerHandlerAdapter,\
        org.springframework.web.servlet.mvc.annotation.AnnotationMethodHandlerAdapter
    // 异常处理器（共 3 个）
```

```
    org.springframework.web.servlet.HandlerExceptionResolver=org.springframework.we
b.servlet mvc.annotation.AnnotationMethodHandlerExceptionResolver,\
        org.springframework.web.servlet.mvc.annotation.ResponseStatusExceptionResolver,\
        org.springframework.web.servlet.mvc.support.DefaultHandlerExceptionResolver
    // 视图名称解析器
    org.springframework.web.servlet.RequestToViewNameTranslator=org.springframework
.web.servlet.view.DefaultRequestToViewNameTranslator
    // 视图解析器
    org.springframework.web.servlet.ViewResolver=org.springframework.web.servlet.vi
ew.InternalResourceViewResolver
    // FlashMap 映射管理器
    org.springframework.web.servlet.FlashMapManager=org.springframework.web.servlet
.support.SessionFlashMapManager
```

如果开发者希望使用自定义类型的组件，则只需要在 Spring 配置文件中配置自定义的Bean组件印可。Spring MVC如果发现上下文中有用户自定义的组件，就不会使用默认的组件。

以下是 DispatcherServlet 装配每种组件的细节。

➢ **本地化解析器**。只允许一个实例
- 查找名为 localeResolver、类型为.LocaleResolver 的 Bean 作为该类型组件。
- 如果没有找到，则使用默认的实现类 AcceptHeaderLocaleResolver 作为该类型组件。

➢ **主题解析器**。只允许一个实例
- 查找名为 themeResolver、类型为 ThemeResolver 的 Bean 作为该类型组件。
- 如果没有找到，则使用默认的实现类 FixedThemeResolve 作为该类型组件。

➢ **处理器映射器**。允许多个实例
- 如果 detectAllHandlerMappings 的属性为 true（默认为 true），则根据类型匹配机制查找上下文以及 Spring 容器中所有类型为 HandlerMapping 的 Bean，将它们作为该类型组件。
- 如果 detectAllHandlerMappings 的属性为 false，则查找名为 handlerMapping、类型为 HandlerMapping 的 Bean 作为该类型组件。
- 如果通过以上两种方式都没有找到，则使用 BeanNameUrlHandlerMapping 实现类创建该类型的组件。

➢ **处理器适配器**。允许多个实例
- 如果 detectAllHandlerAdapters 的属性为 true（默认为 true），则根据类型匹配机制查找上下文以及 Spring 容器中所有类型为 HandlerAdapter 的 Bean，将它们作为该类型组件。
- 如果 detectAllHandlerAdapters 的属性为 false，则查找名为 handlerAdapter、类型为 HandlerAdapter 的 Bean 作为该类型组件。
- 如果通过以上两种方式都没有找到，则使用 DispatcherServlet.properties 配置文件中指定的三个实现类分别创建一个适配器，并将其添加到适配器列表中。

➢ **处理器异常解析器**。允许多个实例
- 如果 detectAllHandlerExceptionResolvers 的属性为 true（默认为 true），则根据类型匹配机制查找上下文以及 Spring 容器中所有类型为 HandlerExceptionResolver 的 Bean，将它们作为该类型组件。
- 如果 detectAllHandlerExceptionResolvers 的属性为 false，则查找名为 handlerException-Resolver、类型为 HandlerExceptionResolver 的 Bean 作为该类型组件。
- 如果通过以上两种方式都没有找到，则查找 DispatcherServlet.properties 配置文件中定

义的默认实现类。注意，如果该文件中没有对应处理器异常解析器的默认实现类，用户可以自定义处理器异常解析器的实现类，将之添加到 DispatcherServlet.properties 配置文件当中。

➢ **视图名称解析器**。只允许一个实例
- 查找名为 viewNameTranslator、类型为 RequestToViewNameTranslator 的 Bean 作为该类型组件。
- 如果没有找到，则使用默认的实现类 DefaultRequestToViewNameTranslator 作为该类型的组件。

➢ **视图解析器**。允许多个实例
- 如果 detectAllViewResolvers 的属性为 true（默认为 true），则根据类型匹配机制查找上下文以及 Spring 容器中所有类型为 ViewResolver 的 Bean，将它们作为该类型组件。
- 如果 detectAllViewResolvers 的属性为 false，则查找名为 viewResolvers、类型为 ViewResolver 的 Bean 作为该类型组件。
- 如果通过以上两种方式都没有找到，则查找 DispatcherServlet.properties 配置文件中定义的默认实现类 InternalResourceViewResolver 作为该类型的组件。

➢ **文件上传解析器**。只允许一个实例
- 查找名为 muliipartResolver、类型为 MuliipartResolver 的 Bean 作为该类型组件。
- 如果用户没有在上下文中显式定义 MuliipartResolver 类型的组件，则 DispatcherServlet 将不会加载该类型的组件。

➢ **FlashMap 映射管理器**。
- 查找名为 FlashMapManager、类型为 SessionFlashMapManager 的 Bean 作为该类型组件，用于管理 FlashMap，即数据默认存储在 HttpSession 中。

DispatcherServlet 装配的各种组件，有些只允许一个实例，比如文件上传解析器 MuliipartResolver、本地化解析器 LocaleResolver 等；有些则允许多个实例，如处理器映射器 HandlerMapping、处理器适配器 HandlerAdapter 等，读者需要注意这一点。

如果同一类型的组件存在多个，那么它们之间的优先级顺序如何确定呢？因为这些组件都实现了 org.springframework.core.Ordered 接口，所以可以通过 Order 属性确定优先级的顺序，值越小的优先级越高。

2.5 Spring MVC 执行的流程

下面将对开发 Spring MVC 应用的过程进行总结，以让读者对 Spring MVC 有一个大致的了解。

2.5.1 Spring MVC 应用的开发步骤

下面简单介绍 Spring MVC 应用的开发步骤。

① 在 web.xml 文件中定义前端控制器 DispatcherServlet 来拦截用户请求。

由于 Web 应用是基于请求/响应架构的应用，所以不管哪个 MVC Web 框架，都需要在 web.xml 中配置该框架的核心 Servlet 或 Filter，这样才可以让该框架介入 Web 应用中。

例如，开发 Spring MVC 应用的第 1 步就是在 web.xml 文件中增加如下配置片段：

```
<!-- 定义 Spring MVC 的前端控制器 -->
  <servlet>
    <servlet-name>springmvc</servlet-name>
    <servlet-class>
       org.springframework.web.servlet.DispatcherServlet
    </servlet-class>
    <init-param>
      <param-name>contextConfigLocation</param-name>
      <param-value>/WEB-INF/springmvc-config.xml</param-value>
    </init-param>
    <load-on-startup>1</load-on-startup>
  </servlet>
<!-- 让 Spring MVC 的前端控制器拦截所有请求 -->
  <servlet-mapping>
    <servlet-name>springmvc</servlet-name>
    <url-pattern>/</url-pattern>
  </servlet-mapping>
```

② 如果需要以 POST 方式提交请求，则定义包含表单数据的 JSP 页面。如果仅仅只是以 GET 方式发送请求，则无须经过这一步。

③ 定义处理用户请求的 Handle 类，可以实现 Controller 接口或使用@Controller 注解。

这一步也是所有 MVC 框架中必不可少的，因为这个 DispatcherServlet 就是 MVC 中的 C，也就是前端控制器，该控制器负责接收请求，并将请求分发给对应的 Handle，即实现 Controller 接口的 Java 类，而该 Java 类负责调用后台业务逻辑代码来处理请求。

提示

可能有读者会产生疑问：Controller 并未接收到用户请求，它怎么能够处理用户的请求呢？MVC 框架的底层机制是：前端 Servlet 接收到用户请求后，通常会对用户请求进行简单预处理，例如解析、封装参数等，然后通过反射来创建 Controller 实例，并调用 Controller 的指定方法（实现 Controller 接口的是 handleRequest 方法，而使用基于注解的控制器可以是任意方法）来处理用户请求。

这里又产生了一个问题：当 Servlet 拦截用户请求后，它如何知道创建哪个 Controller 接口的实例呢？有两种解决方案。

利用 xml 配置文件：例如在 xml 配置文件中描述 hello 请求对应使用 HelloController 类。这就可以让 MVC 框架知道创建哪个 Controller 接口的实例。

利用注解：例如使用注解@Controller 描述一个类，并使用注解@RequestMapping(value="/hello ")描述 hello 请求对应的方法。这样也可以让 MVC 框架知道创建哪个 Controller 接口的实例并调用哪个方法处理请求。

根据上面的介绍不难发现，在 Spring MVC 框架中，控制器实际上由两个部分组成，即拦截所有用户请求和处理请求的通用代码都由前端控制器 DispatcherServlet 完成，而实际的业务控制（诸如调用后台业务逻辑代码，返回处理结果等）则由 Controller 处理。

④ 配置 Handle。Java 领域的绝大部分 MVC 框架都非常喜欢使用 xml 文件来进行配置管理，这在以前是一种思维定势。即配置哪个请求对应哪个 Controller，从而让前端控制器根据该配置来创建合适的 Controller 实例，并调用该 Controller 的业务控制方法。

例如，可以采用如下代码片段来配置 Handle：

```
<!-- 配置 Handle，映射"/hello"请求 -->
   <bean name="/hello" class="org.fkit.controller.HelloController"/>
```

在 Spring 2.5 之后，推荐使用注解来配置 Handle：

```
@Controller
public class HelloController{
    @RequestMapping(value="/hello")
    public ModelAndView hello(){
       …
    }
}
```

上面的配置片段指定如果用户请求 URL 为 hello，则使用 org.fkit.controller.HelloController 来处理。现在几乎所有的 MVC 框架都使用“约定优于配置”的思想，也就是采用约定方式来规定用户请求地址和 Handle 之间的对应关系。

⑤ 编写视图资源。当 Handle 处理用户请求结束后，通常会返回一个 ModelAndView 对象，该对象中应该包含返回的视图名或视图名和模型，这个视图名就代表需要显示的物理视图资源。如果 Handle 需要把一些数据传给视图资源，则可以通过模型对象。

经过上面 5 个步骤，即可基本完成一个 Spring MVC 处理流程的开发，也就是可以执行一次完整的请求→响应过程。

2.5.2　Spring MVC 执行的流程

上一节所介绍的 Spring MVC 应用的开发流程实际上是按请求→响应的流程来开发的，下面通过一个流程图来介绍请求→响应的完整流程。图 2.5 显示了一次请求→响应的完整流程。

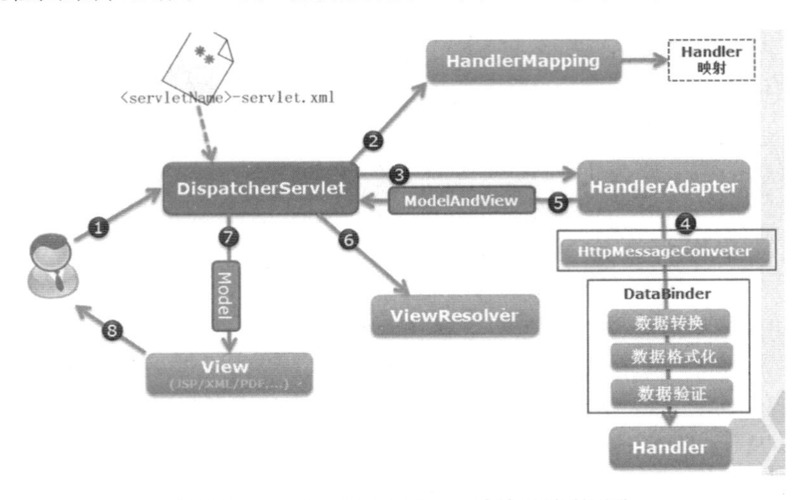

图 2.5　Spring MVC 请求→响应的完整流程

按照图 2.5 中所标识的序号，Spring MVC 请求→响应的完整工作流程如下：

① 用户向服务器发送请求，请求被 Spring 的前端控制器 DispatcherServlet 截获。

② DispatcherServlet 对请求 URL（统一资源定位符）进行解析，得到 URI（请求资源标识符）。然后根据该 URI，调用 HandlerMapping 获得该 Handler 配置的所有相关的对象，包括 Handler 对象以及 Handler 对象对应的拦截器，这些对象会被封装到一个 HandlerExecutionChain 对象当中返回。

③ DispatcherServlet 根据获得的 Handler，选择一个合适的 HandlerAdapter。HandlerAdapter 的设计符合面向对象中的单一职责原则，代码架构清晰，便于维护，最重要的是，代码可复用性高。HandlerAdapter 会被用于处理多种 Handler，调用 Handler 实际处理请求的方法，例如 hello 方法。

④ 提取请求中的模型数据，开始执行 Handler（Controller）。在填充 Handler 的入参过程

中，根据配置，Spring 将帮你做一些额外的工作。

- **消息转换**。将请求消息（如 JSON、XML 等数据）转换成一个对象，将对象转换为指定的响应信息。
- **数据转换**。对请求消息进行数据转换，如 String 转换成 Integer、Double 等。
- **数据格式化**。对请求消息进行数据格式化，如将字符串转换成格式化数字或格式化日期等。
- **数据验证**。验证数据的有效性（长度、格式等），验证结果存储到 BindingResult 或 Error 中。

⑤ Handler 执行完成后，向 DispatcherServlet 返回一个 ModelAndView 对象，ModelAndView 对象中应该包含视图名或视图名和模型。

⑥ 根据返回的 ModelAndView 对象，选择一个合适的 ViewResolver（视图解析器）返回给 DispatcherServlet。

⑦ ViewResolver 结合 Model 和 View 来渲染视图。

⑧ 将视图渲染结果返回给客户端。

以上 8 个步骤，DispatcherServlet、HandlerMapping、HandlerAdapter 和 ViewResolver 等对象协同工作，完成 Spring MVC 请求→响应的整个工作流程，这些对象所完成的工作对于开发者来说都是不可见的，开发者并不需要关心这些对象是如何工作的，开发者只需要在 Handler（Controller）当中完成对请求的业务处理。

提示

关于 DispatcherServlet、HandlerMapping、HandlerAdapter 和 ViewResolver 等对象协同工作的知识，对于还不了解 Spring MVC 的读者来说，可能会感觉很难理解，因为这一节的知识涉及 Spring MVC 的源代码和 Spring MVC 的设计思想，建议初学者学习完 Spring MVC 的知识之后再回过头来看本节的内容。

2.6 本章小结

本章介绍了 Spring MVC 的入门知识，包括如何使用 Spring MVC 开发一个简单的 Web 应用。在 Spring MVC 中，开发者无须编写自己的前端控制器，使用 Spring 提供的 DispatcherServlet 就可以分派请求。Spring MVC 传统风格的控制器开发方式是实现 Controller 接口，从 Spring 2.5 开始，提供了一个更好的控制器开发方式，即采用注解类型。最后，详细分析了 Spring MVC 请求→响应的完整工作流程。

第 3 章将重点介绍 Spring MVC 常用的注解。

CHAPTER

3

第3章 Spring MVC的常用注解

本章要点

- @Controller 注解
- @RequestMapping 注解
- @GetMapping 注解
- @PostMapping 注解
- @RequestParam 注解
- @PathVariable 注解
- @MatrixVariable 注解
- @CrossOrigin 注解
- @RequestHeader 注解
- @CookieValue 注解
- @RequestAttribute 注解
- @SessionAttribute 注解
- @SessionAttributes 注解
- @ModelAttribute 注解
- @RequestBody 注解
- @ResponseBody 注解
- @RestController 注解
- @ResponseStatus 注解
- @ExceptionHandle 注解
- @ ControllerAdvice 注解
- @ RestControllerAdvice 注解

Spring 从 2.5 版开始引入注解，用户可以在 Spring MVC 中使用@Controller、@RequestMapping、@RequestParam、@ModelAttribute 等类似的注解。到目前为止，Spring 的版本虽然发生了很大的变化，但注解的特性却一直延续下来，并不断扩展，让广大开发者的工作变得更轻松。这都离不开注解的强大作用，本章将重点讲解 Spring MVC 中常用的注解。

3.1 @Controller 注解

org.springframework.stereotype.Controller 注解用于指示 Spring 类的实例是一个控制器，使用@Controller 注解的类不需要继承特定的父类或者实现特定的接口，相对之前的版本实现 Controller 接口变得更加简单。而且 Controller 接口的实现类只能处理一个单一请求动作，而@Controller 注解的控制器可以支持同时处理多个请求动作，更加灵活。

@Controller 用于标记一个类，使用它标记的类就是一个 Spring MVC Controller 对象，即一个控制器类。Spring 使用扫描机制查找应用程序中所有基于注解的控制器类。分发处理器会扫描使用了该注解的类的方法，并检测该方法是否使用了@RequestMapping 注解，而使用@RequestMapping 注解的方法才是真正处理请求的处理器。为了保证 Spring 能找到控制器，需要完成两件事情：

- 在 Spring MVC 的配置文件的头文件中引入 spring-context。
- 使用<context:component-scan/>元素，该元素的功能为：启动包扫描功能，以便注册带有@Controller、@Service、@repository、@Component 等注解的类成为 Spring 的 Bean。base-package 属性指定了需要扫描的类包，类包及其递归子包中所有的类都会被处理。

配置文件如下所示：

```
<context:component-scan base-package="org.fkit.controller"/>
```

应该将所有控制器类都放在基本包下，并且指定扫描该包，即 org.fkit.controller，而不应该指定扫描 org.fkit 包，以免 Spring MVC 扫描了无关的包。

示例：@Controller 注解的使用

新建一个项目 ControllerTest，加入所需的 jar 文件，示例代码如下：

程序清单：codes/03/ControllerTest/src/org/fkit/controller/HelloWorldController

```
package org.fkit.controller;
import org.springframework.stereotype.Controller;
import org.springframework.ui.Model;
import org.springframework.web.bind.annotation.RequestMapping;
@Controller
public class HelloWorldController{
    @RequestMapping("/helloWorld")
    public String helloWorld(Model model) {
        model.addAttribute("message", "Hello World!");
        return "helloWorld";
    }
}
```

HelloWorldController 是一个基于@Controller 注解的控制器，@RequestMapping 注解用来映射一个请求，value="/helloWorld"表示请求由 helloWorld 方法进行处理。helloWorld 方法接收一个 org.springframework.ui.Model 类型的参数，本例在 model 中添加了一个名为“message”的字符串对象，该对象可以在返回的视图当中通过 request 对象获取。最后，方法返回一个字符串“helloWorld”作为视图名称。

程序清单：codes/03/ControllerTest/WebContent/WEB-INF/springmvc-config.xml

```
<?xml version="1.0" encoding="UTF-8"?>
<beans xmlns="http://www.springframework.org/schema/beans"
    xmlns:xsi="http://www.w3.org/2001/XMLSchema-instance"
    xmlns:p="http://www.springframework.org/schema/p"
    xmlns:mvc="http://www.springframework.org/schema/mvc"
    xmlns:context="http://www.springframework.org/schema/context"
    xsi:schemaLocation="
        http://www.springframework.org/schema/beans
        http://www.springframework.org/schema/beans/spring-beans.xsd
        http://www.springframework.org/schema/mvc
        http://www.springframework.org/schema/mvc/spring-mvc.xsd
        http://www.springframework.org/schema/context
        http://www.springframework.org/schema/context/spring-context.xsd">

    <!-- spring 可以自动去扫描 base-pack 下面的包或者子包下面的 java 文件，
        如果扫描到有 Spring 的相关注解的类，则把这些类注册为 Spring 的 bean -->
    <context:component-scan base-package="org.fkit.controller"/>
     <!-- 默认装配方案 -->
     <mvc:annotation-driven/>
       <!-- 静态资源处理 -->
     <mvc:default-servlet-handler/>

    <!-- 视图解析器  p:prefix 属性表示前缀  p:suffix 表示后缀  -->
     <bean id="viewResolver"
       class="org.springframework.web.servlet.view.InternalResourceViewResolver"
         p:prefix="/WEB-INF/content/" p:suffix=".jsp"/>
</beans>
```

springmvc-config.xml 文件配置信息解释如下：

（1）由于使用了@Controller 注解，因此不需要再在配置文件中使用 XML 描述 Bean。<context:component-scan base-package="org.fkit.controller"/>指定需要 Spring 扫描 org.fkit.controller 包及其子包下面的所有 java 文件。

（2）<mvc:annotation-driven> 是一种简写形式，可以让初学者快速应用默认配置方案。<mvc:annotation-driven> 会自动注册 RequestMappingHandlerMapping 与 Request MappingHandlerAdapter 两个 Bean，这是 Spring MVC 为@Controllers 分发请求所必需的，并提供了数据绑定支持、@NumberFormatannotation 支持、@DateTimeFormat 支持、@Valid 支持、读写 XML 的支持（JAXB）和读写 JSON 的支持（默认为 Jackson）等功能。本例处理 Ajax 请求时，就使用到了对 JSON 的支持功能。

（3）<mvc:default-servlet-handler/>是 Spring MVC 的静态资源处理器，在 web.xml 中，如果将 DispatcherServlet 请求映射配置为“/”，则 Spring MVC 将捕获 Web 容器所有的请求，包括静态资源的请求，而引入类似<script type="text/javascript"src= "js/jquery-1.11.0.min.js"/>这种静态资源文件的时候，DispatcherServlet 会将“/”看成请求路径，找不到它的时候会导致提示 404 错误。当在 springmvc-config.xml 中配置<mvc:default-servlet-handler />后，会在 Spring MVC 上下文中定义一个 org.springframework.web.servlet.resource.DefaultServletHttpRequestHandler，它就像一个检查员，对进入 DispatcherServlet 的 URL 进行筛查，如果发现是静态资源的请求，就将该请求转由 Web 应用服务器默认的 Servlet 处理；如果不是静态资源的请求，才由 DispatcherServlet 继续处理。

（4）最后配置了视图解析器 InternalResourceViewResolver 来解析视图，将 View 呈现给用户。视图解析器中配置的 prefix 属性表示视图的前缀，suffix 表示视图的后缀，返回的视图字符串是“helloWorld”，经过视图解析器解析之后，视图的完整路径为：/WEB-INF/

content/helloWorld.jsp。需要注意的是，此处没有配置处理器映射器和处理器适配器，当用户没有配置这两项时，Spring会使用默认的处理器映射器和处理器适配器处理请求。

此外，还需要在web.xml文件中配置Spring MVC的前端控制器DispatcherServlet，因为每次配置基本一致，此处不再赘述，读者可自行配置。

部署ControllerTest这个Web应用，在浏览器中输入如下URL来测试应用：

```
http://localhost:8080/ControllerTest/helloWorld
```

之后，会看到如图3.1所示的界面，表示Spring MVC访问成功。

图3.1　测试@Controller注解

Spring MVC中用于参数绑定的注解有很多，都在org.springframework.web.bind.annotation包中，根据它们处理的request的不同内容部分可以分为六类。

- **处理请求参数和内容部分的注解：**@RequestMapping、@RequestParam、@GetMapping、@PostMapping、@PutMapping、@DeleteMapping、@PatchMapping、@RequestBody、@ResponseBody、@RequestPart、@RestController。
- **处理请求URL部分的注解：**@PathVariable 、@MatrixVariable、@CrossOrigin。
- **处理请求头部分的注解：**@RequestHeader、@CookieValue。
- **处理属性类型的注解：**@RequestAttribute、@SessionAttribute、@SessionAttributes、@ModelAttribute。
- **处理异常类型的注解：**@ExceptionHandler、@ControllerAdvice、@RestControllerAdvice、@ResponseStatus。
- **绑定表单数据的注解：**@InitBinder。

接下来重点讲解常用的Spring MVC注解。@InitBinder注解用于解决类型转换问题，在第6章中讲解；@RequestPart注解用于绑定“multipart/form-data”参数，常用于文件上传，在第7章中讲解。

3.2　@RequestMapping注解

3.2.1　@RequestMapping注解简介

开发者需要在控制器内部为每一个请求动作开发相应的处理方法。org.springframework.web.bind.annotation.RequestMapping注解指示Spring用哪一个类或方法来处理请求动作，该注解可用于类或方法。

提示

@RequestMapping注解虽然也包含在org.springframework.web.bind.annotation里面，但是严格来说，它并不属于参数绑定注解。

@RequestMapping注解可以用来注释一个控制器类，在这种情况下，所有方法都将映射为相对于类级别的请求，表示该控制器处理的所有请求都被映射到value属性所指示的路径下。

示例代码如下：

```
@Controller
@RequestMapping(value="/user")
public class UserController {
    @RequestMapping(value="/register")
    public String register(){
        return "register";
    }
    @RequestMapping(value="/login")
    public String login(){
        return "login";
    }
}
```

由于 UserController 类中添加了 value="/user"的@RequestMapping 注解，因此所有相关路径都要加上"/user"，此时方法被映射到如下请求 URL：

```
http://localhost:8080/user/register
http://localhost:8080/user/login
```

使用@RequestMapping 注解可指定如表 3.1 所示的属性。

表 3.1 @RequestMapping 注解支持的属性

属性	类型	是否必要	说明
value	String[]	否	用于将指定请求的实际地址映射到方法上
name	String	否	给映射地址指定一个别名
method	RequestMethod[]	否	映射指定请求的方法类型，包括 GET、POST、HEAD、OPTIONS、PUT、PATCH、DELETE、TRACE
consumes	String[]	否	指定处理请求的提交内容类型（Content-Type），例如 application/json、text/html 等
produces	String[]	否	指定返回的内容类型，返回的内容类型必须是 request 请求头（Accept）中所包含的类型
params	String[]	否	指定 request 中必须包含某些参数值时，才让该方法处理
headers	String[]	否	指定 request 中必须包含某些指定的 header 值时，才能让该方法处理请求
Path	String[]	否	在 Servlet 环境中只有:uri 路径映射（例如“/ myPath.do”）。也支持如 ant 的基于路径模式（例如“/ myPath / *,”）。在方法层面上，支持相对路径（例如“edit.do”）

@RequestMapping 注解支持的常用属性示例如下。

1. value 属性

@RequestMapping 是一个用来处理请求地址映射的注解，可以使用@RequestMapping 注释一个方法或类。

一个采用@RequestMapping 注释的方法将成为一个请求处理方法，例如：

```
@RequestMapping(value="/hello")
public ModelAndView hello(){
    return …;
}
```

该示例使用@RequestMapping 注释的 value 属性将 URL 映射到方法上。在这个例子中，将 hello 映射到 hello 方法上，使用如下 URL 访问应用时将由 hello 方法进行处理。

```
http://localhost:8080/context/hello
```

由于 value 属性是@RequestMapping 注释的默认属性，因此，如果只有唯一的属性，则可以省略属性名，即如下两个注解含义相同。

```
@RequestMapping(value="/hello")
@RequestMapping("/hello")
```

但如果有超过一个属性，就必须写上 value 属性名称。

value 属性的值也可以是一个空字符串，此时该方法被映射到如下请求 URL：

```
http://localhost:8080/context
```

2. method 属性

该属性用来指示该方法仅处理哪些 HTTP 请求方式。

```
@RequestMapping(value="/hello",method=RequestMethod.POST)
```

以上代码 method=RequestMethod.POST 表示该方法只支持 POST 请求。

也可以同时支持多个 HTTP 请求方式。如：

```
@RequestMapping(value="/hello",
          method={RequestMethod.POST,RequestMethod.GET})
```

如果没有指定 method 属性值，则请求处理方法可以处理任意的 HTTP 请求方式。

提示

Spring 4.3 之后，新增了@GetMapping、@PostMapping、@PutMapping、@DeleteMapping、@PatchMapping 注解，这几个注解可以指定的属性和@RequestMapping 注解类似，区别在于@GetMapping 注解只支持 GET 方式的请求；@PostMapping 注解只支持 POST 方式的请求；@PutMapping、@DeleteMapping、@PatchMapping 分别对应 PUT 请求、DELETE 请求和 PATCH 请求，使用比较少。

3. consumes 属性

该属性指定处理请求的提交内容类型（Content-Type）。

```
    @RequestMapping(value="/hello",method=RequestMethod.POST, consumes=
"application/json")
```

表示方法仅处理 request Content-Type 为“application/json”类型的请求。

4. produces 属性

该属性指定返回的内容类型，返回的内容类型必须是 request 请求头（Accept）中所包含的类型。

```
    @RequestMapping(value="/hello",method=RequestMethod.POST, produces=
"application/json")
```

方法仅处理 request 请求中 Accept 头中包含了"application/json"的请求，同时指明了返回的内容类型为 application/json。

5. params 属性

该属性指定 request 中必须包含某些参数值时，才让该方法处理。

```
    @RequestMapping(value="/hello",method=RequestMethod.POST,
params="myParam=myValue")
```

方法仅处理其中名为“myParam”、值为“myValue”的请求。

6. headers 属性

该属性指定 request 中必须包含某些指定的 header 值时，才能让该方法处理请求。

```
    @RequestMapping(value="/hello",method=RequestMethod.POST,
headers="Referer=http://www.fkit.org/")
```

方法仅处理 header 中包含了指定“Referer”请求头和对应值为“http://www.fkit.org/”的请求。

3.2.2 请求处理方法中可出现的参数类型

每个请求处理方法可以有多个不同类型的参数。

如果需要访问 HttpServletRequest 对象，则可以添加 HttpServletRequest 作为参数，Spring 会将对象正确地传递给方法：

```
@RequestMapping(value="/login")
   public String login(HttpServletRequest request){
      return "login";
   }
```

如果需要访问 HttpSession 对象，则可以添加 HttpSession 作为参数，Spring 会将对象正确地传递给方法：

```
@RequestMapping(value="/login")
   public String login(HttpSession session){
      return "login";
   }
```

下面是可以在请求处理方法中出现的参数类型：

- javax.servlet.ServletRequest 或 javax.servlet.http. HttpServletRequest
- javax.servlet.ServletResponse 或 javax.servlet.http. HttpServletResponse
- javax.servlet.http.HttpSession
- org.springframework.web.context.request.WebRequest 或 org.springframework.web.context.request.NativeWebRequest
- java.util.Locale
- java.io.InputStream 或 java.io.Reader
- java.io.OutputStream 或 java.io.Writer
- java.security.Principal
- HttpEntity<?>
- java.util.Map
- org.springframework.ui.Model
- org.springframework.ui.ModelMap
- org.springframework.web.servlet.mvc.support.RedirectAttributes
- org.springframework.validation.Errors
- org.springframework.validation.BindingResult
- org.springframework.web.bind.support.SessionStatus
- org.springframework.web.util.UriComponentsBuilder
- @PathVariable、@MatrixVariable 注解
- @RequestParam、@RequestHeader、@RequestBody、@RequestPart 注解

其中最重要的是 org.springframework.ui.Model 类型。这不是一个 Servlet API 类型，而是一个 Spring MVC 类型，其中包含了 Map 对象用来存储数据。如果方法中添加了 Model 参数，则每次调用请求处理方法时，Spring MVC 都会创建 Model 对象，并将其作为参数传递给方法。

还要重点说明的一个对象是 org.springframework.web.context.request.WebRequest。WebRequest 是 Spring MVC 提供的统一请求访问接口，其不仅可以访问请求相关数据，如参数

数据、请求头数据，但访问不到 Cookie 区数据，而且还可以访问请求作用域和会话作用域中的数据；NativeWebRequest 继承了 WebRequest，并提供访问本地 Servlet API 的方法。

WebRequest 中有如下这些重点方法：

- getParameter(String name)。根据请求参数名获取请求参数值，等同于 Servlet API 的 request.getParameter(String name)。
- getHeader(String name)。根据请求头参数名获取请求头参数值，等同于 Servlet API 的 request. getHeader (String name)。
- setAttribute(String name, Object value,int scope)。保存数据到指定的作用域，等同于 Servlet API 的 scope(request, session).setAttribute(String name, Object value)方法。参数中的 int scope 是 WebRequest 对象中定义的两个常量 SCOPE_REQUEST（请求作用域）、SCOPE_SESSION（会话作用域）。

```
    // 保存名称为 username 的值到 request 作用域
    webRequest.setAttribute("username", "疯狂软件", WebRequest.SCOPE_REQUEST);
    // 保存名称为 bookname 的值到 request 作用域
    webRequest.setAttribute("bookname", "Spring+MyBatis 企业应用实战",
WebRequest.SCOPE_SESSION);
```

- getAttribute(String name,int scope)。从指定的作用域中提取数据，等同于 Servlet API 的 scope(request,session).getAttribute(String name)方法。

```
    // 从 request 作用范围域中取出名称为 username 的值
    webRequest.getAttribute("username", WebRequest.SCOPE_REQUEST);
    // 从 session 作用范围域中取出名称为 bookname 的值
    webRequest.getAttribute("bookname", WebRequest.SCOPE_SESSION);
```

提示

从框架的无侵入性来说，建议使用 WebRequest 对象替代 Servlet API 的 HttpServletRequest 对象和 HttpSession 对象。

3.2.3 请求处理方法可返回的类型

每个请求处理方法可以返回以下类型的返回结果：

- org.springframework.web.portlet.ModelAndView
- org.springframework.ui.Model
- java.util.Map<k,v>
- org.springframework.web.servlet.View
- java.lang.String
- HttpEntity 或 ResponseEntity
- java.util.concurrent.Callable
- org.springframework.web.context.request.async.DeferredResult
- void

3.2.4 Model 和 ModelMap

在请求处理方法中可出现和返回的参数类型中，最重要的就是 Model 和 ModelAndView 了。对于 MVC 框架，控制器（Controller）执行业务逻辑，用于产生模型数据（Model），而视图（View）则用于渲染模型数据。

如何将模型数据传递给视图是 Spring MVC 框架的一项重要工作，Spring MVC 提供了多种途径输出模型数据，如：

- Model 和 ModelMap
- ModelAndView
- @ModelAttribute
- @SessionAttributes

下面将重点介绍 Model、ModelMap 以及 ModelAndView，@ SessionAttributes 将在 3.11 节中重点介绍；@ModelAttribute 将在 3.12 节中重点介绍。

Spring MVC 在内部使用了一个 org.springframework.ui.Model 接口存储模型数据，它的功能类似 java.util.Map 接口，但是比 Map 易于使用。org.springframework.ui.ModelMap 接口实现了 Map 接口。

Spring MVC 在调用处理方法之前会创建一个隐含的模型对象，作为模型数据的存储容器。如果处理方法的参数为 Model 或 ModelMap 类型，则 Spring MVC 会将隐含模型的引用传递给这些参数。在处理方法内部，开发者可以通过这个参数对象访问模型中的所有数据，也可以向模型中添加新的属性数据。

在处理方法中，Model 和 ModelMap 对象都可以使用如下方法添加模型数据：

- addAttribute (String attributeName,Object attributeValue)

示例：Model 和 ModelMap 的使用

程序清单：codes/03/ModelTest/src/org/fkit/controller/ModelController

```
import org.fkit.domain.User;
import org.springframework.stereotype.Controller;
import org.springframework.ui.Model;
import org.springframework.ui.ModelMap;
import org.springframework.web.bind.annotation.RequestMapping;

@Controller
public class ModelController{

    @RequestMapping(value="/modelTest")
     public String modelTest(Model model){
        System.out.println("modelTest");
        User user = new User();
        // 设置 user 对象的 username 属性
        user.setUsername("疯狂软件");
        // 将 User 对象添加到 Model 当中
        model.addAttribute("user", user);
        return "result1";
    }

    @RequestMapping(value="/ModelMapTest")
     public String ModelMapTest(ModelMap modelMap){
        System.out.println("ModelMapTest");
        User user = new User();
        // 设置 user 对象的 username 属性
        user.setUsername("疯狂软件");
        // 将 User 对象添加到 ModelMap 当中
        modelMap.addAttribute("user", user);
        return "result2";
    }


}
```

在 modelTest 方法中创建了一个自定义的 User 对象，并且给 username 属性赋值。使用 Model 对象的 addAttribute("user", user)方法将 User 对象添加到 Model 当中，即 JSP 的 requestScope 当中。

ModelMapTest 方法的代码功能与此类似，只是存储对象由 Model 改成了 ModelMap。

3.2.5 ModelAndView

控制器处理方法的返回值如果是 ModelAndView，则其既包含模型数据信息，也包含视图信息，这样 Spring MVC 将使用包含的视图对模型数据进行渲染。可以简单地将模型数据看成一个 Map<String,Object>对象。

在处理方法中可以使用 ModelAndView 对象的如下方法添加模型数据：

```
addObject(String attributeName,Object attributeValue);
```

可以通过如下方法设置视图：

```
setViewName(String viewName)
```

示例：ModelAndView 的使用

程序清单：codes/03/ModelAndViewTest/src/org/fkit/controller/ModelAndViewController

```
import org.fkit.domain.User;
import org.springframework.stereotype.Controller;
import org.springframework.ui.Model;
import org.springframework.ui.ModelMap;
import org.springframework.web.bind.annotation.RequestMapping;
import org.springframework.web.servlet.ModelAndView;
import org.springframework.web.servlet.View;
import org.springframework.web.servlet.view.RedirectView;

@Controller
public class ModelAndViewController{

    @RequestMapping(value="/ModelAndViewTest")
     public ModelAndView ModelAndViewTest(ModelAndView mv){
        System.out.println("ModelAndViewTest");
        User user = new User();
        // 设置 user 对象的 username 属性
        user.setUsername("疯狂软件");
        // 将 User 对象添加到 ModelAndView 当中
        mv.addObject("user", user);
        // 设置要转发的页面
        mv.setViewName("result");
        return mv;
    }

}
```

在 ModelAndViewTest 方法中创建了一个自定义的 User 对象，并且给 username 属性赋值。使用 ModelAndView 对象的 addObject ("user", user)方法将 User 对象添加到 ModelAndView 当中，即 JSP 的 requestScope 当中。同时调用 setViewName("result")方法设置要转发的页面。

此处需要注意的是，方法的返回值必须是 ModelAndView，方法的返回结果必须是 ModelAndView 对象，否则保存在 ModelAndView 对象中的“user”在 result.jsp 页面中获取不到。

3.2.6 页面转发

在 JSP 中页面转发有两种情况：服务器内部跳转（forward）和客户端重定向（redirect）。

Spring MVC 选择页面转发的方式也非常简单。

1. 转发到 JSP 页面。

若请求处理方法返回字符串，默认使用服务器内部跳转（forward）。示例代码如下：

```
return "main";
```

如果希望客户端重定向（redirect），在返回的字符串前面加上"redirect:/"。示例代码如下：

```
return "redirect:/main.jsp";
```

若请求处理方法返回 ModelAndView，使用 setViewName 方法设置转发的页面，默认使用服务器内部跳转（forward）。示例代码如下：

```
mv.setViewName("main");
```

若请求处理方法返回 ModelAndView，使用 setViewName 方法设置转发的页面，例如"mv.setViewName("redirect:/main.jsp");"，则客户端重定向到名为main.jsp的页面。示例代码如下：

```
mv.setViewName("redirect:/main.jsp");
```

需要注意的是，由于客户端重定向（redirect）相当于在浏览器重新发请求，所以不能访问 WEB-INF 下的资源文件，而且也必须写资源文件后缀名，即.jsp，因为此时 springmvc-config.xml 文件的视图解析器设置的前缀和后缀都将无效。

2. 转发到控制器的请求处理方法

若请求处理方法返回字符串，在返回的字符串前面加上"forward:/"，例如"forward:/main"，则服务器内部跳转到名为 main 的请求处理方法。示例代码如下：

```
return "forward:/main";
```

若请求处理方法返回字符串，在返回的字符串前面加上"redirect:/"，例如" redirect:/main"，则客户端重定向到名为 main 的请求处理方法。示例代码如下：

```
return " redirect:/main";
```

若请求处理方法返回 ModelAndView，调用 setViewName 方法设置需要转发的控制器，例如"mv.setViewName("forward:/main");"，则服务器内部跳转到名为 main 的请求处理方法。示例代码如下：

```
mv.setViewName("forward:/main");
```

若请求处理方法返回 ModelAndView，调用 setViewName 方法设置需要转发的控制器，例如"mv.setViewName("redirect:/main");"，则客户端重定向到名为 main 的请求处理方法。示例代码如下：

```
mv.setViewName("redirect:/main");
```

3.3 @RequestParam 注解

org.springframework.web.bind.annotation.RequestParam 注解用于将指定的请求参数赋值给方法中的形参。

使用@RequestParam 注解可指定如表 3.2 所示的属性。

表 3.2　@RequestParam 注解支持的属性

属性	类型	是否必要	说明
name	String	否	指定请求参数绑定的名称

续表

属性	类型	是否必要	说明
value	String	否	name 属性的别名
required	boolean	否	指示参数是否必须绑定
defaultValue	String	否	如果没有传递参数而使用的默认值

请求处理方法参数的可选类型为 Java 基本数据类型和 String。示例代码如下：

```
@RequestMapping(value="/login")
    public ModelAndView login(
            @RequestParam("loginname") String loginname,
            @RequestParam("password") String password){
        return ……;
    }
```

假设请求如下：

```
http://localhost:8080/context/login?loginname=jack&password=123456
```

以上代码会将请求中的 loginname 参数的值“jack”赋给 loginname 变量，password 参数的值“123456”赋给 password 变量。

需要注意的是，如果请求中不包含“loginname”参数，则将产生异常！ 因此，如果不能保证存在“loginname”参数，建议使用：

```
@RequestParam(value = "loginname", required = false)
```

@RequestParam 注解还有如下写法：

```
@RequestParam(value="loginname", required=true, defaultValue="admin")
```

其中 required 参数不是必需的，默认值为 true。

示例：@RequestMapping 和@RequestParam 注解的使用

新建一个项目 RequestMappingTest，加入所需的 jar 文件，示例代码如下：

程序清单：codes/03/RequestMappingTest/src/org/fkit/domain/User

```
// 域对象，实现序列化接口
public class User implements Serializable{
    // 私有字段
    private String loginname;
    private String password;
    private String username;
    // 公共构造器
    public User() {
        super();
    }
    // set/get 方法
    public String getLoginname() {
        return loginname;
    }
    public void setLoginname(String loginname) {
        this.loginname = loginname;
    }
    public String getPassword() {
        return password;
    }
    public void setPassword(String password) {
        this.password = password;
    }
    public String getUsername() {
        return username;
```

```
    }
    public void setUsername(String username) {
        this.username = username;
    }
}
```

User 是一个域对象，用来接收并封装从前台页面传递过来的数据。

程序清单：codes/03/RequestMappingTest/src/org/fkit/controller/UserController

```
import java.util.ArrayList;
import java.util.List;
import org.fkit.domain.User;
import org.springframework.stereotype.Controller;
import org.springframework.ui.Model;
import org.springframework.web.bind.annotation.GetMapping;
import org.springframework.web.bind.annotation.PostMapping;
import org.springframework.web.bind.annotation.RequestMapping;
import org.springframework.web.bind.annotation.RequestParam;
// Controller 注解用于指示该类是一个控制器，可以同时处理多个请求动作
@Controller
// RequestMapping 可以用来注释一个控制器类，此时，所有方法都将映射为相对于类级别的请求
// 表示该控制器处理的所有请求都被映射到 value 属性所指示的路径下
@RequestMapping(value="/user")
public class UserController{
    // 静态 List<User>集合，其代替数据库用来保存注册的用户信息
    private static List<User> userList;
    // UserController 类的构造器，初始化 List<User>集合
    public UserController() {
        super();
        userList = new ArrayList<User>();
    }

    // 该方法映射的请求为 http://localhost:8080/RequestMappingTest /user/register，该
// 方法支持 GET 请求
    @GetMapping(value="/register")
     public String registerForm() {
         logger.info("register GET 方法被调用...");
         // 跳转到注册页面
         return "registerForm";
     }
    // 该方法映射的请求为 http://localhost:8080/ RequestMappingTest /user/register，
// 该方法支持 POST 请求
     @PostMapping(value="/register")
     // 将请求中的 loginname 参数的值赋给 loginname 变量,password 和 username 同样处理
     public String register(
             @RequestParam("loginname") String loginname,
             @RequestParam("password") String password,
             @RequestParam("username") String username) {
         System.out.println("register POST 方法被调用...");
         // 创建 User 对象
         User user = new User();
         user.setLoginname(loginname);
         user.setPassword(password);
         user.setUsername(username);
         // 模拟数据库存储 User 信息
         userList.add(user);
         // 跳转到登录页面
         return "loginForm";
     }
    // 该方法映射的请求为 http://localhost:8080/RequestMappingTest /user/login
     @RequestMapping("/login")
     public String login(
```

```
            // 将请求中的 loginname 参数的值赋给 loginname 变量,password 同样处理
            @RequestParam("loginname") String loginname,
            @RequestParam("password") String password,
            Model model) {
        System.out.println("登录名:"+loginname + " 密码:" + password);
        // 到集合中查找用户是否存在，此处用来模拟数据库验证
        for(User user : userList){
            if(user.getLoginname().equals(loginname)
                    && user.getPassword().equals(password)){
                model.addAttribute("user",user);
                return "welcome";
            }
        }
        return "loginForm";
    }
}
```

UserController 类的代码解释如下：

（1）UserController 类使用了@Controller 注解，是一个控制器类。

（2）在 UserController 类上面使用了@RequestMapping(value="/user")注解，表示该控制器处理的所有请求都被映射到 user 路径下。

（3）本例没有使用数据库存储用户注册信息，所以定义了一个静态的 List 集合 userList 用来代替数据库存储用户数据。

（4）registerForm 方法使用了@GetMapping(value="/register")注解，表示该方法映射的请求为 http://localhost:8080/RequestMappingTest /user/register，并且只支持 GET 请求。该方法返回字符串“registerForm”，参考 springmvc-config.xml 中的配置信息，可以知道该方法只是跳转到 registerForm.jsp 注册页面。

（5）register 方法使用了@PostMapping(value="/register")注解，表示该方法映射的请求为 http://localhost:8080/RequestMappingTest/user/register，并且只支持 POST 请求。该方法使用@RequestParam 注解将指定的请求参数赋值给方法中的形参，之后创建了一个 User 对象保存用户传递的注册信息，最后将 User 对象存储到 userList 集合中，然后登录页面就可以到 userList 集合中进行用户登录业务逻辑的判断。该方法返回字符串“loginForm”，并跳转到 loginForm.jsp 登录页面。

提示

registerForm 方法和 register 方法虽然映射的请求一样，但是 registerForm 方法支持的是 GET 请求，而 register 方法支持的是 POST 请求。

（6）login 方法使用了@RequestMapping("/login")注解，表示该方法映射的请求为 http://localhost:8080/RequestMappingTest/user/login，这里没有设置 method 属性，表示支持所有方式的请求。该方法也使用@RequestParam 注解将指定的请求参数赋值给方法中的形参。之后到集合中查找用户是否存在，此处用来模拟数据库验证。login 方法中还有一个参数 Model 对象，调用该对象的 addAttribute 方法可以将数据添加到 request 当中。最后，如果用户登录成功则返回字符串“welcome”，并跳转到 welcome.jsp 欢迎页面；登录失败则返回字符串“loginForm”，并跳转到 loginForm.jsp 登录页面。

程序清单：codes/03/RequestMappingTest/WebContent/WEB-INF/content/registerForm.jsp

```
<h3>注册页面</h3>
<br>
<form action="register" method="post">
```

```
        <table>
            <tr>
                <td><label>登录名：</label></td>
                <td><input type="text" id="loginname" name="loginname" ></td>
            </tr>
            <tr>
                <td><label>密码：</label></td>
                <td><input type="password" id="password" name="password"></td>
            </tr>
            <tr>
                <td><label>真实姓名：</label></td>
                <td><input type="text" id="username" name="username" ></td>
            </tr>
            <tr>
                <td><input id="submit" type="submit" value="注册"></td>
            </tr>
        </table>
    </form>
```

registerForm.jsp 是一个注册页面，用户可以在这里输入登录名、密码和真实姓名，该表单被提交到 register 请求。注意，这里使用的是 POST 方式，响应请求的是 UserController 类的 register 方法。

程序清单：codes/03/RequestMappingTest/WebContent/WEB-INF/content/loginForm.jsp

```
<h3>登录页面</h3>
<br>
<form action="login" method="post">
    <table>
        <tr>
            <td><label>登录名：</label></td>
            <td><input type="text" id="loginname" name="loginname" ></td>
        </tr>
        <tr>
            <td><label>密码：</label></td>
            <td><input type="password" id="password" name="password"></td>
        </tr>
        <tr>
            <td><input id="submit" type="submit" value="登录"></td>
        </tr>
    </table>
</form>
```

loginForm.jsp 是一个登录页面，用户可以在这里输入登录名和密码，该表单被提交到 login 请求。这里使用的是 POST 方式，响应请求的是 UserController 类的 login 方法。

程序清单：codes/03/RequestMappingTest/WebContent/WEB-INF/content/welcome.jsp

```
<!-- 页面可以访问 Controller 传递出来的模型 User 对象 -->
<h3>欢迎[${requestScope.user.username }]登录</h3>
<br>
```

welcome.jsp 是一个欢迎页面，用户登录成功后跳转到该页面，该页面使用了 EL 表达式访问 request 当中的 user 对象的 username 属性。

此外，还需要在 web.xml 文件中配置 Spring MVC 的前端控制器 DispatcherServlet，因为每次配置基本相同，此处不再赘述，读者可自行配置。

同时 Spring MVC 还需要 springmvc-config.xml 配置文件，该文件内容和 ControllerTest 项目中的 springmvc-config.xml 文件内容一致，读者可自行配置。

部署 RequestMappingTest 这个 Web 应用，在浏览器中输入如下 URL 来测试应用：

```
http://localhost:8080/RequestMappingTest/user/register
```

可看到如图 3.2 所示的页面，表示 Spring MVC 成功跳转到注册页面 registerForm.jsp。

图 3.2 注册页面

输入登录名“test”，密码“123456”，真实姓名“测试用户”，单击“注册”按钮。请求将会被提交到 UserController 类的 register 方法进行注册，注册的用户信息会被保存到 UserController 类的 userList 静态集合中。若注册成功，将会跳转到如图 3.3 所示的登录页面。

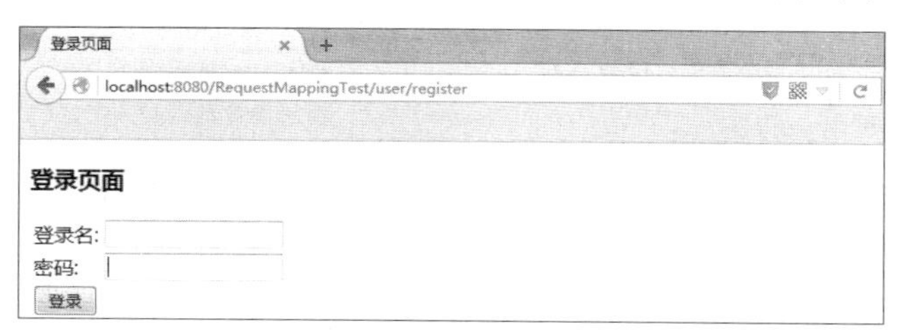

图 3.3 登录页面

输入登录名“test”，密码“123456”，单击“登录”按钮。请求将会被提交到 UserController 类的 login 方法进行登录验证，若验证成功，将会跳转到如图 3.4 所示的欢迎页面。

图 3.4 欢迎页面

3.4 @PathVariable 注解

org.springframework.web.bind.annotation.PathVariable 注解可以非常方便地获得请求 URL 中的动态参数。

使用@PathVariable 注解可指定如表 3.3 所示的属性。

表 3.3 @PathVariable 注解支持的属性

属性	类型	是否必要	说明
name	String	否	指定请求参数绑定的名称，如果省略则绑定同名参数
value	String	否	name 属性的别名
required	boolean	否	指示参数是否必须绑定

@PathVariable 注解示例代码如下：

```
@RequestMapping(value="/pathVariableTest/{userId}")
public void pathVariableTest(@PathVariable Integer userId)
```

假如请求的 URL 为“http://localhost:8080/VariableTest/pathVariableTest/1”，则自动将 URL

中模板变量{userId}绑定到通过@PathVariable注解的同名参数上，即userId变量将被赋值为1。

3.5　@MatrixVariable 注解

org.springframework.web.bind.annotation.MatrixVariable 注解拓展了 URL 请求地址的功能。使用@MatrixVariable 注解时多个变量可以使用“;”（分号）分隔，该注解允许开发者进行多条件组合查询。

使用@ MatrixVariable 注解可指定如表 3.4 所示的属性。

表 3.4　@MatrixVariable 注解支持的属性

属性	类型	是否必要	说明
name	String	否	指定请求参数绑定的名称，如果省略则绑定同名参数
value	String	否	name 属性的别名
pathVar	String	否	matrix variable 所在路径的 url path 变量的名称
required	boolean	否	指示参数是否必须绑定
defaultValue	String	否	如果没有传递参数而使用的默认值

@MatrixVariable 注解示例代码如下：

```
// 映射的请求为/VariableTest /matrixVariableTest/{userId};name=jack;age=23
@GetMapping(value="/matrixVariableTest/{userId}")
public void matrixVariableTest(
        @PathVariable Integer userId,
        @MatrixVariable(value="name",pathVar="userId") String name,
        @MatrixVariable(value="age",pathVar="userId") Integer age)
```

假如请求的 URL 为“http://localhost:8080/VariableTest /matrixVariableTest/1;name=jack;age=23”，则自动将 URL 中模板变量{userId}绑定到通过@PathVariable 注解的同名参数上，即 userId 变量将被赋值为 1；通过@MatrixVariable 注解绑定 name 变量和 age 变量，即 name 变量将被赋值为“jack”，age 变量将被赋值为“23”。

需要注意的是，@MatrixVariable 注解功能在 Spring MVC 中默认是不启用的，启用它需要设置 enable-matrix-variables="true"。示例代码如下：

```
<mvc:annotation-driven enable-matrix-variables="true"/>
```

3.6　@CrossOrigin 注解

org.springframework.web.bind.annotation.CrossOrigin 注解用于在 Spring MVC 中处理跨域请求。使用@CrossOrigin 注解可指定如表 3.5 所示的属性。

表 3.5　@CrossOrigin 注解支持的属性

属性	类型	是否必要	说明
allowCredentials	String	否	包含与请求的域相关的 Cookie，使用时必须指定具体的域
allowedHeaders	String[]	否	请求头中的请求列表
exposedHeaders	String[]	否	客户端允许访问的响应头列表
maxAge	long	否	响应前的缓存最大有效时间，单位是秒
methods	RequestMethod[]	否	请求支持的方法，默认支持 RequestMapping 中设置的方法
origins	String[]	否	所有支持域的集合，如果没有定义，默认支持所有域
value	String[]	否	同 origins 属性一样

@CrossOrigin 注解示例代码如下：

```
@CrossOrigin(maxAge=3600)
@Controller
public class CrossOriginController
```

表示 CrossOriginController 控制器的所有方法可以处理所有域上的请求：

```
@CrossOrigin(origins="http://www.fkit.org",maxAge=3600)
@Controller
public class CrossOriginController
```

表示 CrossOriginController 控制器的所有方法可以处理 http://www.fkit.org 域上的请求：

```
@CrossOrigin(maxAge=3600)
@Controller
public class CrossOriginController{
    @CrossOrigin(origins="http://www.fkit.org")
    @GetMapping(value="/login")
     public String login() {
        // ......
     }
}
```

CrossOriginController 类上有@CrossOrigin 注解，login 方法上也有@CrossOrigin 注解，Spring MVC 会合并两个注解的属性一起使用。

示例：@PathVariable 注解和@MatrixVariable 注解的使用

新建一个项目 VariableTest，加入所需的 jar 文件，示例代码如下：

程序清单：codes/03/VariableTest/WebContent/index.jsp

```
<%@ page language="java" contentType="text/html; charset=UTF-8"
    pageEncoding="UTF-8"%>
<!DOCTYPE html PUBLIC "-//W3C//DTD HTML 4.01 Transitional//EN" "http://www.
w3.org/TR/html4/loose.dtd">
<html>
<head>
<meta http-equiv="Content-Type" content="text/html; charset=UTF-8">
<title>处理请求 URL 注解测试</title>
</head>
<body>
<h2>处理请求 URL 注解测试</h2>
<a href="pathVariableTest/1">测试@PathVariable 注解</a><br><br>
<a href="matrixVariableTest/1;name=jack;age=23">测试@MatrixVariable 注解</a><br>
<br>
<a href="productTest/computer;brand=apple,acer;low=1000;height=10000">条件查询（品
牌，价格区间）</a><br><br>
<a href="http://localhost:8080/CrossOriginTest/welcome">测试@CrossOrigin 注解</a>
<br><br>
</body>
</html>
```

程序清单：codes/03/VariableTest/src/org/fkit/controller/VariableController

```
import java.util.List;
import org.apache.commons.logging.Log;
import org.apache.commons.logging.LogFactory;
import org.springframework.stereotype.Controller;
import org.springframework.web.bind.annotation.GetMapping;
import org.springframework.web.bind.annotation.MatrixVariable;
import org.springframework.web.bind.annotation.PathVariable;

@Controller
```

```
public class VariableController{

    // 测试@PathVariable 注解
    // 该方法映射的请求为/VariableTest/pathVariableTest/1
    @GetMapping(value="/pathVariableTest/{userId}")
     public void pathVariableTest(
            @PathVariable Integer userId) {
        System.out.println("通过@PathVariable 获得数据： userId=" + userId);
     }

    // 测试@MatrixVariable 注解
    // 该方法映射的请求为/VariableTest/matrixVariableTest/1;name=jack;age=23
    @GetMapping(value="/matrixVariableTest/{userId}")
    public void matrixVariableTest(
            @PathVariable Integer userId,
            @MatrixVariable(value="name",pathVar="userId") String name,
            @MatrixVariable(value="age",pathVar="userId") Integer age){
        System.out.println("通过@PathVariable 获得数据： userId=" + userId);
        System.out.println("通过@MatrixVariable 获得数据： name=" +name + " age="+ age);
    }

    // 测试@MatrixVariable 注解的复杂例子
    // 该方法映射的请求为/VariableTest/productTest /computer;brand=apple,acer;low=
1000;height=5000
    @GetMapping(value="/productTest/{goods}")
    public void productTest(
            @PathVariable String goods,
            @MatrixVariable(value="brand",pathVar="goods") List<String> brand,
            @MatrixVariable(value="low",pathVar="goods") Integer low,
            @MatrixVariable(value="height",pathVar="goods") Integer height){
        System.out.println("通过@PathVariable 获得数据： " + goods);
        System.out.println("通过@MatrixVariable 获得数据： " + brand);
        System.out.println("通过@MatrixVariable 获得数据： " + low + " "+ height);
    }
}
```

程序清单：codes/03/VariableTest/WebContent/WEB-INF/springmvc-config.xml

```
<mvc:annotation-driven enable-matrix-variables="true"/>
```

此外，还需要在 web.xml 文件中配置 Spring MVC 的前端控制器 DispatcherServlet，因为每次配置基本相同，此处不再赘述，读者可自行配置。

部署 VariableTest 这个 Web 应用，在浏览器中输入如下 URL 来测试应用：

```
http://localhost:8080/VariableTest/
```

会看到如图 3.5 所示的页面，表示 Spring MVC 成功跳转到 index.jsp。

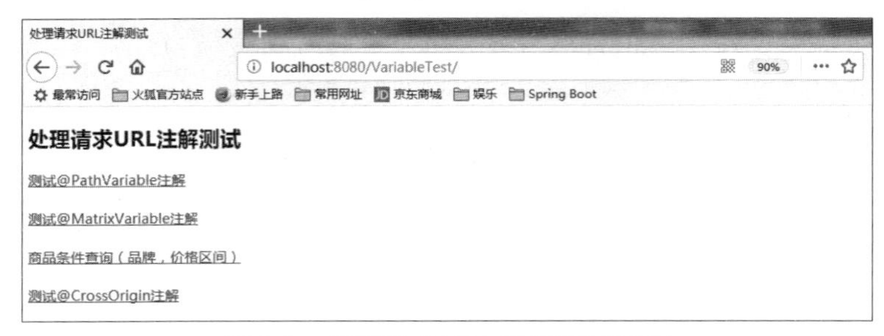

图 3.5　处理请求 URL 注解测试页面

测试：

（1）VariableController 类的 pathVariableTest 方法用于测试@PathVariable 注解，它会将请

求路径“/pathVariableTest/1”中 userId 的值“1”赋给方法参数的 userId 变量。

单击“测试@PathVariable 注解”超链接发送请求，将调用 pathVariableTest 方法，控制台输出结果如下：

```
通过@PathVariable 获得数据： userId=1
```

可以看到，<a href="pathVariableTest/1">测试@PathVariable 注解</a>的数据 1 被传递到方法的变量 userId 并输出在控制台。

（2）VariableController 类的 matrixVariableTest 方法用于测试@MatrixVariable 注解，它会将请求路径“/matrixVariableTest/1;name=jack;age=23”中 name 参数的值 “jack” 赋给方法参数的 name 变量，将 age 参数的值“23” 赋给方法参数的 age 变量。

单击“测试@MatrixVariable 注解”超链接发送请求，将调用 matrixVariableTest 方法，控制台输出结果如下：

```
通过@PathVariable 获得数据： userId=1
通过@MatrixVariable 获得数据： name=jack age=23
```

可以看到，<a href="matrixVariableTest/1;name=jack;age=23">测试@MatrixVariable 注解</a>的参数 name 的值“jack”被传递到方法的 name 变量，参数 age 的值“23”被传递到方法的 age 变量，并输出打印在控制台。

（3）@MatrixVariable 注解还可以完成复杂的参数注入，非常方便地进行多条件组合查询。本例以商品查询为例，详细介绍 matrixVariable 的使用。

VariableController 类的 productTest 方法用于商品条件查询，传递的参数包括商品、品牌和价格区间，它会将请求路径“/productTest/computer;brand=apple,acer;low=2000;height=10000”中 brand 参数的值“apple,acer”赋给方法参数的 brand 变量，该变量是一个 List 集合；将 low 参数的值“2000”赋给方法参数的 low 变量；将 height 参数的值“10000”赋给方法参数的 height 变量。该请求表示一个商品的条件组合查询，商品名称是 computer，查询的品牌是 apple 和 acer，价格区间是从 2000 到 10000。

单击“商品条件查询（品牌，价格区间）”超链接发送请求，将调用 productTest 方法，控制台输出结果如下：

```
通过@PathVariable 获得数据： goods=computer
通过@MatrixVariable 获得数据：brand=[apple, acer]
通过@MatrixVariable 获得数据： low=2000 height=10000
```

可以看到，<a href="productTest/computer;brand=apple,acer;low=2000;height=10000">商品条件查询（品牌，价格区间）</a>的参数 brand 的值“apple,acer”被传递到方法的 brand 集合变量，参数 low 的值“2000”被传递到方法的 low 变量，参数 height 的值“10000”被传递到方法的 height 变量，并输出打印在控制台。

示例：@CrossOrigin 注解的使用

接下来测试跨域发送请求，再新建一个项目 CrossOriginTest，加入所需的 jar 文件，示例代码如下：

程序清单：codes/03/CrossOriginTest/src/org/fkit/controller/CrossOriginController

```
import org.springframework.stereotype.Controller;
import org.springframework.web.bind.annotation.CrossOrigin;
import org.springframework.web.bind.annotation.GetMapping;

@CrossOrigin(maxAge=3600)
```

```
@Controller
public class CrossOriginController{

    // 测试@CrossOrigin 注解
    @CrossOrigin(origins="http://localhost:8080/VariableTest")
    @GetMapping(value="/welcome")
     public String welcome() {
         System.out.println("处理跨域请求");
         return "welcome";
     }

}
```

CrossOriginController 类和 welcome 方法上都使用了@CrossOrigin 注解。welcome 方法接收到跨域请求进行简单处理后，跳转到 welcome.jsp。

程序清单：codes/03/CrossOriginTest/WebContent/WEB-INF/content/welcome.jsp

```
<%@ page language="java" contentType="text/html; charset=UTF-8"
    pageEncoding="UTF-8"%>
<!DOCTYPE    html    PUBLIC    "-//W3C//DTD    HTML    4.01    Transitional//EN"
"http://www.w3.org/TR/html4/loose.dtd">
<html>
<head>
<meta http-equiv="Content-Type" content="text/html; charset=UTF-8">
<title>测试@CrossOrigin 注解</title>
</head>
<body>
恭喜您，测试跨域访问成功！
<br>
</body>
</html>
```

此外，还需要在 web.xml 文件中配置 Spring MVC 的前端控制器 DispatcherServlet，因为每次配置基本一致，此处不再赘述，读者可自行配置。

同时部署 VariableTest 和 CrossOriginTest 两个 Web 应用，在浏览器中输入如下 URL 来测试应用：

```
http://localhost:8080/VariableTest/
```

会看到如图 3.5 所示的界面，单击"<a href="http://localhost:8080/CrossOriginTest/welcome">测试@CrossOrigin 注解</a>"超链接发送跨域请求，请求被发送到 CrossOriginTest Web 应用，调用 CrossOriginController 控制器的 welcome 方法进行处理，控制台输出结果如下：

```
处理跨域请求
```

同时跳转到 welcome.jsp 页面，如图 3.6 所示。

图 3.6 跨域访问成功

3.7 @RequestHeader 注解

org.springframework.web.bind.annotation.RequestHeader 注解用于将请求的头信息数据映射到功能处理方法的参数上。

使用@RequestHeader注解可指定如表3.6所示的属性。

表3.6 @RequestHeader注解支持的属性

属性	类型	是否必要	说明
name	String	否	指定请求参数绑定的名称
value	String	否	name属性的别名
required	boolean	否	指示参数是否必须绑定
defaultValue	String	否	如果没有传递参数而使用的默认值

@RequestHeader注解示例代码如下：

```
@RequestMapping(value="/requestHeaderTest")
    public void requestHeaderTest(
           @RequestHeader("User-Agent") String userAgent,
           @RequestHeader(value="Accept") String[] accepts)
```

以上配置自动将请求头“User-Agent”的值赋到userAgent变量上，并将请求头“Accept”的值赋到accepts变量上。

示例：@RequestHeader注解的使用

新建一个项目RequestHeaderTest，加入所需的jar文件，示例代码如下。

程序清单：codes/03/RequestHeaderTest/src/org/fkit/controller/RequestHeaderController

```
import org.springframework.stereotype.Controller;
import org.springframework.web.bind.annotation.GetMapping;
import org.springframework.web.bind.annotation.RequestHeader;

@Controller
public class RequestHeaderController{

    // 测试@RequestHeader注解
    // 该方法映射的请求为 /requestHeaderTest
    @GetMapping(value="/requestHeaderTest")
     public void requestHeaderTest(
            @RequestHeader("User-Agent") String userAgent,
            @RequestHeader(value="Accept") String[] accepts) {
        System.out.println("通过@requestHeaderTest获得数据： " + userAgent);
System.out.println("通过@requestHeaderTest获得Accept： ");
        // 循环输出头信息
        for(String accept : accepts){
            System.out.println (accept);
        }
     }

}
```

在requestHeaderTest方法中使用@RequestHeader注解获取请求头“User-Agent”的值并赋给userAgent变量，获取请求头“Accept”的值并赋给accepts变量。

程序清单：codes/03/RequestHeaderTest/WebContent/index.jsp

```
<%@ page language="java" contentType="text/html; charset=UTF-8"
    pageEncoding="UTF-8"%>
<!DOCTYPE html PUBLIC "-//W3C//DTD HTML 4.01 Transitional//EN" "http://www.
w3.org/TR/html4/loose.dtd">
<html>
<head>
<meta http-equiv="Content-Type" content="text/html; charset=UTF-8">
<title>@RequestHeader测试</title>
```

```
</head>
<body>
<h2>@RequestHeader 测试</h2>
<a href="requestHeaderTest">测试@RequestHeader 注解</a><br><br>
</body>
</html>
```

此外，还需要在 web.xml 文件中配置 Spring MVC 的前端控制器 DispatcherServlet，因为每次配置基本一致，此处不再赘述，读者可自行配置。

部署 RequestHeaderTest 这个 Web 应用，在浏览器中输入如下 URL 来测试应用：

```
http://localhost:8080/RequestHeaderTest/
```

会看到如图 3.7 所示的页面，表示 Spring MVC 成功跳转到 index.jsp。

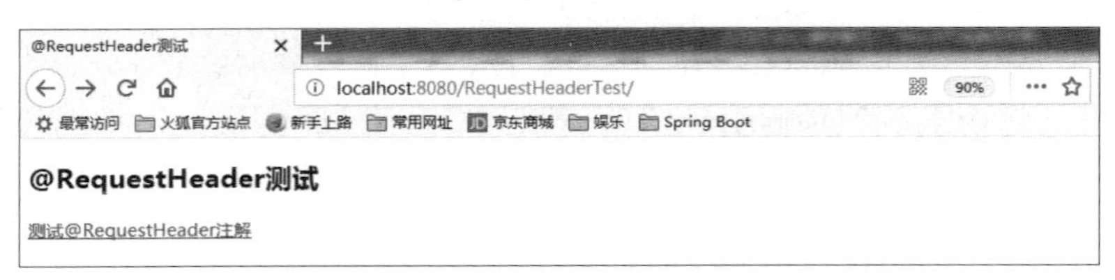

图 3.7 测试@RequestHeader 注解页面

单击“测试@RequestHeader 注解”超链接发送请求，将调用 requestHeaderTest 方法，控制台输出结果如下：

```
通过@requestHeaderTest 获得 User-Agent: Mozilla/5.0 (Windows NT 6.1; Win64; x64; rv:57.0) Gecko/20100101 Firefox/57.0
通过@requestHeaderTest 获得 Accept:
text/html
application/xhtml+xml
application/xml;q=0.9
*/*;q=0.8
```

可以看到，请求头“User-Agent”的值赋给了 userAgent 变量，请求头“Accept”的值赋给了 accepts 变量，并输出打印在控制台。读者请注意，不同的浏览器请求头数据略有不同。

3.8 @CookieValue 注解

org.springframework.web.bind.annotation.CookieValue 注解用于将请求的 Cookie 数据映射到功能处理方法的参数上。

使用@CookieValue 注解可指定如表 3.7 所示的属性。

表 3.7 @CookieValue 注解支持的属性

属性	类型	是否必要	说明
name	String	否	指定请求参数绑定的名称
value	String	否	name 属性的别名
required	boolean	否	指示参数是否必须绑定
defaultValue	String	否	如果没有传递参数而使用的默认值

@CookieValue 注解示例代码如下：

```
@RequestMapping(value="/cookieValueTest")
public void cookieValueTest(
        @CookieValue(value="JSESSIONID", defaultValue="") String sessionId)
```

以上配置会自动将 Cookie 中 JSESSIONID 的值设置到 sessionId 参数上，defaultValue 表示

Cookie 中没有 JSESSIONID 时默认为空。

示例：@CookieValue 注解的使用

新建一个项目 CookieValueTest，加入所需的 jar 文件，示例代码如下。

程序清单：codes/03/CookieValueTest/src/org/fkit/controller/CookieValueController

```
import org.springframework.stereotype.Controller;
import org.springframework.web.bind.annotation.CookieValue;
import org.springframework.web.bind.annotation.GetMapping;

@Controller
public class CookieValueController{

    // 测试@CookieValue 注解
    // 该方法映射的请求为 /cookieValueTest
    @GetMapping(value="/cookieValueTest")
     public void cookieValueTest(
        @CookieValue(value="JSESSIONID",defaultValue="")String sessionId) {
         System.out.println("通过@CookieValue 获得 JSESSIONID： " + sessionId);
    }
}
```

cookieValueTest(@CookieValue(value="JSESSIONID", defaultValue="") String sessionId)方法会自动将 Cookie 中 JSESSIONID 的值入参到 sessionId 参数上，defaultValue 表示 Cookie 中没有 JSESSIONID 时默认为空。

程序清单：codes/03/CookieValueTest/WebContent/index.jsp

```
<%@ page language="java" contentType="text/html; charset=UTF-8"
    pageEncoding="UTF-8"%>
<!DOCTYPE html PUBLIC "-//W3C//DTD HTML 4.01 Transitional//EN" "http://www.
w3.org/TR/html4/loose.dtd">
<html>
<head>
<meta http-equiv="Content-Type" content="text/html; charset=UTF-8">
<title>@CookieValue 测试</title>
</head>
<body>
<h2>@CookieValue 测试</h2>
<a href="cookieValueTest">测试@CookieValue 注解</a><br><br>
</body>
</html>
```

此外，还需要在 web.xml 文件中配置 Spring MVC 的前端控制器 DispatcherServlet，因为每次配置基本一致，此处不再赘述，读者可自行配置。

同时 Spring MVC 还需要 springmvc-config.xml 配置文件，该文件内容和 ControllerTest 项目中的 springmvc-config.xml 文件一致，读者可自行配置。

部署 CookieValueTest 这个 Web 应用，在浏览器中输入如下 URL 来测试应用：

```
http://localhost:8080/CookieValueTest/index.jsp
```

会看到如图 3.8 所示的页面，表示 Spring MVC 成功跳转到初始页面 index.jsp。

单击“测试@CookieValue 注解”超链接发送请求，将调用 cookieValueTest 方法，控制台输出结果如下：

```
通过@requestHeaderTest 获得 JSESSIONID： A1DE158AE570CD41EFAB2D9BC881BCEA
```

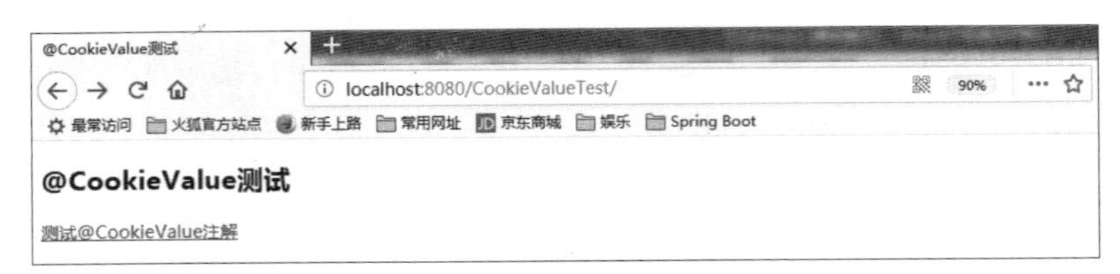

图 3.8　测试@CookieValue 注解页面

可以看到，请求中 sessionId 的值被传递到方法的 sessionId 变量并输出打印在控制台。

3.9　@RequestAttribute 注解

org.springframework.web.bind.annotation.RequestAttribute 注解用于访问由请求处理方法、过滤器或拦截器创建的、预先存在于 request 作用域中的属性，将该属性转换到目标方法的参数。

使用@RequestAttribute 注解可指定如表 3.8 所示的属性。

表 3.8　@RequestAttribute 注解支持的属性

属性	类型	是否必要	说明
name	String	否	指定请求参数绑定的名称
value	String	否	name 属性的别名
required	boolean	否	指示参数是否必须绑定

@RequestAttribute 注解示例代码如下：

```
@RequestMapping(value="/arrtibuteTest")
public void arrtibuteTest(
        @RequestAttribute(value="username") String username)
```

以上代码会自动将 request 作用域中名为 username 的属性的值设置到 username 参数上。

3.10　@SessionAttribute 注解

org.springframework.web.bind.annotation.SessionAttribute 注解用于访问由请求处理方法、过滤器或拦截器创建的、预先存在于 session 作用域中的属性，将该属性转换到目标方法的参数。

使用@SessionAttribute 注解可指定如表 3.9 所示的属性。

表 3.9　@SessionAttribute 注解支持的属性

属性	类型	是否必要	说明
name	String	否	指定请求参数绑定的名称
value	String	否	name 属性的别名
required	boolean	否	指示参数是否必须绑定

@SessionAttribute 注解示例代码如下：

```
@RequestMapping(value="/arrtibuteTest")
public void arrtibuteTest(
        @SessionAttribute(value="username") String username)
```

以上代码会自动将 session 作用域中名为 username 的属性的值设置到 username 参数上。

示例：@RequestAttribute注解和@SessionAttribute注解的使用

新建一个项目AttributeTest，加入所需的jar文件，示例代码如下：

程序清单：codes/03/AttributeTest/WebContent/index.jsp

```
<%@ page language="java" contentType="text/html; charset=UTF-8"
    pageEncoding="UTF-8"%>
<!DOCTYPE html PUBLIC "-//W3C//DTD HTML 4.01 Transitional//EN" "http://www.w3.org/TR/html4/loose.dtd">
<html>
<head>
<meta http-equiv="Content-Type" content="text/html; charset=UTF-8">
<title>@RequestAttribute和@SessionAttribute测试</title>
</head>
<body>
<h2>@RequestAttribute和@SessionAttribute测试</h2>
<a href="attributeTest">测试@RequestAttribute和@SessionAttribute</a><br><br>
</body>
</html>
```

程序清单：codes/03/AttributeTest/src/org/fkit/controller/AttributeController

```
import org.springframework.stereotype.Controller;
import org.springframework.web.bind.annotation.GetMapping;
import org.springframework.web.bind.annotation.RequestAttribute;
import org.springframework.web.bind.annotation.RequestMapping;
import org.springframework.web.bind.annotation.SessionAttribute;
import org.springframework.web.servlet.ModelAndView;

@Controller
public class AttributeController {

    // 该方法映射的请求为/attributeTest
     @GetMapping("/attributeTest")
     public ModelAndView attributeTest(ModelAndView mv) {
         System.out.println("attributeTest方法被调用...");
         // 客户端重定向到main请求，其会被自定义过滤器拦截
         mv.setViewName("redirect:main");;
         return mv;

     }
    /*
     * 该方法映射的请求为/main
     * @RequestAttribute("school")会获取request作用域中名为"school"的属性的值赋给
     * school变量
     * @SessionAttribute("author")会获取session作用域中名为"author"的属性的值赋给
     * author变量
     */
     @RequestMapping("/main")
     public String main(
             @RequestAttribute("school") String school,
             @SessionAttribute("author") String author) {
         System.out.println("main方法被调用...");
         // 输出@RequestAttribute获得的school
         System.out.println("school: " + school);
         // 输出@SessionAttribute获得的author
         System.out.println("author: " + author);
```

```
        return "welcome";
    }

}
```

attributeTest 方法处理请求后重定向到 main 请求，main 请求会被自定义过滤器拦截，在过滤器中会分别设置两个属性到 request 作用域和 session 作用域。在 main 方法中使用@RequestAttribute 和@SessionAttribute 进行赋值。

程序清单：codes/03/AttributeTest/src/org/fkit/filter/TestAttributeFilter

```
import java.io.IOException;
import javax.servlet.Filter;
import javax.servlet.FilterChain;
import javax.servlet.FilterConfig;
import javax.servlet.ServletException;
import javax.servlet.ServletRequest;
import javax.servlet.ServletResponse;
import javax.servlet.annotation.WebFilter;
import javax.servlet.http.HttpServletRequest;


// 过滤器拦截/main 请求
@WebFilter(value="/main")
public class TestAttributeFilter implements Filter{

    @Override
    public void destroy() {
    }

    @Override
    public void doFilter(ServletRequest arg0, ServletResponse arg1, FilterChain chain)
          throws IOException, ServletException {

        System.out.println("进入 AuthFilter 过滤器的 doFilter 方法");
        // 将 ServletRequest 对象强转成 HttpServletRequest 对象
        HttpServletRequest request = (HttpServletRequest)arg0;
        // 在 request 作用域中设置一个 school 属性
        request.setAttribute("school", "疯狂软件");
        // 在 session 作用域中设置一个 author 属性
        request.getSession().setAttribute("author", "肖文吉");
        // 如果还有过滤器执行过滤器，否则进入请求处理方法
        chain.doFilter(request, arg1);

    }

    @Override
    public void init(FilterConfig arg0) throws ServletException {
    }

}
```

TestAttributeFilter 过滤器拦截“main”请求，在 doFilter 方法中分别设置两个属性到 request 作用域和 session 作用域。

此外，还需要在 web.xml 文件中配置 Spring MVC 的前端控制器 DispatcherServlet，因为每次配置基本一致，此处不再赘述，读者可自行配置。

同时 Spring MVC 还需要 springmvc-config.xml 配置文件，该文件内容和 ControllerTest 项目中的 springmvc-config.xml 文件一致，读者可自行配置。

部署 AttributeTest 这个 Web 应用，在浏览器中输入如下 URL 来测试应用：

```
http://localhost:8080/AttributeTest/index.jsp
```

会看到如图 3.9 所示的页面，表示 Spring MVC 成功跳转到初始页面 index.jsp。

图 3.9 测试@RequestAttribut 和@SessionAttribute 注解页面

单击“测试@RequestAttribut 和@SessionAttribute”超链接发送请求，将调用 attributeTest 方法，经过过滤器，重定向到 main 方法，控制台输出结果如下：

```
attributeTest 方法被调用...
进入 AuthFilter 过滤器的 doFilter 方法
main 方法被调用...
school: 疯狂软件
author: 肖文吉
```

可以看到，request 作用域中的 school 的值被传递到方法的 school 变量，session 作用域中的 author 的值被传递到方法的 author 变量，并输出打印在控制台。

3.11 @SessionAttributes 注解

org.springframework.web.bind.annotation.SessionAttributes 注解允许我们有选择地指定 Model 中的哪些属性转存到 HttpSession 对象当中。

使用@SessionAttributes 注解可指定如表 3.10 所示的属性。

表 3.10 @SessionAttributes 注解支持的属性

属性	类型	是否必要	说明
names	String[]	否	Model 中属性的名称，即存储在 HttpSession 当中的属性名称
value	String[]	否	names 属性的别名
types	Class<?>[]	否	指示参数是否必须绑定

@SessionAttributes 注释只能声明在类上，而不能声明在方法上。

示例：@SessionAttributes 注解的使用

程序清单：codes/03/SessionAttributesTest/src/org/fkit/controller/SessionAttributesController

```
import org.fkit.domain.User;
import org.springframework.stereotype.Controller;
import org.springframework.ui.Model;
import org.springframework.web.bind.annotation.PathVariable;
import org.springframework.web.bind.annotation.RequestMapping;
import org.springframework.web.bind.annotation.RequestParam;
import org.springframework.web.bind.annotation.SessionAttributes;

@Controller
// 将 Model 中的 user 属性转存到 HttpSession 对象当中
@SessionAttributes("user")
public class SessionAttributesController{
```

```
    // 该方法映射的请求为 /login
    @RequestMapping(value="/login")
     public String login(
            @RequestParam("loginname") String loginname,
            @RequestParam("password") String password,
            Model model ) {
        // 创建 User 对象，装载用户信息
        User user = new User();
        user.setLoginname(loginname);
        user.setPassword(password);
        user.setUsername("admin");
        // 将 user 对象添加到 Model 当中
        model.addAttribute("user",user);
        return "welcome";
    }
}
```

login 方法中使用的 User 对象和 3.3 节中 RequestMappingTest 项目中的 User 对象一致，读者可以自行参考，此处不再赘述。

接下来创建一个登录页面 index.jsp。

程序清单：codes/03/SessionAttributesTest/ WebContent/index.jsp

```
    <%@ page language="java" contentType="text/html; charset=UTF-8"
       pageEncoding="UTF-8"%>
    <!DOCTYPE html PUBLIC "-//W3C//DTD HTML 4.01 Transitional//EN" "http://www.
w3.org/TR/html4/loose.dtd">
    <html>
    <head>
    <meta http-equiv="Content-Type" content="text/html; charset=UTF-8">
    <title>登录页面</title>
    </head>
    <body>
    <h3>测试@SessionAttributes 注解</h3>
    <form action="login" method="post">
        <table>
            <tr>
                <td><label>登录名: </label></td>
                <td><input type="text" id="loginname" name="loginname" ></td>
            </tr>
            <tr>
                <td><label>密码: </label></td>
                <td><input type="password" id="password" name="password"></td>
            </tr>
            <tr>
                <td><input id="submit" type="submit" value="登录"></td>
            </tr>
        </table>
    </form>
    </body>
    </html>
```

最后创建一个查看作用域中属性数据的页面 welcome.jsp。

程序清单：codes/03/SessionAttributesTes/WebContent/WEB-INF/content/welcome.jsp

```
    ${requestScope.user.username }<br>     // 访问 request 作用域中的 user 对象:
    ${sessionScope.user.username }<br>     // 访问 session 作用域中的 user 对象:
```

在浏览器中输入如下 URL 来测试应用：

```
    http://localhost:8080/SessionAttributesTes
```

会看到如图 3.10 所示的界面。

图 3.10 测试@SessionAttributes 注解页面

输入登录名“test”，密码“123456”，单击“登录”按钮。请求将会被提交到 SessionAttributesController 类的 login 方法，该方法将会创建 User 对象来保存数据，并将其设置到 Model 当中。因为类上面使用了@SessionAttributes 注解，所以 User 同时也会被设置到 HttpSession 作用域当中。方法执行完跳转到如图 3.11 所示的欢迎界面。

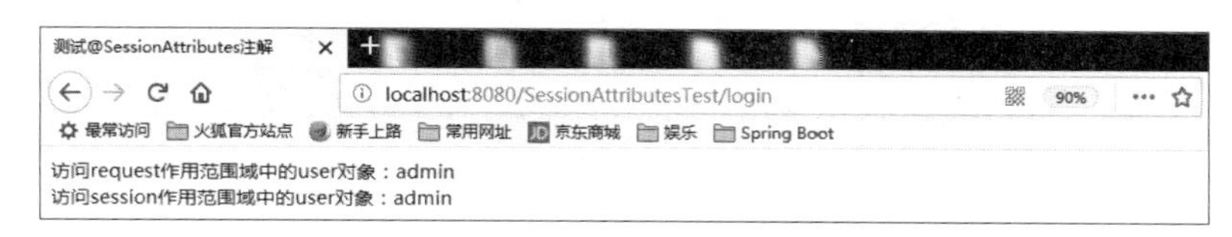

图 3.11 @SessionAttributes 注解测试结果

可以看到，User 对象被成功设置到 HttpSession 作用域当中。

@SessionAttributes 还有如下写法：

```
@SessionAttributes(types={User.class}, value="user")
```

还可以设置多个对象到 HttpSession 当中：

```
@SessionAttributes(types={User.class,Dept.class},value={"user","dept"})
```

types 属性用来指定放入 HttpSession 当中的对象类型。

3.12 @ModelAttribute 注解

org.springframework.web.bind.annotation.ModelAttribute 注解用于将请求参数绑定到对象。

@ModelAttribute 注解只支持一个属性 value，类型为 String，表示绑定的属性名称。

提示

被@ModelAttribute 注释的方法会在 Controller 每个方法执行前被执行，因此在一个 Controller 被映射到多个 URL 时，要谨慎使用。

@ModelAttribute 注解的使用方式有很多种，下面为读者逐一介绍。

示例：@ModelAttribute 注解的使用

程序清单：codes/03/ModelAttributeTest/WebContent/index.jsp

```
<%@ page language="java" contentType="text/html; charset=UTF-8"
    pageEncoding="UTF-8"%>
<!DOCTYPE html PUBLIC "-//W3C//DTD HTML 4.01 Transitional//EN" "http://www.w3.org/TR/html4/loose.dtd">
<html>
<head>
<meta http-equiv="Content-Type" content="text/html; charset=UTF-8">
<title>测试@ModelAttribute</title>
</head>
<body>
```

```
<h3>测试@ModelAttribute 的不同用法</h3>
<a href="loginForm1">测试@ModelAttribute(value="")注释返回具体类的方法 </a><br><br>
<a href="loginForm2">测试@ModelAttribute 注释 void 返回值的方法</a><br><br>
<a href="loginForm3">测试@ModelAttribute 注释返回具体类的方法</a><br><br>
<a href="loginForm4">测试@ModelAttribute 和@RequestMapping 同时注释一个方法
</a><br><br>
<a href="loginForm5">测试@ModelAttribute 注释一个方法的参数 </a><br><br>
</body>
</html>
```

程序清单：codes/03/ModelAttributeTest/src/org/fkit/controller/FormController

```
@Controller
public class FormController{
   // 该方法映射的请求为 http://localhost:8080/ModelAttributeTest/{formName}
   @RequestMapping(value="/{formName}")
    public String loginForm(@PathVariable String formName){
       // 动态跳转页面
       return formName;
    }
}
```

1. 测试@ModelAttribute(value="")注释返回具体类的方法

程序清单：codes/03/ModelAttributeTest/WebContent/WEB-INF/content/loginForm1.jsp

```
<h3>测试@ModelAttribute(value="")注释返回具体类的方法</h3>
<form action="login1" method="post">
    <table>
       <tr>
          <td><label>登录名: </label></td>
          <td><input type="text" id="loginname" name="loginname" ></td>
       </tr>
       <tr>
          <td><input id="submit" type="submit" value="登录"></td>
       </tr>
    </table>
</form>
```

程序清单：codes/03/ModelAttributeTest/src/org/fkit/controller/ModelAttribute1Controller

```
import org.springframework.stereotype.Controller;
import org.springframework.web.bind.annotation.ModelAttribute;
import org.springframework.web.bind.annotation.RequestMapping;
import org.springframework.web.bind.annotation.RequestParam;
@Controller
public class ModelAttribute1Controller{
   // 使用@ModelAttribute 注释的 value 属性，来指定 model 属性的名称,model 属性的值就是方法
的返回值
   @ModelAttribute("loginname")
   public String userModel1(
         @RequestParam("loginname") String loginname){
      return loginname;
   }
   @RequestMapping(value="/login1")
    public String login1() {
      return "result1";
    }
}
```

在 ModelAttribute1Controller 类中除了@RequestMapping 映射的 login1 方法之外，还提供了一个 userModel1 方法，该方法上有一个@ModelAttribute 注解。此处@ModelAttribute 注解默认的 value 值为“loginname”，用来指定 model 属性的名称，而 model 属性的值就是 userModel1

方法的返回值。被@ModelAttribute注解的userModel1方法会先于login1调用，它把请求参数loginname的值赋给loginname变量，并设置了一个属性loginname到Model当中，而属性的值就是loginame变量的值。

程序清单：codes/03/ModelAttributeTest /WebContent/WEB-INF/content/result1.jsp

```
访问 request 作用域中的 loginname 对象：${requestScope.loginname }<br>
```

在跳转的result1.jsp中可以访问到由@ModelAttribute设置的loginname的值。

此外，还需要在web.xml文件中配置Spring MVC的前端控制器DispatcherServlet。因为每次配置基本一致，此处不再赘述，读者可自行配置。

同时，Spring MVC还需要springmvc-config.xml配置文件，该文件内容和ControllerTest项目中的springmvc-config.xml文件一致，读者可自行配置。

部署ModelAttributeTest这个Web应用，在浏览器中输入如下URL来测试应用：

```
http://localhost:8080/ModelAttributeTest/index.jsp
```

会看到如图3.12所示的页面，表示Spring MVC成功跳转到初始页面index.jsp。

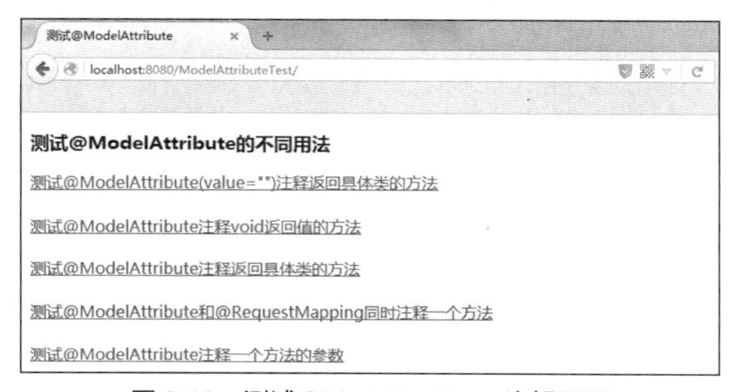

图3.12 测试@ModelAttribute注解页面

单击“测试@ModelAttribute(value="")注释返回具体类的方法”超链接发送请求，跳转到loginForm1.jsp页面，如图3.13所示。

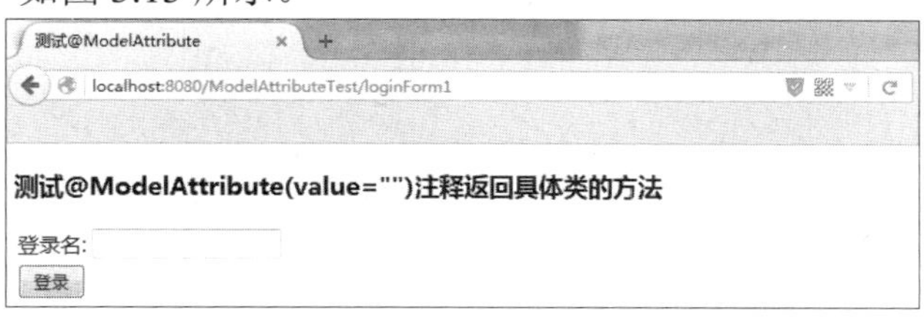

图3.13 测试@ModelAttribute(value="")页面

输入登录名“test”，单击“登录”按钮发送请求，而后将先调用userModel1方法，再调用login1方法，并跳转到result1.jsp页面，如图3.14所示。

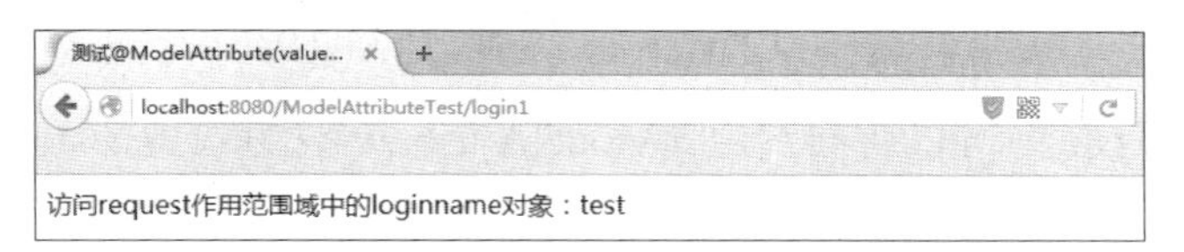

图3.14 测试@ModelAttribute(value="")结果页面

可以看到，在request作用域中访问到了Model的值。

2. 测试@ModelAttribute 注释 void 返回值的方法

程序清单：codes/03/ModelAttributeTest/WebContent/WEB-INF/content/loginForm2.jsp

```
<h3>测试@ModelAttribute 注释 void 返回值的方法</h3>
<form action="login2" method="post">
    <table>
        <tr>
            <td><label>登录名: </label></td>
            <td><input type="text" id="loginname" name="loginname" ></td>
        </tr>
        <tr>
            <td><label>密码: </label></td>
            <td><input type="password" id="password" name="password"></td>
        </tr>
        <tr>
            <td><input id="submit" type="submit" value="登录"></td>
        </tr>
    </table>
</form>
```

程序清单：codes/03/ModelAttributeTest/src/org/fkit/controller/ModelAttribute2Controller

```
import org.springframework.stereotype.Controller;
import org.springframework.ui.Model;
import org.springframework.web.bind.annotation.ModelAttribute;
import org.springframework.web.bind.annotation.RequestMapping;
import org.springframework.web.bind.annotation.RequestParam;
@Controller
public class ModelAttribute2Controller{
// model 属性名称和值由 model.addAttribute()实现，前提是要在方法中加入一个 Model 类型的参数
    @ModelAttribute
    public void userModel2(
            @RequestParam("loginname") String loginname,
            @RequestParam("password") String password,
            Model model){
        model.addAttribute("loginname", loginname);
        model.addAttribute("password", password);
    }
    @RequestMapping(value="/login2")
     public String login2() {
        return "result2";
     }
}
```

在 ModelAttribute2Controller 类中除了@RequestMapping 映射的 login2 方法之外，还提供了一个 userModel2 方法，该方法上有一个@ModelAttribute 注解。userModel2 方法会先于 login2 方法被调用，它把请求参数值赋给对应变量，model 属性名称和值由 model.addAttribute()方法实现，前提是要在方法中加入一个 Model 类型的参数。

程序清单：codes/03/ModelAttributeTest /WebContent/WEB-INF/content/result2.jsp

```
${requestScope.loginname }<br>    // 访问 request 作用域中的 loginname 对象
${requestScope.password }<br>     // 访问 request 作用域中的 password 对象
```

在跳转的 result2.jsp 中可以访问到由@ModelAttribute 注解设置的 loginname 和 password 的值。

在浏览器中输入如下 URL 来测试应用：http://localhost:8080/ModelAttributeTest/index.jsp，将会跳转到如图 3.12 所示页面，单击“测试@ModelAttribute 注释 void 返回值的方法”超链接发送请求，将会跳转到 loginForm2.jsp 页面，如图 3.15 所示。

图 3.15　测试@ModelAttribute 注释 void 返回值的方法

输入登录名“test”，密码“123456”，单击“登录”按钮发送请求，而后将先调用 userModel2 方法，再调用 login2 方法，并跳转到 result2.jsp 页面，如图 3.16 所示。

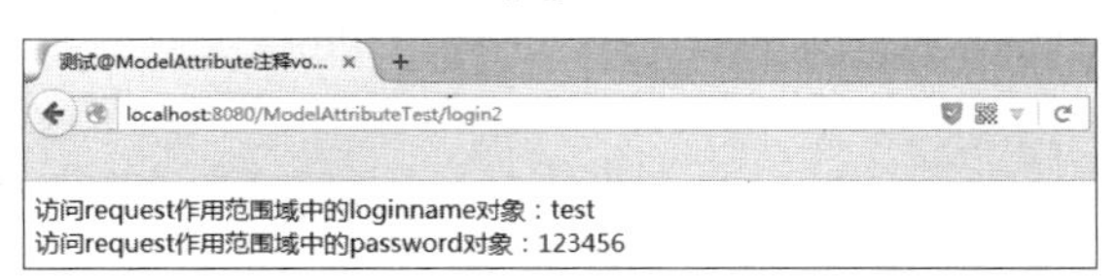

图 3.16　测试@ModelAttribute 注释 void 返回值的方法结果

可以看到，在 request 作用域中访问到了 Model 的值。

3. 测试@ModelAttribute 注释返回具体类的方法

程序清单：codes/03/ModelAttributeTest/WebContent/WEB-INF/content/loginForm3.jsp

loginForm3.jsp 和 loginForm2.jsp 页面内容一致，读者可自行参考，此处不再赘述。

程序清单：codes/03/ModelAttributeTest/src/org/fkit/controller/ModelAttribute3Controller

```
import java.util.ArrayList;
import java.util.List;
import org.fkit.domain.User;
import org.springframework.stereotype.Controller;
import org.springframework.web.bind.annotation.ModelAttribute;
import org.springframework.web.bind.annotation.RequestMapping;
import org.springframework.web.bind.annotation.RequestParam;
@Controller
public class ModelAttribute3Controller{

    // 静态 List<User>集合，其代替数据库用来保存注册的用户信息
    private static List<User> userList;
    // UserController 类的构造器，初始化 List<User>集合
    public ModelAttribute3Controller() {
        super();
        userList = new ArrayList<User>();
        User user1 = new User("test","123456","测试用户");
        User user2 = new User("admin","123456","管理员");
        // 存储 User 用户，用于模拟数据库数据
        userList.add(user1);
        userList.add(user2);
    }
    // 根据登录名和密码查询用户，若用户存在则返回包含用户信息的 User 对象，若不存在则返回 null
    public User find(String loginname,String password){
        for(User user: userList){
            if(user.getLoginname().equals(loginname) && user.getPassword().equals
(password)){
                return user;
            }
        }
        return null;
    }
    //model 属性的名称没有被指定，它由返回类型隐含表示，如这个方法返回 User 类型，那么这个 model
```

```
// 属性的名称是 user
     @ModelAttribute
     public User userModel3(
             @RequestParam("loginname") String loginname,
             @RequestParam("password") String password){
         return find(loginname, password);
     }
     @RequestMapping(value="/login3")
      public String login3() {
          return "result3";
      }
   }
```

在 ModelAttribute3Controller 类中除了@RequestMapping 映射的 login3 方法之外，还提供了一个 userModel3 方法，该方法上有一个@ModelAttribute 注解。userModel3 方法会先于 login3 方法被调用，这里 model 属性的名称没有被指定，它由@ModelAttribute 注解的 userModel3 方法的返回类型隐含表示，如这个方法返回 User 类型，那么这个 model 属性的名称就是 user。此处 find(loginname, password)方法用来模拟数据库根据登录名和密码查询用户的功能实现。

程序清单：codes/03/ModelAttributeTest/WebContent/WEB-INF/content/result3.jsp

```
${requestScope.user.username }<br>    // 访问 request 作用范围域中的 user 对象：
```

在跳转的 result2.jsp 页面中可以访问到由@ModelAttribute 设置的 loginname 和 password 的值。

在浏览器中输入如下 URL 来测试应用：http://localhost:8080/ModelAttributeTest/index.jsp，跳转到如图 3.12 所示页面，单击“测试@ModelAttribute 注释返回具体类的方法”超链接发送请求，将跳转到 loginForm3.jsp 页面，如图 3.15 所示，输入登录名“test”，密码“123456”，单击“登录”按钮发送请求，而后将先调用 userModel3 方法，再调用 login3 方法，并跳转到 result3.jsp 页面，如图 3.17 所示。

图 3.17　测试@ModelAttribute 注释返回具体类的方法结果

可以看到，在 request 作用域中访问到了 User 对象。

4. 测试@ModelAttribute 和@RequestMapping 同时注释一个方法

程序清单：codes/03/ModelAttributeTest/WebContent/WEB-INF/content/loginForm4.jsp

loginForm4.jsp 和 loginForm2.jsp 页面内容一致，读者可自行参考，此处不再赘述。

程序清单：codes/03/ModelAttributeTest/src/org/fkit/controller/ModelAttribute4Controller

```
@Controller
public class ModelAttribute4Controller{
     // 此时 login4 方法的返回值并不是一个视图名称，而是 model 属性的值，视图名称是@RequestMapping
     // 的 value 值"/login4"
     // Model 的属性名称由@ModelAttribute(value="")指定，相当于在 request 中封装了
     // username (key) =admin (value)
    @RequestMapping(value="/login4")
    @ModelAttribute(value="username")
     public String login4() {
         return "admin";
     }
}
```

在 ModelAttribute4Controller 中，@ModelAttribute 和@RequestMapping 同时注释一个方法，此时 login4 方法的返回值并不是一个视图名称，而是 model 属性的值，视图名称是@RequestMapping 的 value 值“/login4”。Model 的属性名称由@ModelAttribute 的 value 值指定，这相当于在 request 中封装了 username（key）=admin（value）。

注意，此处 login4 方法跳转的结果是“/login4”。

程序清单：codes/03/ModelAttributeTest /WebContent/WEB-INF/content/login4.jsp

```
${requestScope.username }<br>    // 访问 request 作用域中的 username 对象的 username 属性
```

在浏览器中输入如下 URL 来测试应用：http://localhost:8080/ModelAttributeTest/index.jsp，跳转到如图 3.12 所示页面，单击“测试@ModelAttribute 和@RequestMapping 同时注释一个方法”超链接发送请求，而后跳转到 loginForm4.jsp 页面，如图 3.15 所示。输入登录名“test”，密码“123456”，单击“登录”按钮发送请求，将调用 login4 方法，跳转到 login4.jsp 页面，如图 3.18 所示。

图 3.18　测试@ModelAttribute 和@RequestMapping 同时注释一个方法结果

可以看到，在 request 作用域中访问到了 username 的值，也就是 login4 方法的返回值“admin”。

5. 测试@ModelAttribute 注释一个方法的参数

程序清单：codes/03/ModelAttributeTest/WebContent/WEB-INF/content/loginForm5.jsp

loginForm5.jsp 和 loginForm2.jsp 页面内容一致，读者可自行参考，此处不再赘述。

程序清单：codes/03/ModelAttributeTest/src/org/fkit/controller/ModelAttribute5Controller

```
@Controller
public class ModelAttribute5Controller{

    /*
     * @ModelAttribute 注释方法参数，会自动入参
     * loginname 控件的值会自动入参到 User 的 loginname 属性
     * password 控件的值会自动入参到 User 的 password 属性
     */
    @RequestMapping(value="/login5")
     public String login5(@ModelAttribute User user) {
        user.setUsername("管理员");
        return "result5";
     }
}
```

ModelAttribute5Controller 类中 login5 方法的参数 User 使用了@ModelAttribute 注解，前台页面的控件的值会自动入参到@ModelAttribute 注解修饰的对象的同名属性当中。这种方式是在实际开发中使用最多的方式。

程序清单：codes/03/ModelAttributeTest /WebContent/WEB-INF/content/result5.jsp

```
访问 request 作用域中的 user 对象的 loginname 属性：${requestScope.user.loginname }<br>
访问 request 作用域中的 user 对象的 password 属性：${requestScope.user.password }<br>
访问 request 作用域中的 user 对象的 username 属性：${requestScope.user.username }<br>
```

在浏览器中输入如下 URL 来测试应用：http://localhost:8080/ModelAttributeTest/index.jsp，跳转到如图 3.12 所示页面，单击“测试@ModelAttribute 注释一个方法的参数”超链接发送请求，跳转到 loginForm5.jsp 页面，如图 3.15 所示。输入登录名“test”，密码“123456”，单击“登录”按钮发送请求，而后将调用 login5 方法，跳转到 result5.jsp 页面，如图 3.19 所示。

图 3.19　测试@ModelAttribute 注释一个方法的参数

可以看到，在 request 作用域中访问到了 User 对象。

提示

@ModelAttribute 注解的使用方法有很多种，非常灵活，读者可以根据业务需求选择使用。

3.13　@RequestBody 注解

org.springframework.web.bind.annotation.RequestBody 注解常用来处理 Content-Type:不是 application/x-www-form-urlencoded 编码的内容，例如 application/json、 application/xml 等。@RequestBody 注解通过使用 HandlerAdapter 配置的 HttpMessageConverters 来解析 JSON 或 XML 数据，然后绑定到相应的 Bean 上。

3.13.1　HttpMessageConverter<T>接口

HttpMessageConverter<T>是 Spring 3.0 之后新增的一个重要接口，它负责将请求信息转换为一个对象（类型为 T），并将对象（类型为 T）绑定到请求方法的参数上或输出为响应信息。

DispatcherServlet 默认已经装配了 RequestMappingHandlerAdapter 作为 HandlerAdapter 组件的实现类，即 HttpMessageConverter 由 RequestMappingHandlerAdapter 使用，将请求信息转换为对象，或将对象转换为响应信息。

HttpMessageConverter<T>接口中定义了以下几个方法：

- boolean canRead(Class<?> clazz, MediaType mediaType)。该方法指定转换器可以读取的对象类型，即转换器可将请求信息转换为 clazz 类型的对象，同时指定支持的 MIME 类型（text/html、application/json 等）。MIME 媒体类型在 RFC2616 中定义，具体请参考 http://tools.ietf.org/html/rfc2616#section-3.7 上的说明。
- boolean canWrite(Class<?> clazz, MediaType mediaType)。该方法指定转换器可以将 clazz 类型的对象写到响应流当中，响应流支持的媒体类型在 mediaType 中定义。
- List<MediaType> getSupportedMediaTypes()。该方法返回当前转换器支持的媒体类型。
- T read(Class<? extends T> clazz,HttpInputMessage inputMessage)。该方法将请求信息流转换为 T 类型的对象。
- void write(T t,MediaType contentType,HttpOutputMessage outputMessage)。该方法将 T 类型的对象写到响应流当中，同时指定响应的媒体类型为 contentType。

Spring 为 HttpMessageConverter<T>提供了多个实现类，这些实现类组成了一个功能强大、用途广泛的信息转换家族。详细说明如下：

- StringHttpMessageConverter。将请求信息转换为字符串。泛型 T 为 String 类型，可以读取所有媒体类型（*/*）的请求信息，可通过设置 supportedMediaTypes 属性指定媒体类型。响应信息的媒体类型为 text/plain（即 Content-Type 的值）。
- FormHttpMessageConverter。将表单数据读取到 MultiValueMap 中。泛型 T 为 org.springframework.util.MultiValueMap<String,?>类型，支持读取 application/x-www-form-urlencoded 类型的信息，但不支持读取 multipart/form-data 类型的信息。可以写 application/x- www-form-urlencoded 及 multipart/form-data 类型的响应信息。
- SourceHttpMessageConverter。如果部分表单属性是 XML 数据，则可用该转换器进行转换。
- ResourceHttpMessageConverter。读写 org.springframework.core.io.Resource 对象。泛型 T 为 org.springframework.core.io.Resource 对象，可以读取所有媒体类型（*/*）的请求信息。如果类路径下提供了 JAF（Java Activation Framework），则根据 Resource 的类型指定响应的类型，否则响应的类型为 application/octet-stream。
- BufferedImageHttpMessageConverter。读写 BufferedImage 对象。泛型 T 为 BufferedImage 对象，可以读取所有类型（*/*）的请求信息，返回 BufferedImage 相应的类型，也可以通过 contentType 显式指定。
- ByteArrayHttpMessageConverter。读写二进制数据。泛型 T 为 byte[]类型，可以读取所有类型（*/*）的请求信息，可以通过设置 supportMediaTypes 属性指定类型，响应信息的媒体类型为 application/octet-stream。
- SourceHttpMessageConverter。读写 javax.xml.transform.Source 类型的数据。泛型 T 为 javax.xml.transform.Source 类型及其扩展类型，包括 javax.xml.transform.dom.DOMSource、javax.xml.transform.sax.SAXSource 及 javax.xml.transform.stream.StreamSource，可以读取 text/xml 和 application/xml 类型请求，响应信息的类型为 text/xml 和 application/xml。
- MarshallingHttpMessageConverter。通过 Spring 的 org.springframework.oxm.Marshalling（将 Java 对象转换成 XML）和 Unmarshaller（将 XML 解析为 Java 对象）读写 XML 消息。泛型 T 为 Object 类型，可以读取 text/xml 和 application/xml 类型请求，响应信息的类型为 text/xml 和 application/xml。
- Jaxb2RootElementHttpMessageConverter。通过 JAXB2 读写 XML 消息，将请求消息转换到注解@XmlRootElement 和 XmlType 作用的类中。泛型 T 为 Object 类型，可以读取 text/xml 和 application/xml 类型请求，响应信息的类型为 text/xml 和 application/xml。
- MappingJackson2HttpMessageConverter。利用 Jackson 开源类包读写 JSON 数据。泛型 T 为 Object 类型，可以读取 application/json 类型的数据，响应信息的类型为 application/json。
- RssChannelHttpMessageConverter。能够读写 RSS 种子消息。泛型 T 为 com.sun.syndication.feed.rss.Channel 类型，可以读取 application/rss+xml 类型的数据，响应信息的类型为 application/rss+xml。
- AtomFeedHttpMessageConverter。能够读写 RSS 种子消息。泛型 T 为 com.sun.syndication.feed.atom.Feed 类型，可以读取 application/atom+xml 类型的数据，响应信息的类型为 application/ atom +xml。

RequestMappingHandlerAdapter 默认已经装配了以下的 HttpMessageConverter:

- StringHttpMessageConverter

➢ ByteArrayHttpMessageConverter

➢ SourceHttpMessageConverter

如果需要装配其他类型的 HttpMessageConverter，则可以在 Spring 的 Web 容器的上下文中自定义一个 RequestMappingHandlerAdapter，如下所示：

```
    <bean class="org.springframework.web.servlet.mvc.method.annotation.
RequestMappingHandlerAdapter">
          <property name="messageConverters">
              <list>
                  <bean class="org.springframework.http.converter.
StringHttpMessageConverter"/>
                  <bean class="org.springframework.http.converter.xml.
SourceHttpMessageConverter"/>
                  <bean class="org.springframework.http.converter.
ByteArrayHttpMessageConverter"/>
                  <bean class="org.springframework.http.converter.
BufferedImageHttpMessageConverter"/>
              </list>
          </property>
      </bean>
```

提示

如果在 Spring Web 容器中显式定义了一个 RequestMappingHandlerAdapter，则 Spring MVC 的 RequestMappingHandlerAdapter 默认装配的 HttpMessageConverter 将不再起作用。

3.13.2 转换 JSON 数据

Spring MVC 提供了处理 JSON 格式请求/响应的 HttpMessageConverter：

➢ MappingJackson2HttpMessageConverter。利用 Jackson 开源类包处理 JSON 格式的请求或响应信息。

因此只需要在 Spring Web 容器中为 RequestMappingHandlerAdapter 装配处理 JSON 的 HttpMessageConverter，并在交互过程中通过请求的 Accept 指定 MIME 类型，Spring MVC 就可以使服务端的处理方法和客户端 JSON 格式的消息进行通信了，开发者几乎不用关心通信层数据格式的问题，可以将精力集中到业务处理上面。

org.springframework.web.bind.annotation.RequestBody 注解用于读取 Request 请求的 body 部分数据，使用系统默认配置的 HttpMessageConverter 进行解析，然后把相应的数据绑定到 Controller 中方法的参数上。

当前台页面使用 GET 或 POST 方式提交数据时，数据编码格式由请求头的 ContentType 指定。可以分为以下几种情况：

➢ application/x-www-form-urlencoded。这种情况的数据@RequestParam、@ModelAttribute 也可以处理，并且很方便，当然@RequestBody 也能处理。

➢ multipart/form-data。@RequestBody 不能处理这种格式的数据。

➢ application/json、application/xml 等格式的数据。必须使用@RequestBody 来处理。

在实际开发工作中使用@RequestBody 注解可以很方便地接收 JSON 格式的数据，并将其转换成对应的数据类型。

Spring 的官方文档说明，Spring MVC 默认使用 MappingJackson2HttpMessageConverter 转换 JSON 格式的数据，Spring 默认使用 Jackson 开源框架作为 JSON 解析框架。Jackson 开源框架可以非常轻松地将 Java 对象转换成 json 对象和 xml 文档，同样也可以将 json 对象、xml 文档转换成 Java 对象。

本书成书时 Jackson 开源框架的最新版本是 2.9.2，jar 包共有 3 个：jackson-annotations-2.9.2.jar、jackson-core-2.9.2.jar、jackson-databind-2.9.2.jar。最新下载地址为：http:// mvnrepository.com/artifact/com.fasterxml.jackson.core，建议读者下载该版本或者更高版本进行测试。

示例：@RequestBody 接收 JSON 格式的数据

创建一个 RequestBodyTest 项目，在 WebContent 目录下创建一个 js 目录，加入 jQuery 和 json2 的 js 文件，在 WEB-INF/lib 目录中加入 Jackson 的 jar 文件。

程序清单：codes/03/RequestBodyTest/WebContent/index.jsp

```
<%@ page language="java" contentType="text/html; charset=UTF-8"
    pageEncoding="UTF-8"%>
<!DOCTYPE html>
<html>
<head>
<meta http-equiv="Content-Type" content="text/html; charset=UTF-8">
<title>测试接收 JSON 格式的数据</title>
<script type="text/javascript" src="js/jquery-1.11.0.min.js"></script>
<script type="text/javascript" src="js/json2.js"></script>
<script type="text/javascript">
$(document).ready(function(){
    testRequestBody();
});
function testRequestBody(){
    $.ajax("${pageContext.request.contextPath}/json/testRequestBody"
    // 发送请求的 URL 字符串
            {
            dataType : "json", // 预期服务器返回的数据类型
               type : "post", //  请求方式为 POST 或 GET
           contentType:"application/json", //  发送信息至服务器时的内容编码格式
           // 发送到服务器的数据
           data:JSON.stringify({id : 1, name : "Spring+MyBatis 企业应用实战"}),
           async:  true , // 默认设置下，所有请求均为异步请求。如果设置为 false，则发送同步请求
           // 请求成功后的回调函数
           success :function(data){
              console.log(data);
             $("#id").html(data.id);
             $("#name").html(data.name);
             $("#author").html(data.author);
           },
           // 请求出错时调用的函数
           error:function(){
              alert("数据发送失败");
           }
    });
}
</script>
</head>
<body>
编号：<span id="id"></span><br>
书名：<span id="name"></span><br>
作者：<span id="author"></span><br>
</body>
</html>
```

index.jsp 页面代码分析如下：

（1）页面使用 jQuery 发送 JSON 数据，在页面的<head>部分，引入了 jQuery 和 json2 的 js 文件。

（2）页面载入时调用 testRequestBody 函数。

（3）testRequestBody 函数发送异步请求到“json/testRequestBody”，注意加粗的代码 contentType:"application/json"，其表示发送的内容编码格式为 JSON；data:JSON.stringify({id : 1, name : "Spring MyBatis 企业应用实战"})，表示发送一个 JSON 数据；请求成功将返回一个 JSON 数据，接到返回的数据后将数据设置到页面的<span>当中。

程序清单：codes/03/RequestBodyTest/src/org/fkit/controller/BookController

```
import javax.servlet.http.HttpServletResponse;
import org.fkit.domain.Book;
import org.springframework.stereotype.Controller;
import org.springframework.web.bind.annotation.RequestBody;
import org.springframework.web.bind.annotation.RequestMapping;
import com.fasterxml.jackson.databind.ObjectMapper;

@Controller
@RequestMapping("/json")
public class BookController {
    private static final Log logger = LogFactory.getLog(BookController.class);
    // @RequestBody 根据 json 数据，转换成对应的 Object
    @RequestMapping(value="/testRequestBody")
    public void setJson(@RequestBody Book book,
            HttpServletResponse response) throws Exception{
        // ObjectMapper 类是 Jackson 库的主要类。它提供一些功能将 Java 对象转换成对应的 JSON
        // 格式的数据
        ObjectMapper mapper = new ObjectMapper();
        // 将 book 对象转换成 json 输出
        System.out.println(mapper.writeValueAsString(book) );
        book.setAuthor("肖文吉");
        response.setContentType("text/html;charset=UTF-8");
        // 将 book 对象转换成 json 写出到客户端
        response.getWriter().println(mapper.writeValueAsString(book));
}
}
```

setJson 方法中的第一个参数@RequestBody Book book 表示，使用@RequestBody 注解获取 JSON 数据后，将 JSON 数据设置到对应的 Book 对象的属性当中。第二个参数是 HttpServletResponse 对象，用来输出响应数据到客户端。

向前台 JSP 页面的 JSON 数据中传入了 id 和 name，为了测试接收数据，使用 logger.info (mapper. writeValueAsString(book))；代码将接收到的 JSON 数据中的 book 对象打印在控制台上。为了测试传递数据到 JSP 页面，在该方法中还给 book 对象的 author 对象设置了一个值，并将其写出到客户端。

程序清单：codes/03/RequestBodyTest/src/org/fkit/domain/Book

```
import java.io.Serializable;
public class Book implements Serializable {
    private Integer id;
    private String name;
    private String author;
public Book() {
        super();
    }
    public Book(Integer id, String name, String author) {
        super();
        this.id = id;
```

```
        this.name = name;
        this.author = author;
    }
    public Integer getId() {
        return id;
    }
    public void setId(Integer id) {
        this.id = id;
    }
    public String getName() {
        return name;
    }
    public void setName(String name) {
        this.name = name;
    }
    public String getAuthor() {
        return author;
    }
    public void setAuthor(String author) {
        this.author = author;
    }
    @Override
    public String toString() {
        return "Book [id=" + id + ", name=" + name + ", author=" + author + "]";
    }
}
```

在 Book 类中定义了 3 个属性：id、name 和 author，用于接收向 JSP 页面传入的 JSON 数据。toString 方法用来输出获取的数据对象信息。

程序清单：codes/03/RequestBodyTest/WebContent/WEB-INF/springmvc-config.xml

```
<!-- Spring 可以自动扫描 base-pack 下面的包或者子包下面的 java 文件，
    如果扫描到有 Spring 相关注解的类，则把这些类注册为 Spring 的 bean -->
    <context:component-scan base-package="org.fkit.controller"/>
<!- 默认配置方案  -->
  <mvc:annotation-driven/>
    <!-- 静态资源处理 -->
  <mvc:default-servlet-handler/>
    <!-- 视图解析器  p:prefix 属性表示前缀  p:suffix 表示后缀  -->
   <bean id="viewResolver"
  class="org.springframework.web.servlet.view.InternalResourceViewResolver"
  p:prefix="/WEB-INF/content/" p:suffix=".jsp"/>
```

提示

在引入静态文件，例如 js 时，需要加入<mvc:default-servlet-handler/>从而使用默认的 Servlet 来响应静态文件。如果没有加入该配置，则执行时页面会报 404 错误，而控制台会提出警告：

No mapping found for HTTP request with URI [/ResponseBodyTest/js/jquery-1.11.0.min.js] in DispatcherServlet with name 'springmvc'

此外，还需要在 web.xml 文件中配置 Spring MVC 的前端控制器 DispatcherServlet，因为每次配置基本一致，此处不再赘述，读者可自行配置。

部署 RequestBodyTest 这个 Web 应用，在浏览器中输入如下 URL 来测试应用：

```
http://localhost:8080/RequestBodyTest/index.jsp
```

载入 index.jsp 页面时会发送 Ajax 请求，传递 JSON 数据，BookController 接收到请求后，@RequestBody 注解会将 JSON 数据设置到 Book 参数对应的属性当中。控制台输出如下：

```
信息：{"id":1,"name":"Spring+MyBatis 企业应用实战"}
```

可以看到，JSON 数据传递的 id 和 name 被赋值到 Book 对象的属性中。接下来，setJson 方法给 Book 对象的 author 属性设置了值，并将 Book 对象转换成 JSON 写出到客户端。

请求响应如图 3.20 所示，表示 Spring MVC 成功将 JSON 数据写出到客户端。

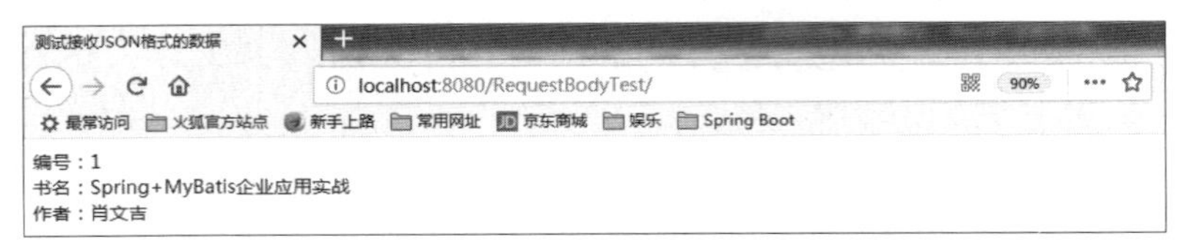

图 3.20　Spring MVC 接收 JSON 数据

可以看到，Book 对象被以 JSON 格式成功写回客户端。

示例：自定义 HttpMessageConverter 接收 JSON 格式的数据

Spring 默认使用 Jackson 处理 JSON 数据。在实际开发中，开发者也可以选择使用其他开源框架处理 JSON 数据。那么，如果使用其他的开源框架处理 JSON 数据，该如何配置 HttpMessageConverter 呢？接下来，我们就使用在业界非常受欢迎的 Fastjson 来接收 JSON 数据。

本书成书时 Fastjson 开源框架的最新版本是 1.2.9，jar 包只有 1 个：fastjson-1.2.9.jar。最新下载地址为：http://mvnrepository.com/artifact/com.alibaba/fastjson，建议读者下载该版本或者更高版本进行测试。

创建一个 FastjsonTest 项目，在 WebContent 目录下创建一个 js 目录，加入 jQuery 和 json2 的 js 文件，在 WEB-INF/lib 目录中加入 Fastjson 的 jar 文件。

程序清单：codes/03/FastjsonTest/src/org/fkit/controller/BookController

```
import javax.servlet.http.HttpServletResponse;
import org.fkit.domain.Book;
import org.springframework.stereotype.Controller;
import org.springframework.web.bind.annotation.RequestBody;
import org.springframework.web.bind.annotation.RequestMapping;
import com.alibaba.fastjson.JSONObject;
@Controller
@RequestMapping("/json")
public class BookController {
    @RequestMapping(value="/testRequestBody")
    public void setJson(@RequestBody Book book,
            HttpServletResponse response) throws Exception{
    // JSONObject-lib 包是一个 beans、collections、maps、java arrays、xml 和 JSON 互相转
    // 换的包
        // 使用 JSONObject 将 book 对象转换成 json 输出
        System.out.println(JSONObject.toJSONString(book));
        book.setAuthor("肖文吉");
        response.setContentType("text/html;charset=UTF-8");
        // 将 book 对象转换成 json 写出到客户端
        response.getWriter().println(JSONObject.toJSONString(book));
    }
}
```

程序清单：codes/03/FastjsonTest/WebContent/WEB-INF/springmvc-config.xml

```
<!-- spring 可以自动扫描 base-pack 下面的包或者子包下面的 java 文件，
        如果扫描到有 Spring 的相关注解的类，则把这些类注册为 Spring 的 bean -->
    <context:component-scan base-package="org.fkit.controller"/>
    <!--静态资源处理 -->
    <mvc:default-servlet-handler/>
```

```
        <!-- 设置配置方案 -->
        <mvc:annotation-driven>
         <!-- 设置不使用默认的消息转换器 -->
         <mvc:message-converters register-defaults="false">
             <!-- 配置 Spring 的转换器 -->
             <bean class="org.springframework.http.converter.
StringHttpMessageConverter"/>
             <bean class="org.springframework.http.converter.xml.
SourceHttpMessageConverter"/>
             <bean class="org.springframework.http.converter.
ByteArrayHttpMessageConverter"/>
             <bean class="org.springframework.http.converter.
BufferedImageHttpMessageConverter"/>
         <!-- 配置 fastjson 中实现 HttpMessageConverter 接口的转换器 -->
             <!-- FastJsonHttpMessageConverter 是 fastjson 中实现了 HttpMessageConverter
接口的类-->
             <bean id="fastJsonHttpMessageConverter"
                 class="com.alibaba.fastjson.support.spring.
FastJsonHttpMessageConverter">
                 <!-- 加入支持的媒体类型：返回 contentType -->
                 <property name="supportedMediaTypes">
                     <list>
                         <!-- 这里顺序不能反，一定先写 text/html，不然 IE 下会出现下载提示 -->
                         <value>text/html;charset=UTF-8</value>
                         <value>application/json;charset=UTF-8</value>
                     </list>
                 </property>
             </bean>
         </mvc:message-converters>
     </mvc:annotation-driven>
     <!-- 视图解析器  p:prefix 属性表示前缀  p:suffix 表示后缀  -->
       <bean id="viewResolver"
      class="org.springframework.web.servlet.view.InternalResourceViewResolver"
      p:prefix="/WEB-INF/content/" p:suffix=".jsp"/>
```

以上配置文件和之前的配置文件主要的区别在于，之前使用的是Spring中默认的Mapping-Jackson2HttpMessageConverter，这样只需要配置默认的<mvc:annotation-driven/>就可以了。而现在使用第三方的开源框架Fastjson处理JSON数据，则需要另行配置HttpMessageConverter。

<mvc:message-converters register-defaults="false">设置不使用默认的消息转换器。在Spring的官方文档中有一段这样的话：

The MappingJackson2JsonView uses the Jackson library's ObjectMapper to render the response content as JSON.

这段话的意思是，Spring MVC默认使用 MappingJackson2JsonView 转换器，所以必须加入 Jackson 这个库的第三方类文件。而在实际开发中，更加受欢迎的是 Fastjson，所以本例并没有使用 Jackson，而是使用了 Fastjson，则转换器需要配置成 com.alibaba.fastjson.support.spring.FastJsonHttpMessageConverter 类型，FastJsonHttpMessageConverter 是 Fastjson 中实现了 HttpMessageConverter 接口的类。

提示

如果加入了Fastjson相关jar文件，但是没有配置FastJsonHttpMessageConverter转换器，则在发送请求时后台会提示错误：

Handler execution resulted in exception: Content type 'application/ json;charset=UTF-8' not supported

此外，其他 JSP 和 Java 文件和之前项目的一致，并且还需要在 web.xml 文件中配置 Spring MVC 的前端控制器 DispatcherServlet，因为每次配置基本一致，此处不再赘述，读者可自行配置。

部署 FastjsonTest 这个 Web 应用，在浏览器中输入如下 URL 来测试应用：

```
http://localhost:8080/FastjsonTest/index.jsp
```

测试结果如图 3.20 所示，由此可知，处理 JSON 格式的开源框架使用 Jackson 和 Fastjson，只是需要使用不同的 HttpMessageConverter。

3.14　@ResponseBody 注解

org.springframework.web.bind.annotation.ResponseBody 注解用于将 Controller 的请求处理方法返回的对象，通过适当的 HttpMessageConverter 转换为指定格式后，写入到 Response 对象的 body 数据区。当返回的数据不是 HTML 标签的页面，而是其他某种格式的数据时（如 JSON、XML 等）使用它。

示例：@ResponseBody 返回 JSON 格式的数据

创建一个 ResponseBodyTest 项目，在 WebContent 目录下创建一个 js 目录，加入 jQuery 和 json2 的 js 文件，在 WEB-INF/lib 目录中加入 Jackson 的 jar 文件。

程序清单：codes/03/ResponseBodyTest/src/org/fkit/controller/BookController

```
import java.util.ArrayList;
import java.util.List;
import org.fkit.domain.Book;
import org.springframework.stereotype.Controller;
import org.springframework.web.bind.annotation.RequestMapping;
import org.springframework.web.bind.annotation.ResponseBody;
@Controller
@RequestMapping("/json")
public class BookController {
    @RequestMapping(value="/testRequestBody")
    // @ResponseBody 会将集合数据转换为 json 格式并将其返回客户端
    @ResponseBody
    public Object getJson() {
        List<Book> list = new ArrayList<Book>();
        list.add(new Book(1,"Spring+MyBatis 企业应用实战","肖文吉"));
        list.add(new Book(2,"轻量级 Java EE 企业应用实战","李刚"));
        return list;
    }
}
```

getJson 方法会将 List 集合数据转换成 JSON 格式，然后将其返回到客户端。

程序清单：codes/03/ResponseBodyTest/WebContent/index.jsp

```
<%@ page language="java" contentType="text/html; charset=UTF-8"
    pageEncoding="UTF-8"%>
<!DOCTYPE html>
<html>
<head>
<meta http-equiv="Content-Type" content="text/html; charset=UTF-8">
<title>测试返回 JSON 格式的数据</title>
```

```
<script type="text/javascript" src="js/jquery-1.11.0.min.js"></script>
<script type="text/javascript" src="js/json2.js"></script>
<script type="text/javascript">
$(document).ready(function(){
    testResponseBody();
});
function testResponseBody(){
    $.post("${pageContext.request.contextPath}/json/testRequestBody",null,
            function(data){
        $.each(data,function(){
            var tr  = $("<tr align='center'/>");
            $("<td/>").html(this.id).appendTo(tr);
            $("<td/>").html(this.name).appendTo(tr);
            $("<td/>").html(this.author).appendTo(tr);
            $("#booktable").append(tr);
        })
    },"json");
}
</script>
</head>
<body>
<table id="booktable" border="1"  style="border-collapse: collapse;">
    <tr align="center">
      <th>编号</th>
      <th>书名</th>
      <th>作者</th>
    </tr>
</table>
</body>
</html>
```

index.jsp 页面代码分析如下：

（1）页面使用 jQuery 发送请求，在页面的<head>部分，引入了 jQuery 和 json2 的 js 文件。

（2）载入页面时调用 testResponseBody 函数。

（3）testResponseBody 函数发送异步请求到"json/testRequestBody"，请求成功将返回一个 JSON 数据，该数据包含多个书籍信息。接到返回的数据后使用 jQuery 将数据设置到页面的<table>表单中。

此外，还需要在 web.xml 文件中配置 Spring MVC 的前端控制器 DispatcherServlet，因为每次配置基本一致，此处不再赘述，读者可自行配置。

同时 Spring MVC 还需要 springmvc-config.xml 配置文件，该文件内容和 RequestBodyTest 项目中的 springmvc-config.xml 文件一致，读者可自行配置。

部署 ResponseBodyTest 这个 Web 应用，在浏览器中输入如下 URL 来测试应用：

```
http://localhost:8080/ResponseBodyTest/index.jsp
```

载入 index.jsp 页面时会发送 Ajax 请求，getJson 方法创建多个 Book 对象并将其封装到 List 集合中返回，方法上的@ResponseBody 注解会将集合数据转换为 JSON 格式数据并将其返回客户端。

请求响应如图 3.21 所示，这表示 Spring MVC 成功将 JSON 数据写到了客户端。

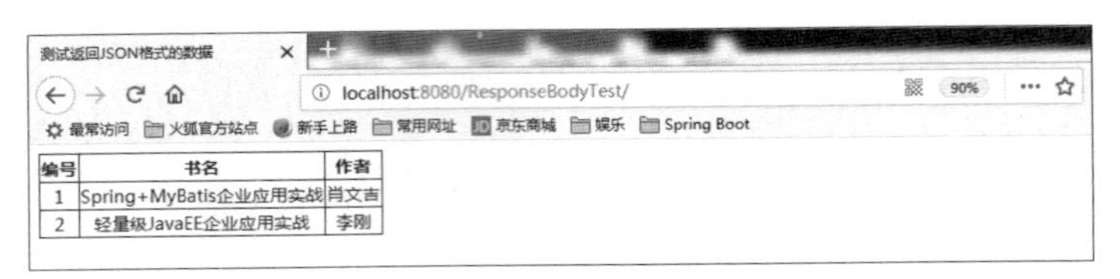

图 3.21　Spring MVC 返回 JSON 数据

可以看到，包含 Book 对象的集合数据被转换成 JSON 格式并被成功写回客户端。

示例：自定义 HttpMessageConverter 返回 JSON 格式的数据

接下来，使用 Fastjson 来返回 JSON 数据。

创建一个 Fastjson2Test 项目，在 WebContent 目录下创建一个 js 目录，加入 jQuery 和 json2 的 js 文件，在 WEB-INF/lib 目录中加入 Fastjson 的 jar 文件。

JsonResponse2Test 项目的所有 JSP 和 Java 文件和 JsonResponseTest 一致，只是在 springmvc-config.xml 中使用了 Fastjson 的 FastJsonHttpMessageConverter。读者可参考配套资源文件中对应的代码，测试结果和 ResponseBodyTest 项目一致，此处不再赘述。

3.15　转换 XML 数据

Spring MVC 提供了处理 XML 格式请求/响应的 HttpMessageConverter，如 Jaxb2RootElement-HttpMessageConverter 通过 JAXB2 读写 XML 消息，并将请求消息转换到注解@XmlRootElement 和@XmlType 作用的类中。

因此只需要在 Spring Web 容器中为 RequestMappingHandlerAdapter 装配处理 XML 的 HttpMessageConverter，并在交互过程中通过请求的 Accept 指定 MIME 类型，Spring MVC 就可以使服务端的处理方法和客户端 XML 格式的消息进行通信了。开发者几乎不用关心通信层数据格式的问题，可以将精力集中到业务处理上面。

在 Spring 的官方文档说明中，Spring MVC 默认使用 Jaxb2RootElementHttpMessage-Converter 转换 XML 格式的数据，JAXB（Java Architecture for XML Binding）可以很方便地生成 XML 格式数据，也能够很方便地生成 JSON 格式数据，这样一来可以更好地在 XML 和 JSON 之间进行转换。

JAXB 是一个业界的标准，是一项可以根据 XML Schema 产生 Java 类的技术。在该过程中，JAXB 提供了将 XML 实例文档反向生成 Java 对象的方法，并能将 Java 对象的内容重新写到 XML 实例文档中，从而使得 Java 开发者在 Java 应用程序中能够很方便地处理 XML 数据。

JAXB 常用的注解包括：@XmlRootElement、@XmlElement，等等。

示例：接收 XML 格式的数据

创建一个 XmlTest 项目，在 WebContent 目录下创建一个 js 目录，加入 jQuery 和 json2 的 js 文件。

程序清单：codes/03/XmlTest/WebContent/sendxml.jsp

```
<%@ page language="java" contentType="text/html; charset=UTF-8"
    pageEncoding="UTF-8"%>
<!DOCTYPE html>
<html>
<head>
<meta http-equiv="Content-Type" content="text/html; charset=UTF-8">
```

```
<title>测试接收 XML 格式的数据</title>
<script type="text/javascript" src="js/jquery-1.11.0.min.js"></script>
<script type="text/javascript" src="js/json2.js"></script>
<script type="text/javascript">
$(document).ready(function(){
    sendxml();
});
function sendxml (){
    var xmlData = "<?xml version=\"1.0\" encoding=\"UTF-8\" standalone=\"yes\"?>
<book><id>1</id><name>疯狂 Java 讲义</name><author>李刚</author></book>";
    $.ajax("${pageContext.request.contextPath}/sendxml",// 发送请求的 URL 字符串
        {
            type : "POST", //  请求方式为 POST 或 GET
        contentType:"application/xml", //  发送信息至服务器时的内容编码类型
        // 发送到服务器的数据
        data: xmlData,
        async:  true , // 在默认设置下，所有请求均为异步请求。如果设置为 false，则发送同步请求
    });
}
</script>
</head>
<body>
</body>
</html>
```

sendxml.jsp 页面代码分析如下：

（1）页面使用 jQuery 发送 JSON 数据，在页面的<head>部分，引入了 jQuery 和 json2 的 js 文件。

（2）载入页面时调用 sendxml 函数。

（3）sendxml 函数发送异步请求到“sendxml”，注意加粗的代码：contentType:"application/xml"，其表示发送的内容编码格式为 XML；data 表示发送一个 XML 数据。

程序清单：codes/03/ XmlTest /src/org/fkit/domain/Book

```
// @XmlRootElement 表示 XML 文档的根元素
@XmlRootElement
public class Book implements Serializable {
    private Integer id;
    private String name;
    private String author;
    public Book() {
        super();
        // TODO Auto-generated constructor stub
    }
    public Book(Integer id, String name, String author) {
        super();
        this.id = id;
        this.name = name;
        this.author = author;
    }
    public Integer getId() {
        return id;
    }
    // 该属性作为 xml 的 element
    @XmlElement
    public void setId(Integer id) {
        this.id = id;
    }
    public String getName() {
        return name;
    }
```

```
    @XmlElement
    public void setName(String name) {
        this.name = name;
    }
    public String getAuthor() {
        return author;
    }
    @XmlElement
    public void setAuthor(String author) {
        this.author = author;
    }
    @Override
    public String toString() {
        return "Book [id=" + id + ", name=" + name + ", author=" + author + "]";
    }
}
```

在 Book 类中定义了 3 个属性：id、name 和 author，分别对应 XML 的元素。toString 方法用来输出获取的数据对象信息。

程序清单：codes/03/ XmlTest /src/org/fkit/controller/BookController

```
import java.io.InputStream;
import javax.xml.bind.JAXBContext;
import javax.xml.bind.Unmarshaller;
import org.fkit.domain.Book;
import org.springframework.stereotype.Controller;
import org.springframework.web.bind.annotation.PostMapping;
import org.springframework.web.bind.annotation.RequestBody;
import org.springframework.web.bind.annotation.ResponseBody;
@Controller
public class BookController {

    // @RequestBody Book book 会将传递的 xml 数据自动绑定到 Book 对象
     @PostMapping(value="/sendxml")
     public void sendxml(@RequestBody Book book) {
         System.out.println(book);
         System.out.println("接收 XML 数据成功");
     }
    // @ResponseBody 会将 Book 自动转成 XML 数据返回
     @PostMapping(value="/readxml")
     @ResponseBody
     public Book readXml()throws Exception {
         // 通过 JAXBContext 的 newInstance 方法，传递一个 class 就可以获得一个上下文
         JAXBContext context = JAXBContext.newInstance(Book.class);
         // 创建一个 Unmarshaller 对象
         Unmarshaller unmar = context.createUnmarshaller();
         InputStream is = this.getClass().getResourceAsStream("/book.xml");
         // Unmarshaller 对象的 unmarshal 方法可以进行 xml 到 Java 对象的转换
         Book book = (Book) unmar.unmarshal(is);
         System.out.println(book);
             return book;
     }
}
```

sendxml 方法中的第一个参数@RequestBody Book book 表示，使用@RequestBody 注解获取到 XML 数据后，将 XML 数据设置到对应的 Book 对象的属性中。为了测试接收数据，使用“logger.info(book);”代码将接收到的 XML 数据的 Book 对象打印在控制台上。

springmvc-config.xml 文件和 JsonRequestTest 项目的一致，重点在于<mvc:annotation-driven/>，该配置默认装配了 Jaxb2RootElementHttpMessageConverter 来处理 XML 数据的转换。

此外，还需要在 web.xml 文件中配置 Spring MVC 的前端控制器 DispatcherServlet，因为

每次配置基本一致，此处不再赘述，读者可自行配置。

部署 XmlTest 这个 Web 应用，在浏览器中输入如下 URL 来测试应用：

```
http://localhost:8080/XmlTest/sendxml.jsp
```

载入 sendxml.jsp 页面时会发送 Ajax 请求，传递 XML 数据。BookController 接收到请求后，@RequestBody 注解会将 XML 数据设置到 Book 参数对应的属性中。控制台输出如下：

```
信息: Book [id=1, name=疯狂 Java 讲义, author=李刚]
信息: 接收 XML 数据成功
```

可以看到，XML 数据传递的 id、name、author 元素被赋值到了 Book 对象对应的属性当中。

示例：返回 XML 格式的数据

程序清单：codes/03/XmlTest/WebContent/readxml.jsp

```
<%@ page language="java" contentType="text/html; charset=UTF-8"
    pageEncoding="UTF-8"%>
<!DOCTYPE html>
<html>
<head>
<meta http-equiv="Content-Type" content="text/html; charset=UTF-8">
<title>测试返回 XML 格式的数据</title>
<script type="text/javascript" src="js/jquery-1.11.0.min.js"></script>
<script type="text/javascript" src="js/json2.js"></script>
<script type="text/javascript">
$(document).ready(function(){
    readxml();
});
function readxml(){
    $.ajax("${pageContext.request.contextPath}/readxml",// 发送请求的 URL 字符串
            {
            dataType : "text", // 预期服务器返回的数据类型
               type : "POST", //  请求方式为 POST 或 GET
           async:  true , // 在默认设置下，所有请求均为异步请求。如果设置为 false，则发送同步请求
           // 请求成功后的回调函数
           success :function(xml){
                  // 获得 xml 数据的 id、name 和 author
                var id = $("id", xml).text();
                var name = $("name", xml).text();
                var author = $("author", xml).text();
                var tr  = $("<tr align='center'/>");
               $("<td/>").html(id).appendTo(tr);
               $("<td/>").html(name).appendTo(tr);
               $("<td/>").html(author).appendTo(tr);
               $("#booktable").append(tr);
           },
           // 请求出错时调用的函数
           error:function(){
               alert("数据接收失败");
           }
    });
}
</script>
</head>
<body>
<table id="booktable" border="1"  style="border-collapse: collapse;">
    <tr align="center">
      <th>编号</th>
      <th>书名</th>
      <th>作者</th>
```

```
    </tr>
</table>
</body>
</html>
```

readxml.jsp 页面代码分析如下：

（1）页面使用 jQuery 发送 XML 数据，在页面的<head>部分，引入了 jQuery 和 json2 的 js 文件。

（2）载入页面时调用 readxml 函数。

（3）readxml 函数发送异步请求到“readxml”，请求成功将返回一个 XML 数据，接到返回的数据后将 XML 数据中的元素读取出来并将其设置到页面的<span>中。

BookController 的 readxml 方法使用 JAXB 读取一个 XML 文件的数据并生成一个 Book 对象返回。@ResponseBody 会将 Book 对象转换成 XML 数据返回到前台 JSP 页面。

在浏览器中输入如下 URL 来测试应用：

```
http://localhost:8080/XmlTest/readxml.jsp
```

请求响应如图 3.22 所示，这表示 Spring MVC 成功将 XML 数据返回到客户端。

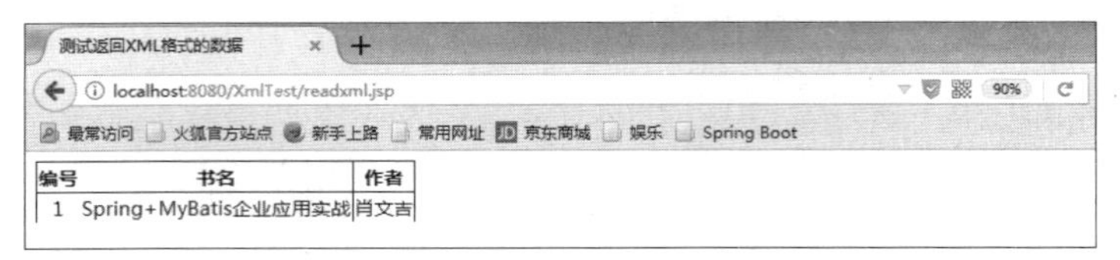

图 3.22　Spring MVC 返回 XML 数据

3.16　@RestController 注解

org.springframework.web.bind.annotation.RestController 注解本身使用@Controller 和@ResponseBody 注解。使用了@RestController 注解的类会被看作一个 Controller，而该类中所有使用@RequestMapping 注解的方法都默认使用了@ResponseBody 注解。

@RestController 注解的源代码如下：

```
@Target(ElementType.TYPE)
@Retention(RetentionPolicy.RUNTIME)
@Documented
@Controller
@ResponseBody
public @interface RestController {

    @AliasFor(annotation = Controller.class)
    String value() default "";

}
```

示例：@RestController 注解的使用

创建一个 RestControllerTest 项目，所有文件和配置基本和 3.14 节的 ResponseBodyTest 项目一致。

程序清单：codes/03/RestControllerTest/src/org/fkit/controller/BookController

```
import java.util.ArrayList;
import java.util.List;
import org.fkit.domain.Book;
import org.springframework.web.bind.annotation.RequestMapping;
```

```
import org.springframework.web.bind.annotation.RestController;

@RestController
@RequestMapping("/json")
public class BookController {

    @RequestMapping(value="/testRequestBody")
    public Object getJson() {
        List<Book> list = new ArrayList<Book>();
        list.add(new Book(1,"Spring+MyBatis 企业应用实战","肖文吉"));
        list.add(new Book(2,"轻量级 Java EE 企业应用实战","李刚"));
        return list;
    }

}
```

BookController 使用了@RestController 注解，该类会被看成一个 Controller，同时该类中所有使用@RequestMapping 注解的方法都默认使用了@ResponseBody 注解，getJson 方法会将 List 集合数据转换成 JSON 格式并返回客户端。

测试结果和 ResponseBodyTest 项目的测试结果一致，此处不再赘述。

3.17 Spring MVC 的异常处理

任何成熟的 MVC 框架都应该提供异常处理机制，当然可以在 Controller 的请求处理方法中手动使用 try...catch 块捕捉异常，当捕捉到特定异常时，返回特定逻辑视图名，但这种处理方式非常烦琐，需要在请求处理方法中书写大量的 catch 块。最大的缺点还在于异常处理与代码耦合，一旦需要改变异常处理方式，必须修改大量代码！这是一种相当糟糕的方式。

```
public class XxxController {
    ...
    @RequestMapping("/hello")
    public String hello() throws Exception{
        try {
            ...
        } catch (异常 1 e) {
            return 结果 1
        }catch (异常 2 e) {
            return 结果 2
        }

    }

}
```

Spring MVC 中提供的异常处理方式有两种：

（1）使用 Spring MVC 提供的简单异常处理器 SimpleMappingExceptionResolver。

（2）使用@ExceptionHandler 注解实现局部异常处理或使用@ControllerAdvice 注解实现统一异常处理。

3.17.1 Spring MVC 异常处理接口以及实现类

- HandlerExceptionResolver 是 Spring 3.0 之后新增的一个重要接口，负责 Spring MVC 的异常处理。

该接口只有一个方法签名：

```
ModelAndView resolveException(HttpServletRequest request,
```

```
                                HttpServletResponse response,
                                @Nullable
                                java.lang.Object handler,
                                java.lang.Exception ex)
```

resolveException 方法处理程序执行期间被抛出的异常，返回一个模型和视图，视图通常是一个特定的错误处理页面。

- AbstractHandlerExceptionResolver 抽象类实现了 HandlerExceptionResolver 接口，重写了 HandlerExceptionResolver 接口的 resolveException 方法用于处理异常。
- AbstractHandlerMethodExceptionResolver 抽象类继承了 AbstractHandlerExceptionResolver 抽象类，该类主要就是为 HandlerMethod 类服务，即 handler 参数是 HandlerMethod 类型。
- ExceptionHandlerExceptionResolver 类继承自 AbstractHandlerMethodExceptionResolver，该类主要处理 Controller 中使用@ExceptionHandler 注解的方法和@ControllerAdvice 注解定义的类。该类也是<mvc:annotation-driven/>配置中定义的 HandlerExceptionResolver 实现类之一，大多数异常处理都由该类操作。
- SimpleMappingExceptionResolver 继承自 AbstractHandlerExceptionResolver 抽象类，是一个根据配置来解析异常的类，包括异常类型、默认的错误视图、默认的响应码及异常映射等配置属性。

因此，在 Spring MVC 中进行异常处理，如果选择 XML 配置，则使用 SimpleMapping-ExceptionResolver 类；如果使用@ExceptionHandler 注解和@ControllerAdvice 注解，则由 ExceptionHandlerExceptionResolver 类进行处理。

提示 两种异常处理方式在项目中不能共存，在开发中通常根据实际情况来选择。

示例：SimpleMappingExceptionResolver 处理异常

新建一个项目 SimpleMappingExceptionResolverTest，加入所需的 jar 文件，示例代码如下：

程序清单：codes/03/SimpleMappingExceptionResolverTest/WebContent/index.jsp

```
<%@ page language="java" contentType="text/html; charset=UTF-8"
    pageEncoding="UTF-8"%>
<!DOCTYPE html PUBLIC "-//W3C//DTD HTML 4.01 Transitional//EN" "http://www.w3.org/TR/html4/loose.dtd">
<html>
<head>
<meta http-equiv="Content-Type" content="text/html; charset=UTF-8">
<title>异常处理示例</title>
</head>
<body>
<a href="hello">没有异常处理</a><br><br>
<a href="test">使用简单异常处理器处理异常</a><br><br>
<a href="find">使用简单异常处理器处理特定异常</a><br><br>
</body>
</html>
```

index.jsp 中有 3 个超链接，分别测试没有异常处理、有异常处理、特定异常处理 3 种情况。

程序清单：codes/03/SimpleMappingExceptionResolverTest/src/org/fkit/controller/TestController

```
import java.sql.SQLException;
import org.springframework.stereotype.Controller;
```

```
import org.springframework.web.bind.annotation.GetMapping;

@Controller
public class TestController{

    @GetMapping("/hello")
    public String hello() throws Exception{
        // 抛出异常
        throw new Exception();
    }
    @GetMapping("/test")
    public String test() throws Exception{
        // 模拟异常
        int i = 5/0;
        return "success";
    }
    @GetMapping("/find")
    public String find() throws Exception{
        try {
            // 模拟异常
            int i = 5/0;
            return "success";
        } catch (Exception e) {
            throw new SQLException("查找数据失败!");
        }
    }
}
```

TestController 中有 3 个方法，分别对应 index.jsp 页面的 3 个请求：

- hello 方法什么都没做，直接抛出一个异常。
- test 方法模拟了一个除数不能为 0 异常。
- find 方法模拟了一个除数不能为 0 异常之后，在 catch 块中抛出了一个 SQLException 异常。

部署 SimpleMappingExceptionResolverTest 这个 Web 应用，在浏览器中输入如下 URL 来测试应用，如图 3.23 所示。

```
http://localhost:8080/SimpleMappingExceptionResolverTest/
```

图 3.23　测试异常处理页面

单击“没有异常处理”超链接，发送“hello”请求，此时没有异常处理程序，异常被直接抛给了浏览器，如图 3.24 所示。

异常被直接抛到浏览器，页面上显示一大堆错误堆栈信息，用户看到这些错误堆栈信息，往往都会一头雾水，抱怨这个设计实在太不友好。而且错误堆栈信息由于暴露了后台方法的调用关系，对应用来说这是存在一定潜在风险的。虽然在 web.xml 中可以配置处理异常的 jsp 页面，但这还是远远不够的。Spring MVC 对错误处理提供了更好的解决方案。

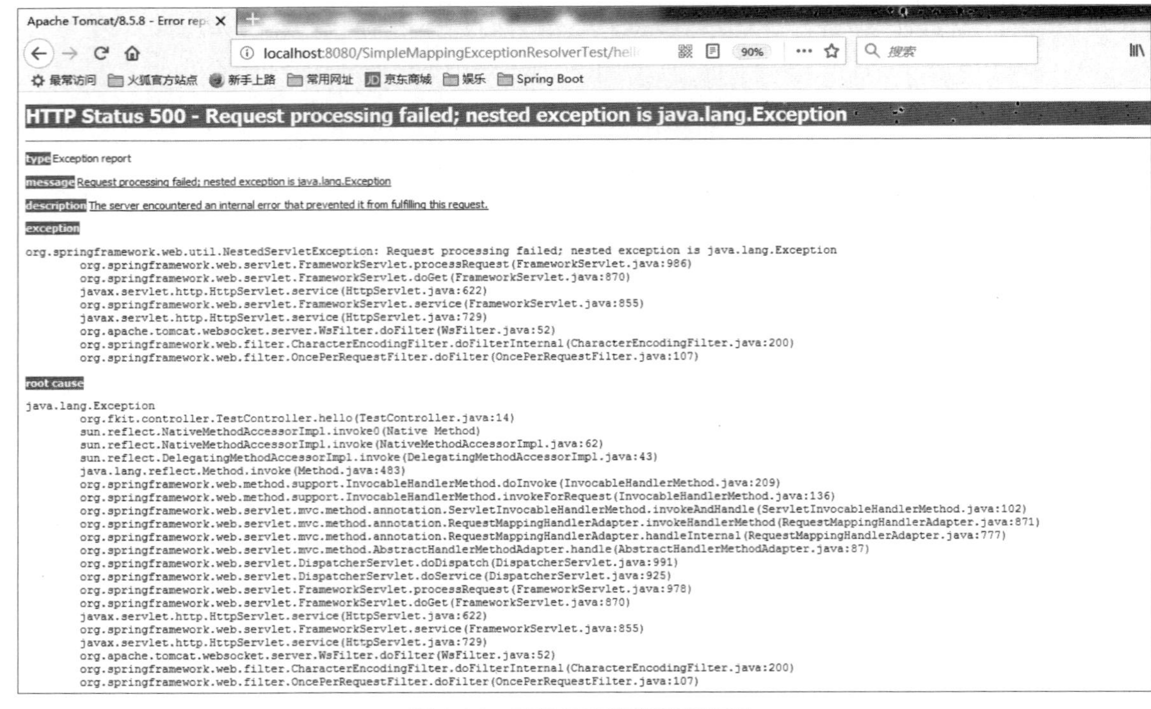

图 3.24　异常被直接抛到浏览器

接下来，在 springmvc-config.xml 中加入异常处理的配置。

程序清单：codes/03/SimpleMappingExceptionResolverTest/WebContent/WEB-INF/springmvc-config.xml

```
<?xml version="1.0" encoding="UTF-8"?>
<beans xmlns="http://www.springframework.org/schema/beans"
    xmlns:xsi="http://www.w3.org/2001/XMLSchema-instance"
    xmlns:p="http://www.springframework.org/schema/p"
    xmlns:mvc="http://www.springframework.org/schema/mvc"
    xmlns:context="http://www.springframework.org/schema/context"
    xsi:schemaLocation="
        http://www.springframework.org/schema/beans
        http://www.springframework.org/schema/beans/spring-beans.xsd
        http://www.springframework.org/schema/mvc
        http://www.springframework.org/schema/mvc/spring-mvc.xsd
        http://www.springframework.org/schema/context
        http://www.springframework.org/schema/context/spring-context.xsd">

    <!-- Spring 可以自动扫描 base-pack 下面的包或者子包下面的 java 文件，
        如果扫描到有 Spring 的相关注解的类，则把这些类注册为 Spring 的 bean -->
    <context:component-scan base-package="org.fkit.controller"/>
    <!-- 默认配置方案 -->
    <mvc:annotation-driven/>
    <!-- 静态资源处理 -->
    <mvc:default-servlet-handler/>

    <!-- 视图解析器  p:prefix 属性表示前缀  p:suffix 表示后缀  -->
     <bean id="viewResolver"

class="org.springframework.web.servlet.view.InternalResourceViewResolver"
          p:prefix="/WEB-INF/content/" p:suffix=".jsp"/>

    <!-- 异常处理
        p:defaultErrorView="error"表示所有没有指定的异常，都跳转到异常处理页面 error
```

```
        p:exceptionAttribute="ex"表示异常处理页面中访问的异常对象变量名是 ex
      -->
      <bean
class="org.springframework.web.servlet.handler.SimpleMappingExceptionResolver"
          p:defaultErrorView="error"
          p:exceptionAttribute="ex">
          <!-- 异常映射
              exceptionMappings 表示映射的异常，接受参数是一个 Properties
              key 是异常类名，value 是处理异常的页面
          -->
          <property name="exceptionMappings">
              <props>
                  <prop key="IOException">ioerror</prop>
                  <prop key="SQLException">sqlerror</prop>
              </props>
          </property>
      </bean>

  </beans>
```

重点是异常处理的配置。SimpleMappingExceptionResolver 是 Spring 提供的处理异常的类，所有抛出的异常都会被该类捕获。p:defaultErrorView="error"表示所有没有指定的异常都跳转到异常处理页面 error，p:exceptionAttribute="ex"表示在异常处理页面中可以访问的异常对象变量名是 ex。如果需要为一些特定的异常指定异常处理页面，可以使用 exceptionMappings 属性，该属性接受的参数是一个 Properties 对象，key 是异常类名或者包名加类名，value 是异常处理页面。例如上面的配置指明，如果是 IOException 则跳转到 ioerror 页面，是 SQLException 则跳转到 sqlerror 页面，是其他异常则全部跳转到 error 页面，在所有异常页面中可以通过 ex 变量访问异常对象 Exception。

程序清单：codes/03/SimpleMappingExceptionResolverTest/WebContent/content/error.jsp

```
  <%@ page language="java" contentType="text/html; charset=UTF-8"
      pageEncoding="UTF-8"%>
  <!DOCTYPE html PUBLIC "-//W3C//DTD HTML 4.01 Transitional//EN" "http://www.
w3.org/TR/html4/loose.dtd">
  <html>
  <head>
  <meta http-equiv="Content-Type" content="text/html; charset=UTF-8">
  <title>异常处理示例</title>
  </head>
  <body>
  <h3>异常处理页面</h3>
  抛出异常信息：${requestScope.ex.message}
  </body>
  </html>
```

程序清单：codes/03/SimpleMappingExceptionResolverTest/WebContent/content/sqlerror.jsp

```
  <%@ page language="java" contentType="text/html; charset=UTF-8"
      pageEncoding="UTF-8"%>
  <!DOCTYPE html PUBLIC "-//W3C//DTD HTML 4.01 Transitional//EN" "http://www.
w3.org/TR/html4/loose.dtd">
  <html>
  <head>
  <meta http-equiv="Content-Type" content="text/html; charset=UTF-8">
  <title>异常处理示例</title>
  </head>
  <body>
  <h3>特定异常处理页面</h3>
  抛出异常信息：${requestScope.ex.message}
  </body>
```

```
</html>
```

再次运行 SimpleMappingExceptionResolverTest 这个 Web 应用，在浏览器中输入如下 URL 来测试应用：

```
http://localhost:8080/SimpleMappingExceptionResolverTest/
```

如图 3.23 所示。单击“使用简单异常处理器处理异常”超链接，发送“test”请求，抛出的异常被 SimpleMappingExceptionResolver 捕获，转发到异常处理页面 error.jsp，如图 3.25 所示。

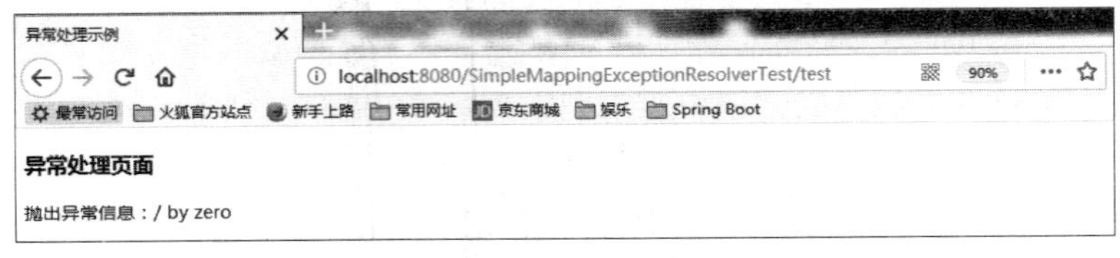

图 3.25　异常处理页面

单击“使用简单异常处理器处理特定异常”超链接，发送“find”请求，请求处理方法抛出的是 SQLException 异常，被 SimpleMappingExceptionResolver 捕获，转发到异常处理页面 sqlerror.jsp，如图 3.26 所示。

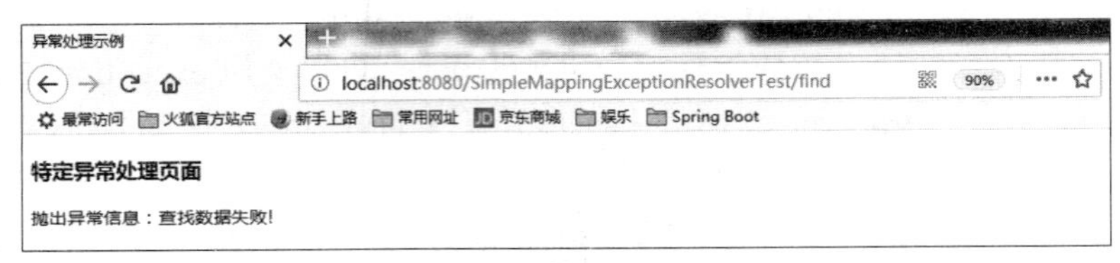

图 3.26　特定异常处理页面

3.17.2　@ResponseStatus 注解

org.springframework.web.bind.annotation.ExceptionHandle.ResponseStatus 注解是处理异常最简单的方式，其可以修饰一个类或者一个方法，当修饰一个类的时候，通常修饰的是一个异常类。

使用@ResponseStatus 注解可指定如表 3.11 所示的属性。

表 3.11　@ResponseStatus 注解支持的属性

属性	类型	是否必要	说明
code	HttpStatus	否	http 状态码，如 HttpStatus.CREATED、HttpStatus.OK 等
value	String	否	同 code 属性
reason	HttpStatus	否	错误信息

使用时，先声明一个自定义异常类，在自定义异常类上面加上@ResponseStatus 注解，就表示在系统运行期间，当抛出自定义异常的时候，使用@ResponseStatus 注解中声明的 value 属性和 reason 属性将异常信息返回给客户端，提高可读性。

示例：@ResponseStatus 处理异常

新建一个项目 ResponseStatusTest，加入所需的 jar 文件，示例代码如下：

程序清单：codes/03/ResponseStatusTest/WebContent/index.jsp

```
<%@ page language="java" contentType="text/html; charset=UTF-8"
    pageEncoding="UTF-8"%>
<!DOCTYPE html PUBLIC "-//W3C//DTD HTML 4.01 Transitional//EN" "http://www.
```

```
w3.org/TR/html4/loose.dtd">
    <html>
    <head>
    <meta http-equiv="Content-Type" content="text/html; charset=UTF-8">
    <title>异常处理示例</title>
    </head>
    <body>
    <br>
    <a href="find">@ResponseStatus 异常处理</a><br><br>
    </body>
    </html>
```

程序清单：codes/03/ResponseStatusTest/src/org/fkit/exception/BookException

```
    import org.springframework.web.bind.annotation.ResponseStatus;
    @ResponseStatus(reason="查询数据失败")
    public class BookException extends RuntimeException {
    }
```

BookException 是自定义异常类，使用了@ResponseStatus 注解修饰，reason 表示抛出异常时显示的错误信息，此处没有使用 value 属性，表示适用于所有 http 状态码。

程序清单：codes/03/ResponseStatusTest/src/org/fkit/controller/BookController

```
    import org.fkjava.exception.BookException;
    import org.springframework.stereotype.Controller;
    import org.springframework.web.bind.annotation.GetMapping;

    @Controller
    public class BookController{

        @GetMapping("/find")
        public String find() throws Exception{
            try {
                int i = 5/0;
                return "success";
            } catch (Exception e) {
                throw new BookException();
            }
        }

    }
```

find 方法模拟了一个异常，发生异常时抛出自定义异常 BookException。

部署 ResponseStatusTest 这个 Web 应用，在浏览器中输入如下 URL 来测试应用，如图 3.27 所示。

```
    http://localhost:8080/ResponseStatusTest/
```

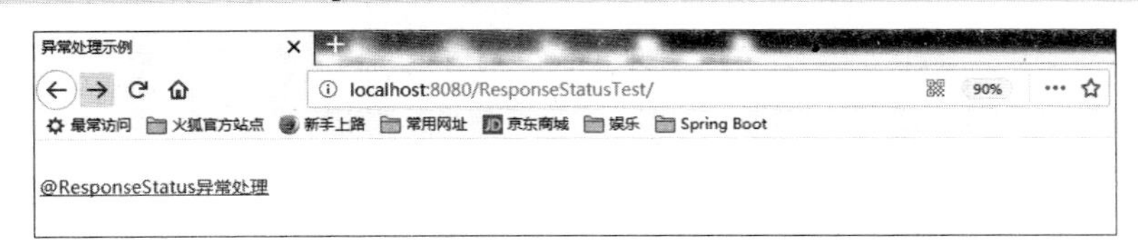

图 3.27　@ResponseStatus 注解测试页面

单击“@ResponseStatus 异常处理”超链接，发送“find”请求，请求处理方法抛出自定义 BookException 异常，显示异常信息，如图 3.28 所示。

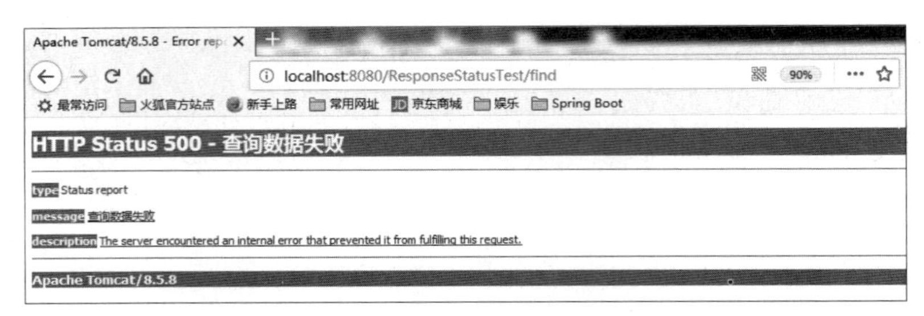

图 3.28　错误信息

3.17.3　@ExceptionHandle 注解

org.springframework.web.bind.annotation.ExceptionHandle 注解的作用对象为方法，并且在运行时有效，value()可以指定异常类。

@ExceptionHandler 注解的源代码如下：

```
@Target(ElementType.METHOD)
@Retention(RetentionPolicy.RUNTIME)
@Documented
public @interface ExceptionHandler {
    Class<? extends Throwable>[] value() default {};
}
```

@ExceptionHandler 注解的方法可以支持的参数除了 HttpServletRequest、HttpServletResponse 等对象之外，还支持一个异常参数，包括一般的异常或自定义异常。如果注解没有指定异常类，会默认进行映射。

示例：@ExceptionHandler 处理异常

新建一个项目 ExceptionHandlerTest，加入所需的 jar 文件，示例代码如下：

程序清单：codes/03/ExceptionHandlerTest/WebContent/index.jsp

```
<%@ page language="java" contentType="text/html; charset=UTF-8"
    pageEncoding="UTF-8"%>
<!DOCTYPE html PUBLIC "-//W3C//DTD HTML 4.01 Transitional//EN" "http://www.
w3.org/TR/html4/loose.dtd">
<html>
<head>
<meta http-equiv="Content-Type" content="text/html; charset=UTF-8">
<title>异常处理示例</title>
</head>
<body>
<a href="test">@ExceptionHandler 处理异常</a><br><br>
<hr>
<a href="login">UserController：父级 Controller 异常处理</a><br><br>
<a href="find">BookController：父级 Controller 异常处理</a><br><br>
</body>
</html>
```

index.jsp 中有 3 个超链接，分别用于测试@ExceptionHandler 异常处理和使用父级 Controller 异常处理。

程序清单：codes/03/ExceptionHandlerTest/src/org/fkit/controller/TestController

```
import javax.servlet.http.HttpServletRequest;
import org.springframework.stereotype.Controller;
import org.springframework.web.bind.annotation.ExceptionHandler;
import org.springframework.web.bind.annotation.GetMapping;
import org.springframework.web.servlet.ModelAndView;
```

```
@Controller
public class TestController{

    @GetMapping("/test")
    public String test() throws Exception{
        // 模拟异常
        int i = 5/0;
        return "success";
    }

    /**
     * 在异常抛出的时候，Controller 会使用@ExceptionHandler 注解的方法去处理异常
     * */
    @ExceptionHandler(value = Exception.class)
    public ModelAndView testErrorHandler(Exception e)  {
        System.out.println("TestController testErrorHandler()......");
        ModelAndView mav = new ModelAndView();
        mav.addObject("ex", e);
        mav.setViewName("error");
        return mav;
    }

}
```

TestController 中 test()方法是 index.jsp 页面的超链接“@ExceptionHandler 处理异常”的请求处理方法，模拟了一个除数不能为 0 的异常。

testErrorHandler()方法使用了@ExceptionHandler 注解，value = Exception.class 表示处理所有的 Exception 类型异常。当 TestController 类抛出异常的时候，会使用@ExceptionHandler 注解的方法去处理异常，而不会直接抛给浏览器。testErrorHandler()方法将捕捉到的异常对象保存到 ModelAndView 当中，传递到 JSP 页面。

程序清单：codes/03/ExceptionHandlerTest/WebContent/WEB-INF/content/error.jsp

```
<html>
<head>
<meta http-equiv="Content-Type" content="text/html; charset=UTF-8">
<title>测试@ExceptionHandler 注解</title>
</head>
<body>
<h3>异常处理页面</h3>
抛出异常信息：${requestScope.ex.message}
</body>
</html>
```

部署 ExceptionHandlerTest 这个 Web 应用，在浏览器中输入如下 URL 来测试应用，如图 3.29 所示。

```
http://localhost:8080/ExceptionHandlerTest/
```

图 3.29 @ExceptionHandler 测试异常处理页面

单击“@ExceptionHandler 处理异常”超链接，发送“test”请求，TestController 的 test()

方法处理请求，抛出异常，异常被@ExceptionHandler 注解修饰的 testErrorHandler 方法捕获，处理之后跳转到 error.jsp 页面，如图 3.30 所示。

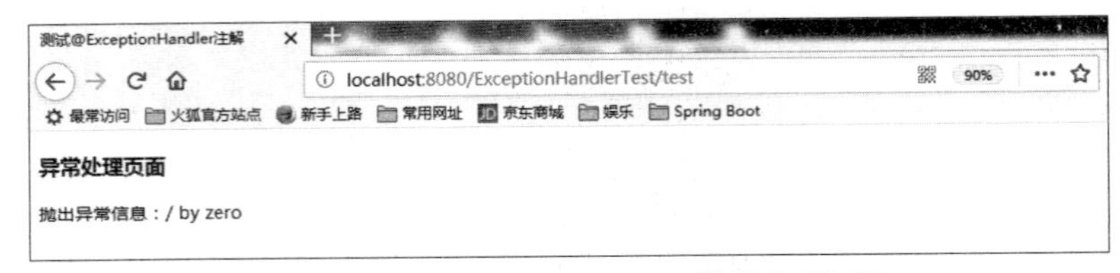

图 3.30　@ExceptionHandler 异常处理页面

基于 Controller 的@ExceptionHandler 注解方法在进行异常处理时，对于每个 Controller 都需要写@ExceptionHandler 注解的异常处理方法，在实际开发当中这非常烦琐。可以写一个父类，在父类中完成@ExceptionHandler 注解的异常处理方法，所有的 Controller 继承这个父类，则所有的 Controller 就都拥有了@ExceptionHandler 注解的异常处理方法。

程序清单：codes/03/ExceptionHandlerTest/src/org/fkit/controller/BaseController

```
import org.springframework.web.bind.annotation.ExceptionHandler;
import org.springframework.web.servlet.ModelAndView;
public class BaseController {
    @ExceptionHandler(value = Exception.class)
    public ModelAndView defaultErrorHandler(Exception e) throws Exception {
        ModelAndView mav = new ModelAndView();
        mav.addObject("ex", e);
        mav.setViewName("error");
        return mav;
    }
}
```

BaseController 作为父类，定义了一个@ExceptionHandler 注解修饰的方法。

程序清单：codes/03/ExceptionHandlerTest/src/org/fkit/controller/UserController

```
import org.springframework.stereotype.Controller;
import org.springframework.web.bind.annotation.GetMapping;
@Controller
public class UserController extends BaseController{
    @GetMapping("/login")
    public String login(String username) throws Exception{
        if(username == null ){
            throw new NullPointerException("用户名不存在!");
        }
        return "success";
    }
}
```

UserController 继承 BaseController，如果抛出异常，将使用父类的@ExceptionHandler 注解修饰的方法处理异常。

程序清单：codes/03/ExceptionHandlerTest/src/org/fkit/controller/BookController

```
import org.springframework.stereotype.Controller;
import org.springframework.web.bind.annotation.GetMapping;
@Controller
public class BookController extends BaseController{
    @GetMapping("/find")
       public String find() throws Exception{
    try {
        int i = 5/0;
        return "success";
    } catch (Exception e) {
```

```
            throw new SQLException("查找图书信息失败!");
        }

    }}
```

BookController 继承 BaseController，如果抛出异常，将使用父类的@ExceptionHandler 注解修饰的方法处理异常。

再次部署 Exception2Test 这个 Web 应用，在浏览器中输入如下 URL 来测试应用，如图 3.29 所示。单击“UserController：父级 Controller 异常处理”超链接，发送“login”请求，异常处理之后跳转到 error.jsp 页面，如图 3.31 所示。

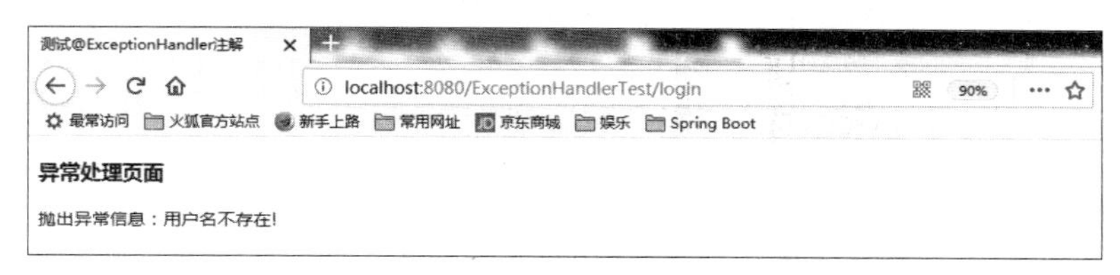

图 3.31　@ExceptionHandle 异常处理页面

再次请求 index.jsp 页面，如图 3.29 所示。单击“BookController：父级 Controller 异常处理”超链接，发送“find”请求，异常处理之后跳转到 error.jsp 页面，如图 3.32 所示。

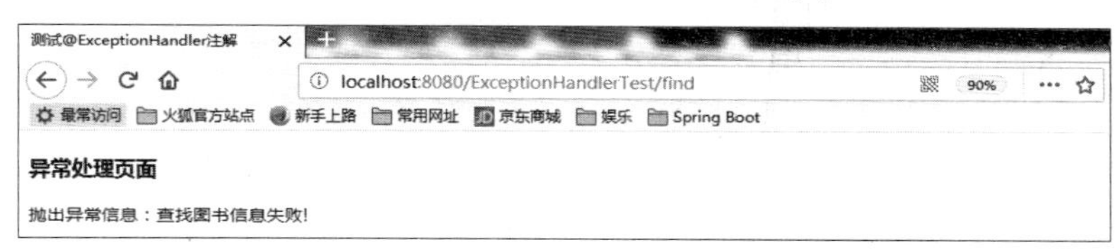

图 3.32　@ExceptionHandle 异常处理页面

3.17.4　@ControllerAdvice 注解

org.springframework.web.bind.annotation.ControllerAdvice 注解是 Spring 3.2 提供的新注解，它是一个控制器增强功能注解。该注解源代码如下：

```
@Target(ElementType.TYPE)
@Retention(RetentionPolicy.RUNTIME)
@Documented
@Component
public @interface ControllerAdvice {
    @AliasFor("basePackages")
    String[] value() default {};
    @AliasFor("value")
    String[] basePackages() default {};
    Class<?>[] basePackageClasses() default {};
    Class<?>[] assignableTypes() default {};
    Class<? extends Annotation>[] annotations() default {};
}
```

该注解使用@Component 注解，也就是说可以使用<context:component-scan>扫描该注解。Spring 官方文档说明，扫描到@ControllerAdvice 注解之后，会将@ControllerAdvice 注解修饰的类的内部使用@ExceptionHandler、@InitBinder、@ModelAttribute 注解的方法应用到所有的请求处理方法上。在实际开发中，@ExceptionHandler 注解的功能最强大，另外两个用处不大。

示例：@ControllerAdvice 处理异常

新建一个项目 ControllerAdviceTest，加入所需的 jar 文件，示例代码如下：

程序清单：codes/03/ControllerAdviceTest/WebContent/index.jsp

```
<%@ page language="java" contentType="text/html; charset=UTF-8"
    pageEncoding="UTF-8"%>
<!DOCTYPE html PUBLIC "-//W3C//DTD HTML 4.01 Transitional//EN" "http://www.w3.org/
TR/html4/loose.dtd">
<html>
<head>
<meta http-equiv="Content-Type" content="text/html; charset=UTF-8">
<title>异常处理示例</title>
<script type="text/javascript" src="js/jquery-1.11.0.min.js"></script>
<script type="text/javascript">
      $(function(){
          $("#search").click(function(){
                $.post("${pageContext.request.contextPath}/search",null,
                        function(data){
                  // 处理异常
                     if (data.message)
                     {
                        alert("与服务器交互出现异常：" + data.message);
                     }
                     else
                     {
                        // 获取服务器响应，显示所有订单信息

                     }
                  },"json");
                });
      })
   </script>
</head>
<body>
<br>
<a href="find">@ControllerAdvice 异常处理</a><br><br>
<hr>
<button id="search">查询订单（返回 JSON）</button>
</body>
</html>
```

index.jsp 中有两个超链接，分别用来测试@ControllerAdvice 异常处理和异常处理时返回 JSON 的两种情况。需要注意 id="search"的按钮，使用 jQuery 进行异步查询订单时，如果抛出异常，则获取返回的 JSON 数据并提示错误。

程序清单：codes/03/ControllerAdviceTest/src/org/fkit/controller/BookController

```
import org.springframework.stereotype.Controller;
import org.springframework.web.bind.annotation.GetMapping;
@Controller
public class BookController{
   @GetMapping("/find")
   public String find() throws Exception{
      int i = 5/0;
      return "success";
   }
}
```

BookController 处理"find"请求，在请求处理方法中简单地模拟了一个异常。BookController 中并没有@ExceptionHandler 注解修饰的方法，抛出的异常会由@ControllerAdvice 注解修饰的类中的@ExceptionHandler 注解修饰的方法进行处理。

程序清单：codes/03/ControllerAdviceTest/src/org/fkit/controller/OrderException

```
public class OrderException extends RuntimeException {
```

```
        public OrderException() {
            super();
        }
        public OrderException(String message, Throwable cause, boolean enableSuppression,
boolean writableStackTrace) {
            super(message, cause, enableSuppression, writableStackTrace);
        }
        public OrderException(String message, Throwable cause) {
            super(message, cause);
        }
        public OrderException(String message) {
            super(message);
        }
        public OrderException(Throwable cause) {
            super(cause);
        }
    }
```

OrderException是一个自定义异常类型，继承自RuntimeException。

程序清单：codes/03/ControllerAdviceTest/src/org/fkit/controller/OrderController

```
    import org.springframework.stereotype.Controller;
    import org.springframework.web.bind.annotation.PostMapping;
    @Controller
    public class OrderController {
        @PostMapping("/search")
        public String search() throws Exception{
            try {
                int i = 5/0;
                return "success";
            } catch (Exception e) {
                e.printStackTrace();
                throw new OrderException("订单查询失败!");
            }

        }
    }
```

OrderController处理“search”请求，在请求处理方法中简单地模拟了一个异常，被捕捉后抛出OrderException自定义异常类型。

程序清单：codes/03/ControllerAdviceTest/src/org/fkit/controller/GlobalExceptionHandler

```
    import java.util.HashMap;
    import java.util.Map;
    import org.springframework.web.bind.annotation.ControllerAdvice;
    import org.springframework.web.bind.annotation.ExceptionHandler;
    import org.springframework.web.bind.annotation.ResponseBody;
    import org.springframework.web.servlet.ModelAndView;
    @ControllerAdvice
    public class GlobalExceptionHandler {
        // 处理Exception类型异常
        @ExceptionHandler(value = Exception.class)
        public ModelAndView globalErrorHandler(Exception e) throws Exception {
            ModelAndView mav = new ModelAndView();
            mav.addObject("ex", e);
            mav.setViewName("error");
            return mav;
        }
        // 处理OrderException自定义异常
        @ExceptionHandler(value = OrderException.class)
        @ResponseBody
        public Object OrderErrorHandler(Exception e) throws Exception {
            // 创建返回对象Map并设置属性，其会被@ResponseBody注解转换为JSON返回
```

```
        Map<String, Object> map = new HashMap<>();
        map.put("code", 100);
        map.put("message", e.getMessage());
        map.put("data", "请求失败");
        return map;
    }
}
```

GlobalExceptionHandler 类使用了@ControllerAdvice 注解来修饰，其会被<context:component-scan>扫描，该类中使用@ExceptionHandler 注解修饰的方法将被应用到所有请求处理方法上。

GlobalExceptionHandler 类中定义了两个方法：第一个方法 globalErrorHandler 使用@ExceptionHandler 注解修饰时 value= Exception.class 表示该方法处理所有 Exception 类型的异常，处理方式和之前一致，将异常信息对象保存到 Model，并返回异常处理页面 error.jsp；第二个方法 OrderErrorHandler 使用@ExceptionHandler 注解修饰时，value = OrderException.class 表示该方法处理 OrderException 自定义类型的异常，此处的处理方式和之前不同，这里创建一个 Map 对象保存信息并返回，由于方法使用了@ResponseBody 注解，返回的 Map 对象会被转成 JSON 数据。

部署 ControllerAdviceTest 这个 Web 应用，在浏览器中输入如下 URL 来测试应用，如图 3.33 所示。

```
http://localhost:8080/ControllerAdviceTest/
```

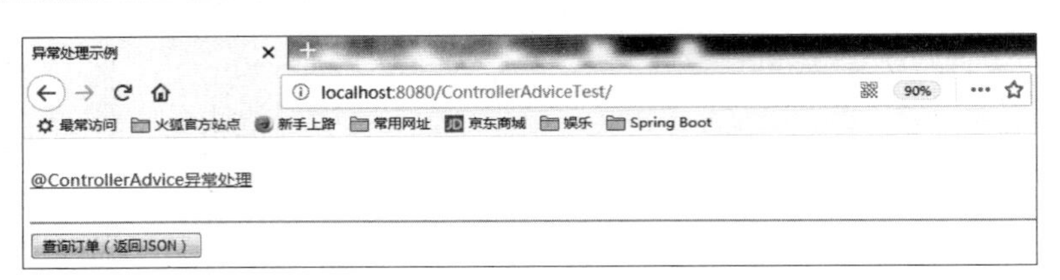

图 3.33 @ControllerAdvice 测试异常处理页面

单击“@ControllerAdvice 异常处理”超链接，发送“find”请求，BookController 的 find()方法处理请求，抛出异常，异常被@ControllerAdvice 注解修饰的 GlobalExceptionHandler 类中@ExceptionHandler(value = Exception.class)注解修饰的 globalErrorHandler 方法捕获，处理之后跳转到 error.jsp 页面，如图 3.34 所示。

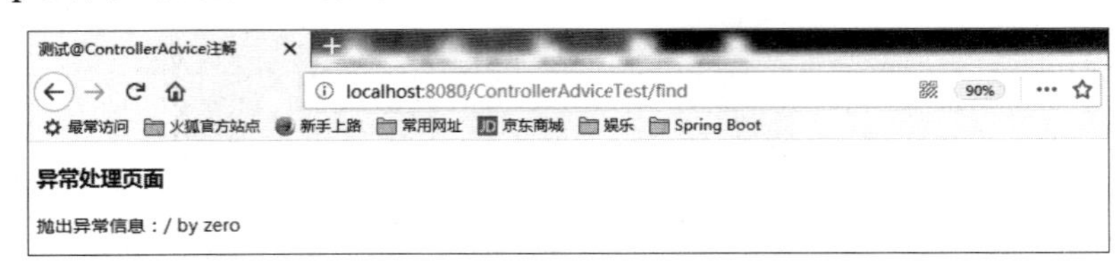

图 3.34 @ControllerAdvice 异常处理页面

再次请求 index.jsp 页面，如图 3.33 所示。单击“查询订单（返回 JSON）”按钮，发送“search”请求，OrderController 的 search()方法处理请求，抛出自定义异常 OrderException，异常被@ControllerAdvice 注解修饰的 GlobalExceptionHandler 类中@ExceptionHandler(value = OrderException.class)注解修饰的 OrderErrorHandler 方法捕获，返回 JSON 信息，如图 3.35 所示。

图 3.35　@ControllerAdvice 异常处理页面

3.17.5　@RestControllerAdvice 注解

org.springframework.web.bind.annotation.RestController 注解本身使用@ControllerAdvice 和@ResponseBody 注解。使用了@RestControllerAdvice 注解的类会被看作一个 ControllerAdvice，而该类中所有使用@ExceptionHandler 注解的方法都默认使用@ResponseBody 注解。

@RestControllerAdvice 注解的源代码如下：

```
@Target(ElementType.TYPE)
@Retention(RetentionPolicy.RUNTIME)
@Documented
@ControllerAdvice
@ResponseBody
public @interface RestControllerAdvice{
    // ......
}
```

示例：@RestControllerAdvice 注解的使用

创建一个 RestControllerAdviceTest 项目，所有文件和配置基本和 3.17.3 节的 ControllerAdviceTest 项目一致。

程序清单：codes/03/RestControllerAdviceTest/src/org/fkit/controller/GlobalExceptionHandler

```
import java.util.HashMap;
import java.util.Map;
import org.springframework.web.bind.annotation.ExceptionHandler;
import org.springframework.web.bind.annotation.RestControllerAdvice;

@RestControllerAdvice
public class GlobalExceptionHandler {

    // 处理 OrderException 自定义异常
    @ExceptionHandler(value = OrderException.class)
    public Object OrderErrorHandler(Exception e) throws Exception {
        // 创建返回对象 Map 并设置属性，其会被@ResponseBody 注解转换为 JSON 返回
        Map<String, Object> map = new HashMap<>();
        map.put("code", 100);
        map.put("message", e.getMessage());
        map.put("data", "请求失败");
        return map;
    }

}
```

GlobalExceptionHandler 使用了@ RestControllerAdvice 注解，该类会被看成一个

ControllerAdvice，同时该类中所有使用@ExceptionHandler 注解的方法都默认使用了@ResponseBody 注解，OrderErrorHandler 方法会将 Map 集合数据转换成 JSON 格式并返回客户端。

测试结果和 ControllerAdviceTest 项目的测试结果一致，此处不再赘述。

3.18 本章小结

本章介绍了 Spring MVC 的常用注解，包括@Controller、@RequestMapping 两个最重要的注解和 Spring MVC 的常用注解。其中，@Controller 注解用于指示 Spring 类的实例是一个控制器；@RequestMapping 注解用来指示一个请求处理方法；@RequestParam 等注解用于参数绑定。

接着介绍了 Spring MVC 中的数据转换，包括重要的数据转换接口 HttpMessageConverter 和 JSON 格式数据以及 XML 格式数据的转换。其中，JSON 格式的数据转换是目前项目开发中最为常用的转换手段，非常重要。

最后，介绍了 Spring MVC 的异常处理。Spring MVC 中的异常处理有两种方式：使用简单异常处理器 SimpleMappingExceptionResolver 和使用@ExceptionHandler 注解和@ControllerAdvice 注解。使用注解处理异常相对接口来说更加灵活。

本章介绍的注解都是在 Spring MVC 中使用较多的注解，读者一定要好好掌握。

第 4 章将重点介绍 Spring MVC 的表单标签库。

CHAPTER

4

第4章 Spring MVC的标签库

本章要点

- form 标签
- input 标签
- password 标签
- hidden 标签
- textarea 标签
- checkbox 标签和 checkboxes 标签
- radiobutton 标签和 radiobuttons 标签
- select 标签
- option 标签和 options 标签
- errors 标签

Spring 从 2.0 版开始，提供了一组功能强大的标签用来在 JSP 和 Spring Web MVC 中处理表单元素。相比其他的标签库，Spring 的标签库集成在 Spring Web MVC 中，因此这里的标签可以访问控制器处理命令对象和绑定数据，这样一来 JSP 更容易开发、阅读和维护。

4.1 表单标签库

表单标签库的实现类在 spring-webmvc.jar 文件当中，标签库描述文件是 spring-form.tld。要使用 Spring MVC 的表单标签库，必须在 JSP 页面的开头处声明 taglib 指令：

```
<%@ taglib prefix="form" uri="http://www.springframework.org/tags/form" %>
```

表 4.1 显示了表单标签库中的所有标签。

表 4.1　表单标签库中的所有标签

标签	描述
form	渲染表单元素
input	渲染<input type= “text” >元素
password	渲染<input type= “password” >元素
hidden	渲染<input type= “hidden” >元素
textarea	渲染 textarea 元素
checkbox	渲染一个<input type= “checkbox” >元素
checkboxes	渲染多个<input type= “checkbox” >元素
radiobutton	渲染一个<input type= “radio” >元素
radiobuttons	渲染多个<input type= “radio” >元素
select	渲染一个选择元素
option	渲染一个可选元素
options	渲染一个可选元素列表
errors	在 span 元素中渲染字段错误

4.1.1 form 标签

Spring MVC 的 form 标签主要有两个作用：

- 自动绑定 Model 中的一个属性值到当前 form 对应的实体对象上，默认为 command 属性，这样我们就可以在 form 表单体里面方便地使用该对象的属性了。
- 支持我们在提交表单时使用除 GET 和 POST 之外的其他方法进行提交，包括 DELETE 和 PUT 等。

form 标签可使用如表 4.2 所示的属性。表 4.2 中列出的只是 Spring MVC 的 form 标签的常用属性，并没有包含 HTML 中如 method 和 action 等属性。

表 4.2　from 标签的属性

属性	描述
modelAttribute	form 绑定的模型属性名称，默认为 command，可自定义修改
commandName	form 绑定的模型属性名称，默认为 command，可自定义修改
acceptCharset	定义服务器接受的字符编码
cssClass	定义要应用到被渲染的 form 元素的 CSS 类
cssStyle	定义要应用到被渲染的 form 元素的 CSS 样式
htmlEscape	boolean 值，表示被渲染的值是否应该进行 HTML 转义

commandName 属性是其中最重要的属性，它定义了模型属性的名称，其中包含了一个绑

定的 JavaBean 对象，该对象的属性将用于填充所生成的表单。如果 commandName 属性存在，则必须在返回包含该表单的视图的请求处理方法中添加响应的模型属性。

通常我们都会指定 commandName 或 modelAttribute 属性，指定绑定到的 JavaBean 的名称，这两个属性功能基本一致。

4.1.2 input 标签

Spring MVC 的 input 标签会被渲染为一个类型为 text 的普通 HTML input 标签。使用 Spring MVC 的 input 标签的唯一目的就是绑定表单数据，通过 path 属性来指定要绑定的 Model 中的值。

Input 标签可使用如表 4.3 所示的属性。表 4.3 中列出的只是 Spring MVC 的 input 标签的常用属性，并没有包含 HTML 的相关属性。

表 4.3 input 标签的属性

属性	描述
cssClass	定义要应用到被渲染的 input 元素的 CSS 类
cssStyle	定义要应用到被渲染的 input 元素的 CSS 样式
cssErrorClass	定义要应用到被渲染 input 元素的 CSS 类，如果 bound 属性中包含错误，则覆盖 cssClass 属性值
htmlEscape	boolean 值，表示被渲染的值是否应该进行 HTML 转义
path	要绑定的属性路径

示例：form 和 input 标签的使用

程序清单：codes/04/FormTest/WebContent/content/registerForm.jsp

```
<%@ taglib prefix="form" uri="http://www.springframework.org/tags/form" %>
<form:form  method ="post" action="register">
    <table>
        <tr>
            <td>姓名:</td>
            <td><form:input path="username"/></td>
        </tr>
        <tr>
            <td>性别:</td>
            <td><form:input path="sex"/></td>
        </tr>
        <tr>
            <td>年龄:</td>
            <td><form:input path="age"/></td>
        </tr>
    </table>
</form:form>
```

如果 Model 中存在一个属性名称为 command 的 JavaBean，而且该 JavaBean 拥有属性 username、sex 和 age，则在渲染上面的代码时就会取 command 的对应属性值赋给对应标签的属性。

程序清单：codes/04/FormTest/src/org/fkit/domain/User

```
import java.io.Serializable;
public class User implements Serializable{
    private String username;
    private String sex;
    private Integer age;
public User() {
        super();
```

```
    }

    public User(String username, String sex, Integer age) {
        super();
        this.username = username;
        this.sex = sex;
        this.age = age;
    }
    // 省略 set 和 get 方法
}
```

程序清单：codes/04/FormTest/src/org/fkit/controller/UserController

```
import org.fkit.domain.User;
import org.springframework.stereotype.Controller;
import org.springframework.ui.Model;
import org.springframework.web.bind.annotation.GetMapping;
@Controller
public class UserController{
     @GetMapping(value="/registerForm")
     public String registerForm(Model model) {
         User user = new User("jack","男",28);
       // 向 model 中添加属性 command，值是 user 对象
         model.addAttribute("command",user);
         return "registerForm";
     }
}
```

注意加粗的代码，该代码将 user 设置到 Model 当中，属性名为“command”。

web.xml 文件和 springmvc-config.xml 文件和之前描述的一致，此处不再赘述。

部署 FormTest 这个 Web 应用，在浏览器中输入如下 URL 来测试应用：

```
http://localhost:8080/FormTest/registerForm
```

会看到如图 4.1 所示的界面。

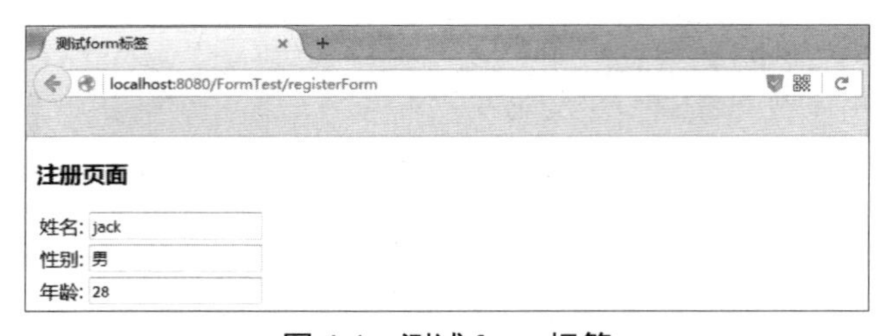

图 4.1　测试 form 标签

在上面的代码中，假设 Model 中存在一个属性名称为 command 的 JavaBean，且它的 username、sex 和 age 属性值分别为“jack”、“男”和“28”，则在浏览器页面单击→右键→查看页面源代码，可以看到 Spring MVC 的 from 标签和 input 标签渲染时生成的 html 代码如下：

```
<form id="command" action="register" method="post">
    <table>
        <tr>
            <td>姓名:</td>
            <td><input id="username" name="username" type="text" value="jack"/></td>
        </tr>
        <tr>
            <td>性别:</td>
            <td><input id="sex" name="sex" type="text" value="男"/></td>
        </tr>
        <tr>
            <td>年龄:</td>
```

```
            <td><input id="age" name="age" type="text" value="28"/></td>
        </tr>
    </table>
</form>
```

从上面生成的代码，我们可以看出，当没有指定form标签的id属性时它会自动获取该form标签绑定的Model中对应属性名称command作为id；而对于input标签，在没有指定id的情况下它会自动获取path指定的属性值作为它的id和name。

Spring MVC指定form标签默认自动绑定的是Model的command属性，那么当form对象对应的属性名称不是command时，应该怎么办呢？对于这种情况，Spring提供了一个commandName属性，可以通过该属性来指定将使用Model中的哪个属性作为form标签需要绑定的command对象。除了commandName属性外，指定modelAttribute属性也可以达到相同的效果。这里假设上面代码存放在Model中的是user对象而不是默认的command对象，那么我们的代码就可以如下定义了。

程序清单：codes/04/FormTest/WebContent/content/registerForm2.jsp

```
<%@ taglib prefix="form" uri="http://www.springframework.org/tags/form" %>
<form:form modelAttribute="user" method="post" action="register" >
    <table>
        <tr>
            <td>姓名:</td>
            <td><form:input path="username"/></td>
        </tr>
        <tr>
            <td>性别:</td>
            <td><form:input path="sex"/></td>
        </tr>
        <tr>
            <td>年龄:</td>
            <td><form:input path="age"/></td>
        </tr>
    </table>
</form:form>
```

注意加粗的代码，这里使用了form表单标签的modelAttribute属性，设置属性值为“user”。

程序清单：codes/04/FormTest/src/org/fkit/controller/UserController

```
@GetMapping (value="/registerForm2")
    public String registerForm2(Model model) {
        User user = new User("jack","男",28);
        // 向model中添加属性user，值是user对象
        model.addAttribute("user",user);
        return "registerForm2";
    }
```

注意加粗的代码，它将user设置到Model当中，属性名不是“command”，而是“user”。

在浏览器中输入如下URL来测试应用：

```
http://localhost:8080/FormTest/registerForm2
```

结果如图4.1所示。

4.1.3 password标签

Spring MVC的password标签会被渲染为一个类型为password的普通HTML input标签。password标签的用法跟input标签相似，也能绑定表单数据，只是它生成的是一个密码框，并且多了一个showPassword属性。

password 标签可使用如表 4.4 所示的属性。表 4.4 中列出的只是 Spring MVC 的 password 标签的常用属性，并没有包含 HTML 的相关属性。

表 4.4　password 标签的属性

属性	描述
cssClass	定义要应用到被渲染的 password 元素的 CSS 类
cssStyle	定义要应用到被渲染的 password 元素的 CSS 样式
cssErrorClass	定义要应用到被渲染的 password 元素的 CSS 类，如果 bound 属性中包含错误，则覆盖 cssClass 属性值
htmlEscape	boolean 值，表示被渲染的值是否应该进行 HTML 转义
path	要绑定的属性路径
showPassword	表示是否应该显示或遮盖密码，默认值为 false

下面是一个 password 标签的例子：

```
<form:password path="password"/>
```

上面的代码运行时 password 标签会被渲染成下面的 HTML 元素：

```
<input id="password" name="password" type="password" value=""/>
```

4.1.4　hidden 标签

Spring MVC 的 hidden 标签会被渲染为一个类型为 hidden 的普通 HTML input 标签。其用法跟 input 标签相似，也能绑定表单数据，只是它生成的是一个隐藏域，没有可视的外观。

hidden 标签可使用如表 4.5 所示的属性。表 4.5 中列出的只是 Spring MVC 的 hidden 标签的常用属性，并没有包含 HTML 的相关属性。

表 4.5　hidden 标签的属性

属性	描述
htmlEscape	boolean 值，表示被渲染的值是否应该进行 HTML 转义
path	要绑定的属性路径

下面是一个 hidden 标签的例子：

```
<form:hidden path="id"/>
```

上面的代码运行时 hidden 标签会被渲染成下面的 HTML 元素：

```
<input id="id" name="id" type="hidden" value=""/>
```

4.1.5　textarea 标签

Spring MVC 的 textarea 标签会被渲染为一个类型为 textarea 的 HTML 标签。textarea 是一个支持多行输入的 HTML 元素。

textarea 标签可使用如表 4.6 所示的属性。表 4.6 中列出的只是 Spring MVC 的 textarea 标签的常用属性，并没有包含 HTML 的相关属性。

表 4.6　textarea 标签的属性

属性	描述
cssClass	定义要应用到被渲染的 textarea 元素的 CSS 类
cssStyle	定义要应用到被渲染的 textarea 元素的 CSS 样式
cssErrorClass	定义要应用到被渲染的 textarea 元素的 CSS 类，如果 bound 属性中包含错误，则覆盖 cssClass 属性值
htmlEscape	boolean 值，表示被渲染的值是否应该进行 HTML 转义
path	要绑定的属性路径

下面是一个 textarea 标签的例子：

```
<form:textarea path="remark" rows="5" cols="20"/>
```

上面的代码运行时 textarea 标签会被渲染成下面的 HTML 元素：

```
<textarea id="remark" name="remark" rows="5" cols="20"></textarea>
```

4.1.6 checkbox 标签

Spring MVC 的 checkbox 标签会被渲染为一个类型为 checkbox 的普通 HTML input 标签。

checkbox 标签可使用如表 4.7 所示的属性。表 4.7 中列出的只是 Spring MVC 的 checkbox 标签的常用属性，并没有包含 HTML 中的相关属性。

表 4.7 checkbox 标签的属性

属性	描述
cssClass	定义要应用到被渲染的 checkbox 元素的 CSS 类
cssStyle	定义要应用到被渲染的 checkbox 元素的 CSS 样式
cssErrorClass	定义要应用到被渲染的 checkbox 元素的 CSS 类，如果 bound 属性中包含错误，则覆盖 cssClass 属性值
htmlEscape	boolean 值，表示被渲染的值是否应该进行 HTML 转义
path	要绑定的属性路径
label	要作为 label 被渲染的复选框的值

1. 绑定 boolean 数据

当 checkbox 绑定的是一个 boolean 数据时，checkbox 的状态跟被绑定的 boolean 数据的状态是一样的，即为 true 时复选框选中，为 false 时复选框不选中。

2. 绑定列表数据

这里的列表数据包括数组、List 和 Set。假设有一个 User 类，User 类有一个类型为 List 的属性 courses。当我们需要显示该 User 的 courses 时，可以使用 checkbox 标签来绑定 courses 数据进行显示。当 checkbox 标签的 value 属性在我们绑定的列表数据中存在时该 checkbox 将为选中状态。

示例：checkbox 标签的使用

程序清单：codes/04/CheckboxTest/src/org/fkit/domain/User

```
import java.io.Serializable;
import java.util.List;
public class User implements Serializable{
    private boolean reader;
    private List<String> courses;
    public User() {
        super();
    }
    public boolean isReader() {
        return reader;
    }
    public void setReader(boolean reader) {
        this.reader = reader;
    }
    public List<String> getCourses() {
        return courses;
    }
    public void setCourses(List<String> courses) {
        this.courses = courses;
    }
}
```

User 类中有一个 boolean 类型的变量 reader 和 List<String>类型的变量 courses，分别用来测试 checkbox 绑定 boolean 数据和绑定列表数据。

程序清单：codes/04/CheckboxTest/src/org/fkit/domain/UserController

```
import java.util.ArrayList;
import java.util.List;
import org.fkit.domain.User;
import org.springframework.stereotype.Controller;
import org.springframework.ui.Model;
import org.springframework.web.bind.annotation.GetMapping;
@Controller
public class UserController{
    @GetMapping(value="/checkboxForm")
    public String registerForm(Model model) {
        User user = new User();
        // 设置 boolean 变量 reader 的值为 true，页面的 checkbox 复选框会被选中
        user.setReader(true);
        // 为集合变量 courses 添加“JAVAEE”和“Spring”，页面的这两项 checkbox 复选框会被选中
        List<String> list = new ArrayList<String>();
        list.add("JAVAEE");
        list.add("Spring");
        user.setCourses(list);
        // 向 model 中添加属性 user，值是 user 对象
            model.addAttribute("user",user);
        return "checkboxForm";
    }
}
```

在 UserController 中创建了 User 对象，并分别设置了变量 reader 和 courses 的值，并将它们添加到 Model 中和页面进行绑定。

程序清单：codes/04/CheckboxTest/WebContent/content/checkboxForm.jsp

```
<%@ taglib prefix="form" uri="http://www.springframework.org/tags/form" %>
<form:form modelAttribute="user" method="post" action="checkboxForm" >
    <table>
        <tr>
            <td>选择课程:</td>
            <td>
                <form:checkbox path="courses" value="JAVAEE" label="JAVAEE"/> 
                <form:checkbox path="courses" value="Mybatis" label="Mybatis"/> 
                <form:checkbox path="courses" value="Spring" label="Spring"/> 
            </td>
        </tr>
    </table>
    <form:checkbox path="reader" value="true"/>已经阅读相关协议
</form:form>
```

web.xml 文件和 springmvc-config.xml 文件与之前讲述的一致，此处不再赘述。

部署 CheckboxTest 这个 Web 应用，在浏览器中输入如下 URL 来测试应用：

```
http://localhost:8080/CheckboxTest/checkboxForm
```

会看到如图 4.2 所示的界面。

图 4.2 测试 checkbox 标签

4.1.7 checkboxes 标签

Spring MVC 的 checkboxes 标签会渲染多个类型为 checkbox 的普通 HTML input 标签。checkboxes 标签可使用如表 4.8 所示的属性。表 4.8 中列出的只是 Spring MVC 的 checkboxes 标签的常用属性，并没有包含 HTML 的相关属性。

表 4.8 checkboxes 标签的属性

属性	描述
cssClass	定义要应用到被渲染的 checkbox 元素的 CSS 类
cssStyle	定义要应用到被渲染的 checkbox 元素的 CSS 样式
cssErrorClass	定义要应用到被渲染的 checkbox 元素的 CSS 类，如果 bound 属性中包含错误，则覆盖 cssClass 属性值
htmlEscape	boolean 值，表示被渲染的值是否应该进行 HTML 转义
path	要绑定的属性路径
items	用于生成 checkbox 元素的对象的 Collection、Map 或者 Array
itemLabel	item 属性中定义的 Collection、Map 或者 Array 中的对象属性，为每个 checkbox 元素提供 label
itemValue	item 属性中定义的 Collection、Map 或者 Array 中的对象属性，为每个 checkbox 元素提供值
delimiter	定义两个 input 元素之间的分隔符，默认没有分隔符

相对于一个 checkbox 标签只能生成一个对应的复选框而言，一个 checkboxes 标签将根据其绑定的数据生成多个复选框。checkboxes 绑定的数据可以是数组、集合和 Map。在使用 checkboxes 标签时有两个属性是必须指定的，一个是 path，另一个是 items。items 表示当前要用来显示的项有哪些，而 path 所绑定的表单对象的属性表示当前表单对象拥有的项，即在 items 所显示的所有项中表单对象拥有的项会被设定为选中状态。

示例：checkboxes 标签的使用

程序清单：codes/04/CheckboxesTest/src/org/fkit/domain/User

```
import java.io.Serializable;
import java.util.List;
public class User implements Serializable{
    private List<String> courses;
    public User() {
        super();
    }
    public List<String> getCourses() {
        return courses;
    }
    public void setCourses(List<String> courses) {
        this.courses = courses;
    }
}
```

程序清单：codes/04/CheckboxesTest/src/org/fkit/domain/UserController

```
import java.util.ArrayList;
import java.util.HashMap;
import java.util.List;
import java.util.Map;
import org.fkit.domain.Dept;
import org.fkit.domain.Employee;
import org.fkit.domain.User;
import org.springframework.stereotype.Controller;
import org.springframework.ui.Model;
import org.springframework.web.bind.annotation.GetMapping;

@Controller
public class UserController{
```

```
@GetMapping(value="/checkboxesForm")
    public String registerForm(Model model) {
        User user = new User();
        // 为集合变量courses添加“JAVAEE”和“Spring”，页面的这两项checkbox复选框会被选中
        List<String> list = new ArrayList<String>();
        list.add("JAVAEE");
        list.add("Spring");
        user.setCourses(list);
        // 页面展现的可供选择的复选框内容courseList
        List<String> courseList = new ArrayList<String>();
        courseList.add("JAVAEE");
        courseList.add("Mybatis");
        courseList.add("Spring");
        // 向model中添加属性user和courseList
        model.addAttribute("user",user);
        model.addAttribute("courseList",courseList);
        return "checkboxesForm";
    }
}
```

在UserController类中创建User对象，并给User对象的courses集合变量添加了“JAVAEE”和“Spring”课程。之后创建了courseList集合变量，该集合变量的内容作为页面显示的可供选择的复选框内容。而页面显示的内容如果在courses中存在，则会被设置为选中状态，即“JAVAEE”和“Spring”内容会默认被选中。

程序清单：codes/04/CheckboxesTest/WebContent/content/checkboxesForm.jsp

```
<%@ taglib prefix="form" uri="http://www.springframework.org/tags/form" %>
<form:form modelAttribute="user" method="post" action="checkboxesForm" >
    <table>
        <tr>
            <td>选择课程:</td>
            <td>
                <form:checkboxes items="${courseList}" path="courses"/>
            </td>
        </tr>
    </table>
</form:form>
```

web.xml文件和springmvc-config.xml文件与之前讲述的一致，此处不再赘述。

部署CheckboxesTest这个Web应用，在浏览器中输入如下URL来测试应用：

```
http://localhost:8080/CheckboxesTest/checkboxesForm
```

会看到如图4.3所示的界面。

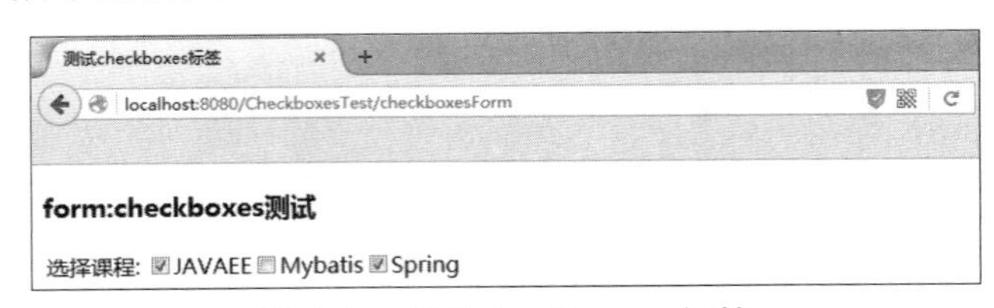

图4.3　测试checkboxes标签

上面介绍的情况是使用List集合作为显示复选框项的数据源，我们可以看到，它所呈现出来的标签label和它的值是一样的。使用Array和Set作为数据源也是这样。

那么如果要让checkboxes呈现出来的label和value不同应该怎么做呢？这时我们可以使用Map作为数据源。使用Map作为checkboxes的items属性的数据源时，Map集合的key将作为真正的复选框的value，而Map集合的value将作为label进行显示。当使用Map作为

checkboxes 的 items 属性的数据源时我们绑定的表单对象属性的类型可以是 Array、集合和 Map，这种情况下就是判断 items Map 中是否含有对应的 key 来决定当前的复选框是否处于选中状态。

程序清单：codes/04/CheckboxesTest/src/org/fkit/domain/UserController

```
@GetMapping(value="/checkboxesForm2")
    public String registerForm2(Model model) {
        User user = new User();
        // 为集合变量 courses 添加“JAVAEE”和“Spring”，页面的这两项 checkbox 复选框会被选中
        List<String> list = new ArrayList<String>();
        list.add("1");
        list.add("3");
        user.setCourses(list);
        // 页面展现的可供选择的复选框内容 courseMap
        Map<String, String> courseMap = new HashMap<String, String>();
        courseMap.put("1","JAVAEE");
        courseMap.put("2","Mybatis");
        courseMap.put("3","Spring");
        // 向 model 中添加属性 user 和 courseList
        model.addAttribute("user",user);
        model.addAttribute("courseMap",courseMap);
        return "checkboxesForm2";
    }
```

在 registerForm2 方法中，提供给页面显示的可供选择的复选框内容 courseMap 是一个 Map，而 user 对象的 courses 集合变量中保存的正是 courseMap 中的 key，它用来决定页面的复选框是否处于选中状态。

程序清单：codes/04/CheckboxesTest/WebContent/content/checkboxesForm2.jsp

```
<form:form modelAttribute="user" method="post" action="checkboxesForm2" >
   <table>
      <tr>
         <td>选择课程:</td>
         <td>
            <form:checkboxes items="${courseMap}" path="courses"/>
         </td>
      </tr>
   </table>
</form:form>
```

在浏览器中输入如下 URL 来测试应用：

```
http://localhost:8080/CheckboxesTest/checkboxesForm2
```

请求的结果和图 4.3 所示的一致，读者可以通过右键→查看源代码，发现 checkbox 的 value 和 label 不同了，value 的值正是 Map 的 key，而 label 的值正是 Map 的 value。

当使用 Array 或者集合作为数据源，且里面的元素都是一个 domain 对象时，还可以使用 checkboxes 标签的 itemLabel 和 itemValue 属性来表示，使用数组或者集合中元素对象的哪一个属性作为需要呈现的单选框的 label 和 value。

程序清单：codes/04/CheckboxesTest/src/org/fkit/domain/Dept

```
import java.io.Serializable;
public class Dept implements Serializable{
   private Integer id;
   private String name;
   public Dept() {
      super();
      // TODO Auto-generated constructor stub
   }
```

```
    public Dept(Integer id, String name) {
        super();
        this.id = id;
        this.name = name;
    }
    public Integer getId() {
        return id;
    }
    public void setId(Integer id) {
        this.id = id;
    }
    public String getName() {
        return name;
    }
    public void setName(String name) {
        this.name = name;
    }
}
```

Dept 类表示一个部门，有 id 和 name 两个属性。之后页面将使用 id 的值作为 value，name 的值作为 label。

程序清单：codes/04/CheckboxesTest/src/org/fkit/domain/Employee

```
import java.io.Serializable;
public class Employee implements Serializable{
    private List<Dept> depts;
    public List<Dept> getDepts() {
        return depts;
    }
    public void setDepts(List<Dept> depts) {
        this.depts = depts;
    }
}
```

Employee 类提供了一个 List 集合属性 depts，用于绑定数据。

程序清单：codes/04/CheckboxesTest/src/org/fkit/domain/UserController

```
@GetMapping(value="/checkboxesForm3")
    public String registerForm3(Model model) {
        Employee employee = new Employee();
        Dept dept = new Dept(1,"开发部");
        // 为集合变量depts添加Dept对象,该对象的id=1,name="开发部",页面的这一项checkbox
        // 复选框会被选中
        List<Dept> list = new ArrayList<Dept>();
        list.add(dept);
        employee.setDepts(list);
        // 页面展现的可供选择的复选框内容deptList
        List<Dept> deptList = new ArrayList<Dept>();
        deptList.add(dept);
        deptList.add(new Dept(2,"销售部"));
        deptList.add(new Dept(3,"财务部"));
        // 向model中添加属性employee和deptList
            model.addAttribute("employee",employee);
            model.addAttribute("deptList",deptList);
        return "checkboxesForm3";
    }
```

程序清单：codes/04/CheckboxesTest/WebContent/content/checkboxesForm3.jsp

```
<form:form modelAttribute="employee" method="post" action="checkboxesForm3" >
    <table>
        <tr>
            <td>选择部门:</td>
```

```
            <td>
                <form:checkboxes items="${deptList}" path="depts"
                    itemLabel="name" itemValue="id"/>
            </td>
        </tr>
    </table>
</form:form>
```

在浏览器中输入如下 URL 来测试应用：

```
http://localhost:8080/CheckboxesTest/checkboxesForm3
```

会看到如图 4.4 所示的界面。

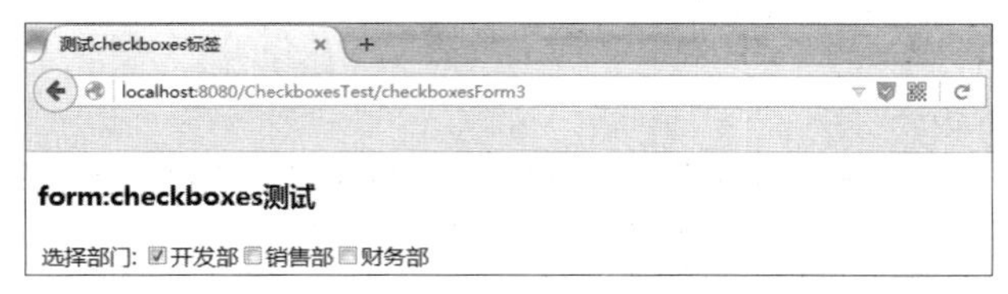

图 4.4　测试 checkboxes 标签

可以看到，从集合中传出来的 Dept 对象的 name 作为 label 显示，id 作为 value 显示。

4.1.8　radiobutton 标签

Spring MVC 的 radiobutton 标签会被渲染为一个类型为 radio 的普通 HTML input 标签。

radiobutton 标签可使用如表 4.9 所示的属性。表 4.9 中列出的只是 Spring MVC 的 radiobutton 标签的常用属性，并没有包含 HTML 的相关属性。

表 4.9　radiobutton 标签的属性

属性	描述
cssClass	定义要应用到被渲染的 radiobutton 元素的 CSS 类
cssStyle	定义要应用到被渲染的 radiobutton 元素的 CSS 样式
cssErrorClass	定义要应用到被渲染的 radiobutton 元素的 CSS 类，如果 bound 属性中包含错误，则覆盖 cssClass 属性值
htmlEscape	boolean 值，表示被渲染的值是否应该进行 HTML 转义
path	要绑定的属性路径
label	要作为 label 被渲染复选框的值

示例：radiobutton 标签的使用

程序清单：codes/04/RadiobuttonTest/src/org/fkit/domain/User

```
import java.io.Serializable;
public class User implements Serializable{
    private String sex;
    public String getSex() {
        return sex;
    }
    public void setSex(String sex) {
        this.sex = sex;
    }
```

User 类中定义了一个属性 sex，用来绑定页面的 radiobutton 标签数据。

程序清单：codes/04/RadiobuttonTest/src/org/fkit/domain/UserController

```
import java.util.ArrayList;
import java.util.HashMap;
import java.util.List;
import java.util.Map;
```

```
import org.fkit.domain.User;
import org.springframework.stereotype.Controller;
import org.springframework.ui.Model;
import org.springframework.web.bind.annotation.GetMapping;
@Controller
public class UserController{
    @GetMapping(value="/radiobuttonForm")
    public String registerForm(Model model) {
        User user = new User();
        // 设置 sex 变量的值为“男”，页面的 radio 单选框的 value=“男”时其会被选中
        user.setSex("男");
        model.addAttribute("user", user);
        return "radiobuttonForm";
    }
}
```

程序清单：codes/04/RadiobuttonTest /WebContent/content/radiobuttonForm.jsp

```
<%@ taglib prefix="form" uri="http://www.springframework.org/tags/form" %>
<form:form modelAttribute="user" method="post" action="radiobuttonForm" >
    <table>
        <tr>
            <td>性别:</td>
            <td>
                <form:radiobutton path="sex" value="男"/>男 
                <form:radiobutton path="sex" value="女"/>女 
            </td>
        </tr>
    </table>
</form:form>
```

web.xml 文件和 springmvc-config.xml 文件与之前讲述的一致，此处不再赘述。

部署 RadiobuttonTest 这个 Web 应用，在浏览器中输入如下 URL 来测试应用：

```
http://localhost:8080/RadiobuttonTest/radiobuttonForm
```

会看到如图 4.5 所示的界面。

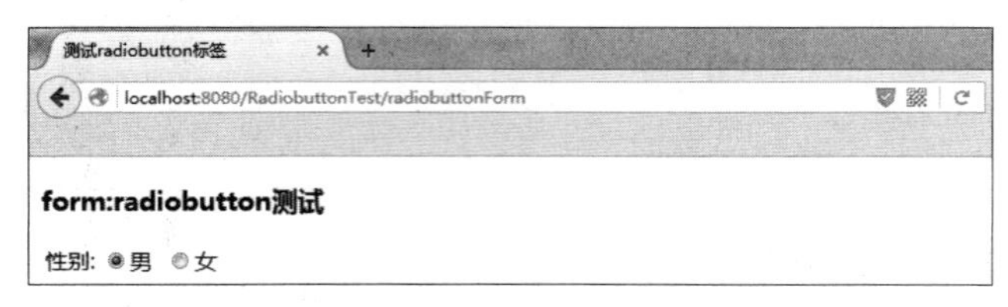

图 4.5 测试 radiobutton 标签

4.1.9 radiobuttons 标签

Spring MVC 的 radiobuttons 标签会渲染多个类型为 radio 的普通 HTML input 标签。radiobuttons 标签可使用如表 4.10 所示的属性。表 4.10 中列出的只是 Spring MVC 的 radiobuttons 标签的常用属性，并没有包含 HTML 的相关属性。

表 4.10 radiobuttons 标签的属性

属性	描述
cssClass	定义要应用到被渲染的 radio 元素的 CSS 类
cssStyle	定义要应用到被渲染的 radio 元素的 CSS 样式
cssErrorClass	定义要应用到被渲染的 radio 元素的 CSS 类，如果 bound 属性中包含错误，则覆盖 cssClass 属性值
htmlEscape	boolean 值，表示被渲染的值是否应该进行 HTML 转义
path	要绑定的属性路径

续表

属性	描述
items	用于生成 radio 元素的对象的 Collection、Map 或者 Array
itemLabel	item 属性中定义的 Collection、Map 或者 Array 中的对象属性，为每个 radio 元素提供 label
itemValue	item 属性中定义的 Collection、Map 或者 Array 中的对象属性，为每个 radio 元素提供值
delimiter	定义两个 input 元素之间的分隔符，默认没有分隔符

相对于一个 radiobutton 标签只能生成一个对应的单选框，一个 radiobuttons 标签将根据其绑定的数据生成多个单选框。radiobuttons 绑定的数据可以是数组、集合和 Map。在使用 radiobuttons 时有两个属性是必须指定的，一个是 path，另一个是 items。items 表示当前要用来显示的项有哪些，而 path 所绑定的表单对象的属性表示当前表单对象拥有的项，即在 items 所显示的所有项中表单对象拥有的项会被设定为选中状态。

示例：radiobuttons 标签的使用

程序清单：codes/04/RadiobuttonTest/src/org/fkit/domain/UserController

```
@GetMapping(value="/radiobuttonsForm")
    public String registerForm2(Model model) {
        User user = new User();
        // 设置 sex 变量的值为“男”，页面的 radio 单选框的 value=“男”时其会被选中
        user.setSex("男");
// 页面展现的可供选择的单选框内容 sexList
        List<String> sexList = new ArrayList<String>();
        sexList.add("男");
        sexList.add("女");
        model.addAttribute("user", user);
        model.addAttribute("sexList", sexList);
        return "radiobuttonsForm";
    }
```

程序清单：codes/04/RadiobuttonTest /WebContent/content/ radiobuttonsForm.jsp

```
<%@ taglib prefix="form" uri="http://www.springframework.org/tags/form" %>
<form:form modelAttribute="user" method="post" action="radiobuttonsForm" >
    <table>
        <tr>
            <td>性别:</td>
            <td>
                <form:radiobuttons path="sex" items="${sexList }"/>
            </td>
        </tr>
    </table>
</form:form>
```

在浏览器中输入如下 URL 来测试应用：

```
http://localhost:8080/RadiobuttonTest/radiobuttonsForm
```

运行结果如图 4.5 所示。内容虽然是一样，这里与 radiobutton 的区别在于，之前的页面有两个 radiobutton 标签，而现在页面只有一个 radiobuttons 标签，单选框的 value 和 label 来自于后台的 List 集合。

我们介绍的这种情况是使用 List 集合作为显示单选框项的数据源，可以看到，它所呈现出来的标签 label 和它的值是一样的。使用 Array 和 Set 作为数据源也是这样。

那么要让 radiobuttons 呈现出来的 label 和 value 不同应该怎么做呢？这时我们可以使用 Map 作为数据源。当使用 Map 作为 radiobuttons 的 items 属性的数据源时，Map 集合的 key 将作为真正的单选框的 value，而 Map 集合的 value 将作为 label 进行显示。当使用 Map 作为

radiobuttons 的 items 属性的数据源时我们绑定的表单对象属性的类型可以是 Array、集合和 Map，这种情况就是判断 items Map 中是否含有对应的 key 来决定当前的单选框是否处于选中状态。

程序清单：codes/04/RadiobuttonTest/src/org/fkit/domain/UserController

```
@GetMapping(value="/radiobuttonsForm2")
    public String registerForm3(Model model) {
        User user = new User();
        // 设置 sex 变量的值为"1"，页面的 radio 单选框的 value=“男”时其会被选中
        user.setSex("1");
         // 页面展现的可供选择的复选框内容 sexMap
        Map<String, String> sexMap = new HashMap<String, String>();
        sexMap.put("1", "男");
        sexMap.put("2", "女");
        model.addAttribute("user", user);
        model.addAttribute("sexMap", sexMap);
        return "radiobuttonsForm2";
    }
```

在 registerForm3 方法中，提供给页面显示的可被选择的单选框内容 sexMap 是一个 Map，而 user 对象的 sex 变量中保存的正是 sexMap 中的 key，其用来决定页面的单选框是否处于选中状态。

程序清单：codes/04/RadiobuttonTest /WebContent/content/ radiobuttonsForm2.jsp

```
<form:form modelAttribute="user" method="post" action="radiobuttonForm2" >
    <table>
        <tr>
            <td>性别:</td>
            <td>
                <form:radiobuttons path="sex" items="${sexMap }"/>
            </td>
        </tr>
    </table>
</form:form>
```

在浏览器中输入如下 URL 来测试应用：

```
http://localhost:8080/RadiobuttonTest/radiobuttonsForm2
```

请求的结果和图 4.5 所示的一致，读者可以通过右键→查看源代码，发现 radiobuttons 的 value 和 label 不同了，value 的值正是 Map 的 key，而 label 的值正是 Map 的 value。

当使用 Array 或者集合作为数据源，且里面的元素都是一个 domain 对象时，我们还可以使用 radiobuttons 标签的 itemLabel 和 itemValue 属性来表示，使用数组或者集合中元素对象的哪一个属性作为需要呈现的单选框的 label 和 value。用法和之前 checkboxes 类似，此处不再赘述。

4.1.10 select 标签

Spring MVC 的 select 标签会渲染一个 HTML 的 select 元素。被渲染元素的选项可能来自其 items 属性的一个 Collectin、Map 及 Array，或者来自一个嵌套的 option 或者 options 标签。select 标签可使用如表 4.11 所示的属性。表 4.11 中列出的只是 Spring MVC 的 select 标签的常用属性，并没有包含 HTML 的相关属性。

表 4.11　select 标签的属性

属性	描述
cssClass	定义要应用到被渲染的 select 元素的 CSS 类
cssStyle	定义要应用到被渲染的 select 元素的 CSS 样式
cssErrorClass	定义要应用到被渲染的 select 元素的 CSS 类，如果 bound 属性中包含错误，则覆盖 cssClass 属性值
htmlEscape	boolean 值，表示被渲染的值是否应该进行 HTML 转义
path	要绑定的属性路径
items	用于生成 select 元素的对象的 Collection、Map 或者 Array
itemLabel	item 属性中定义的 Collection、Map 或者 Array 中的对象属性，为每个 select 元素提供 label
itemValue	item 属性中定义的 Collection、Map 或者 Array 中的对象属性，为每个 select 元素提供值

其中，items 属性特别有用，因为它可以绑定到对象的 Collection、Map、Array 上，为 select 元素生成选项。

4.1.11　option 标签

Spring MVC 的 option 标签会渲染 select 元素中使用的一个 HTML 的 option 元素。option 标签可使用如表 4.12 所示的属性。表 4.12 中列出的只是 Spring MVC 的 option 标签的常用属性，并没有包含 HTML 的相关属性。

表 4.12　option 标签的属性

属性	描述
cssClass	定义要应用到被渲染的 option 元素的 CSS 类
cssStyle	定义要应用到被渲染的 option 元素的 CSS 样式
cssErrorClass	定义要应用到被渲染的 option 元素的 CSS 类，如果 bound 属性中包含错误，则覆盖 cssClass 属性值
htmlEscape	boolean 值，表示被渲染的值是否应该进行 HTML 转义

4.1.12　options 标签

Spring MVC 的 options 标签会渲染 select 元素中使用的一个 HTML 的 option 元素列表。options 标签可使用如表 4.13 所示的属性。表 4.13 中列出的只是 Spring MVC 的 options 标签的常用属性，并没有包含 HTML 的相关属性。

表 4.13　options 标签的属性

属性	描述
cssClass	定义要应用到被渲染的 option 元素的 CSS 类
cssStyle	定义要应用到被渲染的 option 元素的 CSS 样式
cssErrorClass	定义要应用到被渲染的 option 元素的 CSS 类，如果 bound 属性中包含错误，则覆盖 cssClass 属性值
htmlEscape	boolean 值，表示被渲染的值是否应该进行 HTML 转义
items	用于生成 option 列表元素的对象的 Collection、Map 或者 Array
itemLabel	item 属性中定义的 Collection、Map 或者 Array 中的对象属性，为每个 option 元素提供 label
itemValue	item 属性中定义的 Collection、Map 或者 Array 中的对象属性，为每个 option 元素提供值

示例：select、option 和 options 标签的使用

程序清单：codes/04/SelectTest/src/org/fkit/domain/User

```
import java.io.Serializable;
public class User implements Serializable{
    // 部门编号
    private Integer deptId;
```

```
    public User() {
        super();
        // TODO Auto-generated constructor stub
    }
    public Integer getDeptId() {
        return deptId;
    }
    public void setDeptId(Integer deptId) {
        this.deptId = deptId;
    }
}
```

程序清单：codes/04/SelectTest/src/org/fkit/controller/UserController

```
import java.util.ArrayList;
import java.util.HashMap;
import java.util.List;
import java.util.Map;
import org.fkit.domain.Dept;
import org.fkit.domain.User;
import org.springframework.stereotype.Controller;
import org.springframework.ui.Model;
import org.springframework.web.bind.annotation.GetMapping;
@Controller
public class UserController{
@GetMapping (value="/selectForm")
     public String selectForm(Model model) {
         User user = new User();
         // 设置deptId的值，页面的select下拉框对应的option项会被选中
         user.setDeptId(2);
         model.addAttribute("user", user);
         return "selectForm";
     }
}
```

程序清单：codes/04/SelectTest/WebContent/content/selectForm.jsp

```
<h3>form:select 标签直接添加 form:option</h3>
<form:form modelAttribute="user" method="post" action="selectForm" >
    <table>
        <tr>
            <td>部门:</td>
            <td>
                <form:select path="deptId">
                    <form:option value="1">财务部</form:option>
                    <form:option value="2">开发部</form:option>
                    <form:option value="3">销售部</form:option>
                </form:select>
            </td>
        </tr>
    </table>
</form:form>
```

在selectForm.jsp页面中使用Spring MVC的select标签，path绑定Model的deptId属性，select标签里面使用了Spring MVC的option标签来直接添加部门数据。由于被绑定的deptId属性的值是2，所以下拉框的value="2"的"开发部"会默认被选中。

web.xml文件和springmvc-config.xml文件与之前讲述的一致，此处不再赘述。

部署SelectTest这个Web应用，在浏览器中输入如下URL来测试应用：

```
http://localhost:8080/SelectTest/selectForm
```

会看到如图4.6所示的界面。

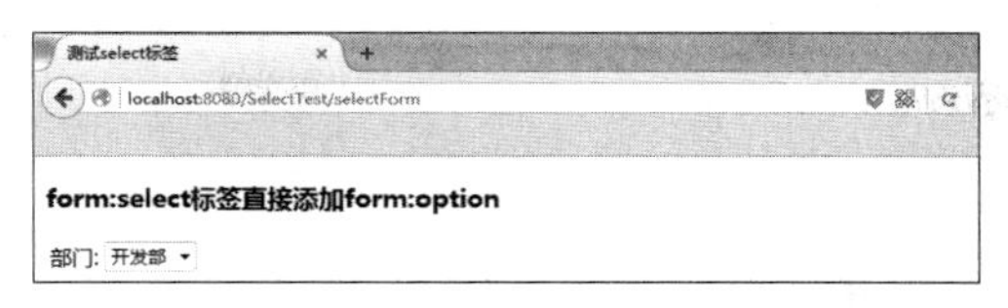

图 4.6　测试 select 标签

在 selectForm.jsp 页面中使用了 Spring MVC 的 option 标签直接添加部门数据，除此之外，还可以使用 select 标签的 items 属性自动加载后台传递出来的数据并将其显示在下拉框当中。

程序清单：codes/04/SelectTest/src/org/fkit/controller/UserController

```
@GetMapping(value="/selectForm2")
    public String selectForm2(Model model) {
        User user = new User();
        user.setDeptId(2);
        // 页面展现的可供选择的 select 下拉框内容 deptMap
        Map<Integer, String> deptMap = new HashMap<Integer, String>();
        deptMap.put(1, "财务部");
        deptMap.put(2, "开发部");
        deptMap.put(3, "销售部");
        model.addAttribute("user", user);
        model.addAttribute("deptMap", deptMap);
        return "selectForm2";
    }
```

程序清单：codes/04/SelectTest /WebContent/content/selectForm2.jsp

```
<h3>form:select 标签 items 属性绑定 Map</h3>
<form:form modelAttribute="user" method="post" action="selectForm2" >
    <table>
        <tr>
            <td>部门:</td>
            <td>
                <form:select path="deptId" items="${deptMap}"/>
            </td>
        </tr>
    </table>
</form:form>
```

selectForm2.jsp 页面中的 items 会加载 Model 中的 deptMap 数据并将其显示在页面上。

在浏览器中输入如下 URL 来测试应用：

```
http://localhost:8080/SelectTest/selectForm2
```

会看到如图 4.7 所示的界面。

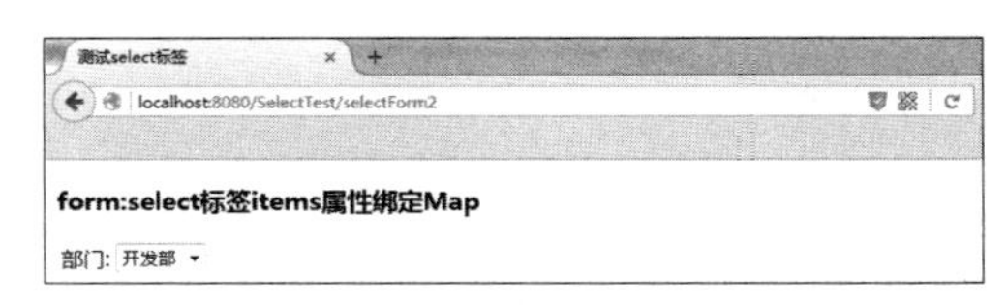

图 4.7　测试 select 标签

还可以使用 options 标签的 items 属性自动加载后台传递出来的数据并将其显示在下拉框当中。

程序清单：codes/04/SelectTest /WebContent/content/selectForm3.jsp

```
<h3>使用 form:options 标签 items 属性绑定 Map</h3>
<form:form modelAttribute="user" method="post" action="selectForm" >
    <table>
        <tr>
            <td>学历:</td>
```

```
            <td>
                <form:select path="deptId">
                    <form:options items="${deptMap}"/>
                </form:select>
            </td>
        </tr>
    </table>
</form:form>
```

selectForm3.jsp 页面中使用了 options 标签的 items 属性加载数据，后台处理请求方法的代码和之前的 selectForm2 方法相同，此处不再赘述。

在浏览器中输入如下 URL 来测试应用：

```
http://localhost:8080/SelectTest/selectForm3
```

会看到如图 4.8 所示的界面。

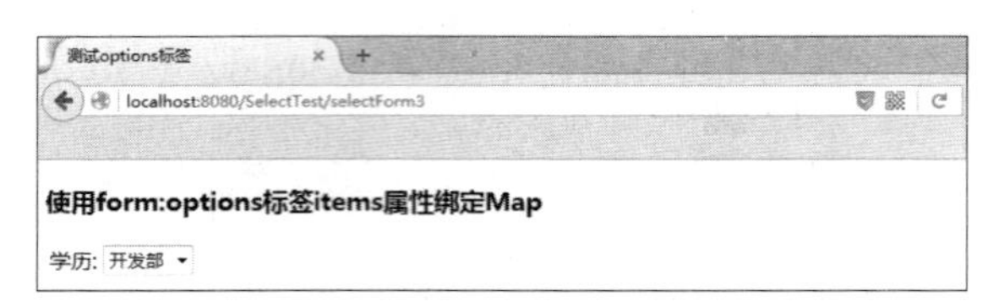

图 4.8　测试 options 标签

在实际开发中，经常会出现一种情况，即 select 下拉框中的数据来自于数据库的表数据，并且获取的数据被封装到 JavaBean 中。此时，就可以使用 select 标签或者 options 标签的 items、itemLabel 和 itemValue 属性来加载数据。

程序清单：codes/04/SelectTest/src/org/fkit/domain/Dept

```
public class Dept {
    private Integer id;
    private String name;
    public Dept() {
        super();
        // TODO Auto-generated constructor stub
    }
    public Dept(Integer id, String name) {
        super();
        this.id = id;
        this.name = name;
    }
    public Integer getId() {
        return id;
    }
    public void setId(Integer id) {
        this.id = id;
    }
    public String getName() {
        return name;
    }
    public void setName(String name) {
        this.name = name;
    }
}
```

程序清单：codes/04/SelectTest/src/org/fkit/controller/UserController

```
@GetMapping (value="/selectForm4")
     public String selectForm4(Model model) {
         User user = new User();
         user.setDeptId(2);
         // 页面展现的可供选择的 select 下拉框内容 deptList，其中的元素为 Dept 对象
```

```
        // 模拟从数据库获取到部门信息并将其封装到对象当中
        List<Dept> deptList = new ArrayList<Dept>();
        deptList.add(new Dept(1, "财务部"));
        deptList.add(new Dept(2, "开发部"));
        deptList.add(new Dept(3, "销售部"));
        model.addAttribute("user", user);
        model.addAttribute("deptList", deptList);
        return "selectForm4";
    }
```

在 selectForm4 方法中模拟从数据库中获取部门信息，并将其封装到 Dept 对象中，且将多个部门信息装载到 List 集合中，最后添加到 Model 当中。

程序清单：codes/04/SelectTest /WebContent/content/selectForm4.jsp

```
<h3>form:select 标签使用 form:options 绑定 Object</h3>
<form:form modelAttribute="user" method="post" action="selectForm" >
    <table>
        <tr>
            <td>学历:</td>
            <td>
                <form:select path="deptId">
                    <form:options items="${deptList}"
                    itemLabel="name" itemValue="id"/>
                </form:select>
            </td>
        </tr>
    </table>
</form:form>
```

selectForm4.jsp 页面的 options 标签的 items 标签加载 Model 中的 deptList，并将集合中的元素及 Dept 对象的 id 属性设置为 option 的 value，name 属性设置为 option 的 label。

在浏览器中输入如下 URL 来测试应用：

```
http://localhost:8080/SelectTest/selectForm4
```

会看到如图 4.9 所示的界面。

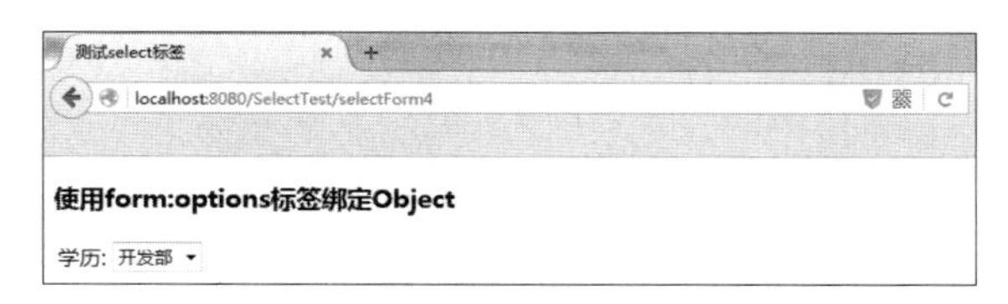

图 4.9 测试 options 标签绑定对象

4.1.13 errors 标签

Spring MVC 的 errors 标签是对应于 Spring MVC 的 Errors 对象的，它的作用就是显示 Errors 对象中包含的错误信息。如果 Errors 不为 null，则会渲染一个 HTML 的 span 元素，用来显示错误信息。

表 4.14 errors 标签的属性

属性	描述
cssClass	定义要应用到被渲染的 errors 元素 CSS 类
cssStyle	定义要应用到被渲染的 errors 元素 CSS 样式
htmlEscape	boolean 值，表示被渲染的值是否应该进行 HTML 转义
delimiter	定义两个 input 元素之间的分隔符，默认没有分隔符
path	要绑定的属性路径

利用errors标签来显示Errors时，是通过errors标签的path属性绑定一个错误信息实现的。可以通过path属性来显示两种类型的错误信息：

- 所有的错误信息，这个时候path的值应该设置为“*”。
- 当前对象的某一个属性的错误信息，这个时候path的值应为所需显示的属性的名称。

示例：errors 标签的使用

提示

errors标签的示例主要是演示errors标签的用法，关于Spring MVC验证的知识，在第6章重点介绍。

程序清单：codes/04/ErrorsTest/src/org/fkit/domain/User

```
import java.io.Serializable;
public class User implements Serializable{
    private String username;
    private String sex;
    private Integer age;
    public User() {
        super();
    }
    public String getUsername() {
        return username;
    }
    public void setUsername(String username) {
        this.username = username;
    }
    public String getSex() {
        return sex;
    }
    public void setSex(String sex) {
        this.sex = sex;
    }
    public Integer getAge() {
        return age;
    }
    public void setAge(Integer age) {
        this.age = age;
    }
}
```

程序清单：codes/04/ErrorsTest/src/org/fkit/validator/UserValidator

```
import org.fkit.domain.User;
import org.springframework.validation.Errors;
import org.springframework.validation.ValidationUtils;
import org.springframework.validation.Validator;
public class UserValidator implements Validator{
    @Override
    public boolean supports(Class<?> clazz) {
        return User.class.equals(clazz);
    }
    @Override
    public void validate(Object object, Errors errors) {
        // 验证 username、sex 和 age 是否为 null
        ValidationUtils.rejectIfEmpty(errors, "username", null, "用户名不能为空");
        ValidationUtils.rejectIfEmpty(errors, "sex", null, "性别不能为空");
        ValidationUtils.rejectIfEmpty(errors, "age", null, "年龄不能为空");
    }
}
```

UserValidator 类实现 org.springframework.validation.Validator 接口，完成验证的功能。

程序清单：codes/04/ErrorsTest/src/org/fkit/controller/UserController

```
import org.fkit.domain.User;
import org.fkit.validator.UserValidator;
import org.springframework.stereotype.Controller;
import org.springframework.ui.Model;
import org.springframework.validation.DataBinder;
import org.springframework.validation.Errors;
import org.springframework.validation.annotation.Validated;
import org.springframework.web.bind.annotation.GetMapping;
import org.springframework.web.bind.annotation.InitBinder;
import org.springframework.web.bind.annotation.PostMapping;
@Controller
public class UserController{
    @GetMapping(value="/registerForm")
    public String registerForm(Model model) {
        User user = new User();
        // 向model中添加属性user，值是user对象
        model.addAttribute("user",user);
        return "registerForm";
    }
    @InitBinder
    public void initBinder(DataBinder binder) {
       // 设置验证的类为UserValidator
       binder.setValidator(new UserValidator());
    }
    @PostMapping(value="/register")
    public String register(@Validated User user, Errors errors) {
        // 如果Errors对象有Field错误，则重新跳回注册页面，否则正常提交
        if (errors.hasFieldErrors())
            return "registerForm";
        return "submit";
    }
}
```

UserController 类使用@InitBinder 注解绑定验证对象。

程序清单：codes/04/ErrorsTest/WebContent/content/registerForm.jsp

```
<form:form modelAttribute="user" method="post" action="register" >
    <table>

        <tr>
            <td>姓名:</td>
            <td><form:input path="username"/></td>
            <td><font color="red"><form:errors path="username"/></font></td>
        </tr>
        <tr>
            <td>性别:</td>
            <td><form:input path="sex"/></td>
            <td><font color="red"><form:errors path="sex"/></font></td>
        </tr>
        <tr>
            <td>年龄:</td>
            <td><form:input path="age"/></td>
            <td><font color="red"><form:errors path="age"/></font></td>
        </tr>
        <tr>
            <td><input type="submit" value="提交"/></td>
        </tr>
    </table>
</form:form>
```

在 registerForm.jsp 页面中，在每个需要输入的控件后面增加了一个 errors 标签，用来显示错误信息。

web.xml 文件和 springmvc-config.xml 文件与之前讲述的一致，此处不再赘述。

部署 ErrorsTest 这个 Web 应用，在浏览器中输入如下 URL 来测试应用：

```
http://localhost:8080/ErrorsTest/registerForm
```

跳转到注册页面，如果不输入任何信息，直接单击“注册”按钮，则会提交请求。因为没有提交任何注册信息，故验证出错。register 请求处理方法会将请求重新转发到注册页面，errors 标签会显示错误提示信息，如图 4.10 所示。

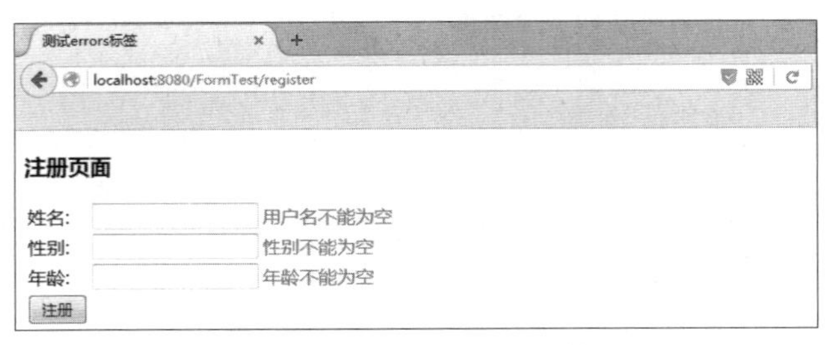

图 4.10　测试 errors 标签

4.2　本章小结

本章介绍了 Spring MVC 的表单标签和如何使用表单标签绑定数据。表单标签的功能非常强大，读者需要好好掌握。

第 5 章将重点介绍 Spring MVC 的国际化知识。

CHAPTER

5

第5章 Spring MVC 的国际化

本章要点

- messageSource 接口
- localeResolver 接口
- message 标签
- AcceptHeaderLocaleResolver 国际化
- SessionLocaleResolver 国际化
- CookieLocaleResolver 国际化

程序国际化是商业系统的一个基本要求，今天的软件系统不再是简单的单机程序，往往都是一个开放的系统，需要面对来自全世界各个地方的访问者，因此，国际化成为商业系统必不可少的一部分。

Spring MVC 的国际化是建立在 Java 国际化的基础之上的，一样也是首先提供不同国家/语言环境的消息资源，然后通过 ResourceBundle 加载指定 Locale 对应的资源文件，再取得该资源文件中指定 key 对应的消息。这整个过程与 Java 程序的国际化完全相同，只是 Spring MVC 框架对 Java 程序国际化进行了进一步的封装，从而简化了应用程序的国际化。

提示： 关于 Java 程序国际化的相关知识，请读者参考“疯狂 Java 体系”的《疯狂 Java 讲义》一书，在该书的 7.6 节有关于 Java 程序国际化的详细介绍。

Spring MVC 的国际化的步骤与 Java 国际化的步骤基本相似，只是实现起来更加简单。Spring MVC 的国际化可按如下步骤进行。

1. 给系统加载国际化资源文件。
2. 输出国际化消息。Spring MVC 输出国际化消息有两种方式：
   - 在视图页面上输出国际化消息，需要使用 Spring MVC 的标签库。
   - 在 Controller 的处理方法中输出国际化消息，需要使用 org.springframework.web.servlet. support.RequestContext 的 getMessage()方法来完成。

5.1 Spring MVC 国际化的相关知识

5.1.1 messageSource 接口

在 Spring MVC 中，不直接使用 java.util.ResourceBundle，而是使用名为 messageSource 的 Bean 告诉 Spring MVC 国际化的属性文件保存在哪里。配置信息代码如下所示：

```
<bean id="messageSource"
      class="org.springframework.context.support.ResourceBundleMessageSource">
      <property name="basenames">
          <list>
              <value>message</value>
              <value>fkit</value>
          </list>
      </property>
</bean>
```

上面的配置使用了 ResourceBundleMessageSource 类作为 messageSource Bean 的实现。basenames 属性用来指定国际化的属性文件名称。如果项目中只有一组属性文件，则可以使用 basenam 来指定国际化的属性文件名称，实例代码如下：

```
<bean id="messageSource"
      class="org.springframework.context.support.ResourceBundleMessageSource"
      p:basenames="message">
</bean>
```

5.1.2 localeResolver 接口

为用户选择语言区域时，最常用的方法是通过读取用户浏览器的 accept-language 标题值，accept-language 标题提供了关于用户浏览器语言的信息。选择语言区域的其他方法还包括读取 HttpSession 或者 Cookie。

在 Spring MVC 中选择语言区域，可以使用语言区域解析器。Spring MVC 提供了一个语言区域解析器接口 LocaleResolver，该接口的常用实现类都在 org.springframework.web.servlet.i18n 包下面，包括：

- AcceptHeaderLocaleResolver
- SessionLocaleResolver
- CookieLocaleResolver

其中，AcceptHeaderLocaleResolver 是默认的，也是最容易使用的语言区域解析器。使用它，Spring MVC 会读取浏览器的 accept-language 标题，来确定使用哪个语言区域。AcceptHeaderLocaleResolver 可以不用显式配置，而 SessionLocaleResolver 和 CookieLocale-Resolver 需要手动显式配置。

5.1.3 message 标签

在 Spring MVC 中显示本地化消息通常使用 Spring 的 message 标签。使用 message 标签，需要在 JSP 页面最前面使用 taglib 指令导入 Spring 的标签库，如下所示：

```
<%@taglib prefix= "spring" uri= "http://www.springframework.org/tags" %>
```

message 标签的属性如表 5.1 所示，所有属性都是可选的。

表 5.1 message 标签的属性

属性	描述
arguments	标签的参数，可以是一个字符串、数组或对象
argumentSeparator	用来分隔该标签参数的字符
code	获取消息的 key
htmlEscape	boolean 值，表示被渲染的值是否应该进行 HTML 转义
javaScriptEscape	boolean 值，表示被渲染的值是否应该进行 JavaScript 转义
message	MessageSourceResolvable 参数
scope	保存 var 属性中定义的变量的作用域
text	如果 code 属性不存在，所显示的默认文本
var	用于保存消息的变量

5.2 Spring MVC 的国际化处理

5.2.1 AcceptHeaderLocaleResolver

AcceptHeaderLocaleResolver 是默认的，也是最容易使用的语言区域解析器。使用它，Spring MVC 会读取浏览器的 accept-language 标题，来确定使用哪个语言区域。AcceptHeaderLocale-Resolver 可以不用显式配置，当然也可以显式配置。

示例：基于浏览器请求的国际化实现

首先，准备两个资源文件，第一个资源文件是 message_en_US.properties，该文件内容如下。

程序清单：codes/05/I18NTest/src/message_en_US.properties

```
loginname= Login name:
password = Password:
submit = Submit
welcome = Welcom {0} access fkit
title = Login Page
username = administrator
```

第二个资源文件的内容如下。

程序清单：codes/05/I18NTest/src/message_zh_CN.properties

```
loginname= 登录名:
password = 密码:
submit = 提交
welcome = 欢迎 {0} 访问 疯狂软件
title = 登录页面
username = 管理员
```

使用 native2ascii 命令处理这个文件，处理后生成的新文件名为 message_zh_CN.properties。

接下来在 JSP 页面中通过<spring:message /> 标签来输出国际化消息。

程序清单：codes/05/I18NTest/WebContent/WEB-INF/content/loginForm.jsp

```
<%@ page language="java" contentType="text/html; charset=UTF-8"
   pageEncoding="UTF-8"%>
<%@taglib prefix= "form" uri= "http://www.springframework.org/tags/form" %>
<%@taglib prefix= "spring" uri= "http://www.springframework.org/tags" %>
<!DOCTYPE html PUBLIC "-//W3C//DTD HTML 4.01 Transitional//EN" "http://www.w3.org/
TR/html4/loose.dtd">
<html>
<head>
<meta http-equiv="Content-Type" content="text/html; charset=UTF-8">
<title>测试基于浏览器请求的国际化</title>
</head>
<body>
<!-- 使用 message 标签来输出国际化信息 -->
<h3><spring:message code="title"/></h3>
<form:form modelAttribute="user" method="post" action="login" >
   <table>
      <tr>
         <td><spring:message code="loginname"/></td>
         <td><form:input path="loginname"/></td>
      </tr>
      <tr>
         <td><spring:message code="password"/></td>
         <td><form:input path="password"/></td>
      </tr>
      <tr>
         <td><input type="submit" value="<spring:message code="submit"/>"/></td>
      </tr>
   </table>
</form:form>
</body>
</html>
```

接下来在 Spring MVC 配置文件中加载国际化资源文件。

程序清单：codes/05/I18NTest/WebContent/WEB-INF/springmvc-config.xml

```
<!-- 国际化 p:basenames 属性用于指定国际化资源文件名 -->
<bean id="messageSource"
      class="org.springframework.context.support.ResourceBundleMessageSource"
     p:basenames="message"/>
<mvc:interceptors>
   <!-- 国际化操作拦截器如果采用基于（Session/Cookie）则必须配置 -->
   <bean class="org.springframework.web.servlet.i18n.LocaleChangeInterceptor" />
</mvc:interceptors>
<!-- AcceptHeaderLocaleResolver 配置，因为 AcceptHeaderLocaleResolver 是默认语言区域
解析器，不配置也可以-->
<bean id="localeResolver" class="org.springframework.web.servlet.i18n.
AcceptHeaderLocaleResolver" />
```

开发处理登录请求的 Controller。

程序清单：codes/05/I18NTest/src/org/fkit/controller/UserController

```
import javax.servlet.http.HttpServletRequest;
import org.fkit.domain.User;
import org.springframework.stereotype.Controller;
import org.springframework.ui.Model;
import org.springframework.web.bind.annotation.ModelAttribute;
import org.springframework.web.bind.annotation.PostMapping;
import org.springframework.web.bind.annotation.RequestMapping;
import org.springframework.web.servlet.support.RequestContext;
@Controller
public class UserController{
    @RequestMapping(value="/loginForm")
     public String loginForm(
            Model model){
        User user = new User();
        model.addAttribute("user",user);
        // 跳转页面
        return " loginForm ";
    }
     @PostMapping (value="/login")
     public String login(
            @ModelAttribute User user,
            Model model,
            HttpServletRequest request) {
        // 如果登录名是 fkit，密码是 123456，则验证通过
        if(user.getLoginname()!=null && user.getLoginname().equals("fkit")
            && user.getPassword()!=null && user.getPassword().equals("123456")){
            //从后台代码获取国际化信息 username
            RequestContext requestContext = new RequestContext(request);
            String username = requestContext.getMessage("username");
            // 将获取的 username 信息设置到 User 对象中，并设置到 Model 当中
            user.setUsername(username);
            model.addAttribute("user", user);
            return "success";
        }
        return "error";
    }
}
```

login()方法接收请求，如果验证通过，则使用 RequestContext 对象的 getMessage()方法来获取国际化消息，跳转到 success.jsp 页面。

程序清单：codes/05/I18NTest/WebContent/WEB-INF/content/success.jsp

```
<body>
<spring:message code="welcome" arguments="${requestScope.user.username}"/>
</body>
```

在 success.jsp 页面中，使用 message 标签读取资源文件中名为 welcome 的消息，并设置了一个参数，参数值为 user 对象的 username 属性。

此外，还需要在 web.xml 文件中配置 Spring MVC 的前端控制器 DispatcherServlet，因为每次配置基本相同，故此处不再赘述，读者可自行配置。

部署 I18NTest 这个 Web 应用，在浏览器中输入如下 URL 来测试应用：

```
http://localhost:8080/I18NTest/loginForm
```

会看到如图 5.1 所示的界面。

输入登录名“fkit”，密码“123456”，单击“提交”按钮，请求将被提交到 Controller。而后跳转到 success.jsp 页面，如图 5.2 所示。

图 5.1　测试基于浏览器请求的国际化

为测试 I18NTest 项目的国际化特性，需要修改浏览器语言的顺序。在 Firefox 中从右上角依次选择菜单→选项→内容→语言，然后单击“选择”按钮，修改浏览器语言的顺序。如图 5.3 所示。

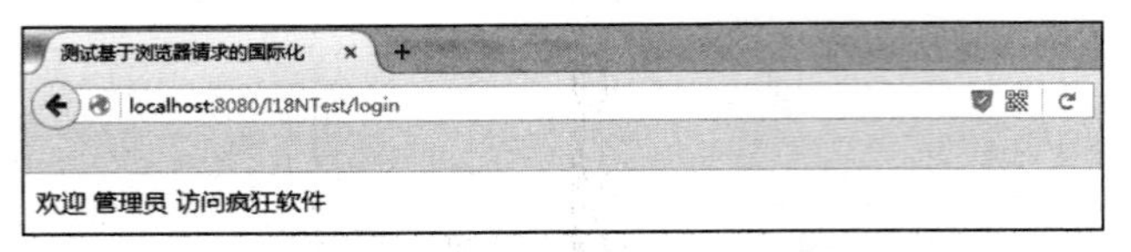

图 5.2　测试基于浏览器请求的国际化

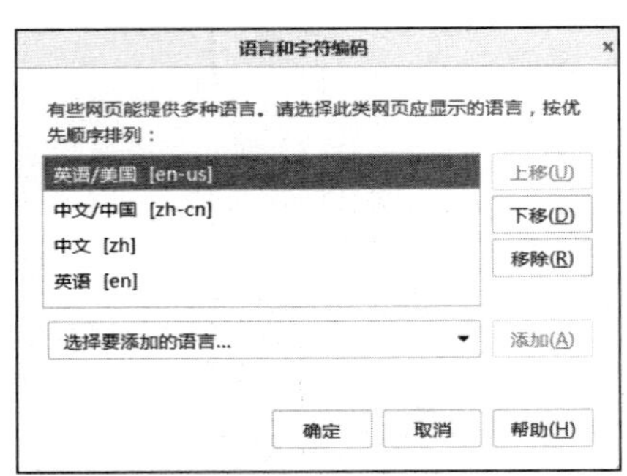

图 5.3　修改浏览器语言顺序

再次在浏览器中输入如下 URL 来测试应用：

```
http://localhost:8080/I18NTest/loginForm
```

会看到如图 5.4 所示的界面。

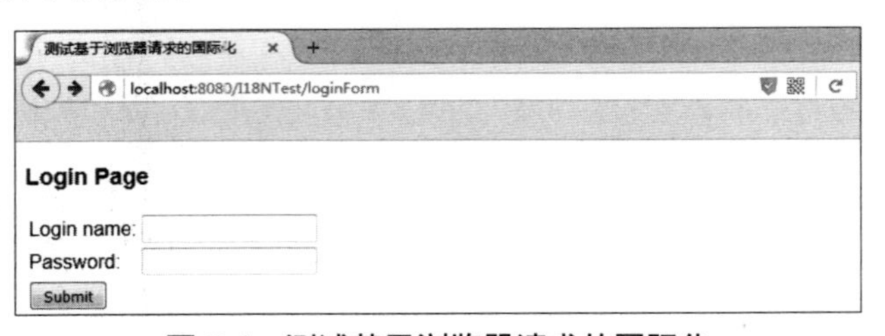

图 5.4　测试基于浏览器请求的国际化

输入登录名“fkit”，密码“123456”，单击“提交”按钮，请求将被提交到 Controller。而后跳转到 success.jsp 页面，如图 5.5 所示。

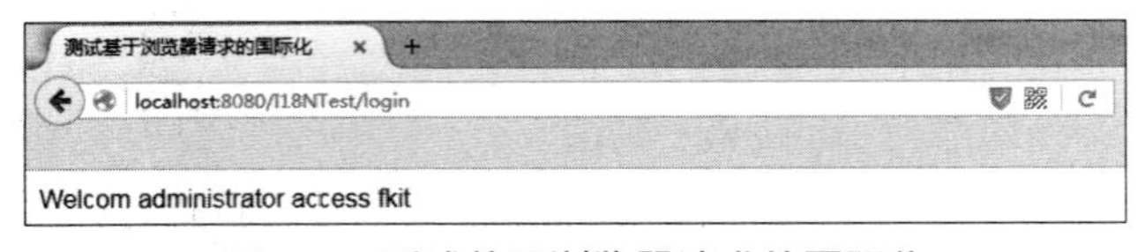

图 5.5　测试基于浏览器请求的国际化

可以看到，页面显示和获取的用户名都变成了英文版，这说明实现了国际化的功能。

5.2.2　SessionLocaleResolver

SessionLocaleResolver 不是默认的语言区域解析器，需要对其进行显式配置。如果使用它，Spring MVC 会从 HttpSession 作用域中获取用户所设置的语言区域，来确定使用哪个语言区域。

示例：基于 HttpSession 的国际化实现

程序清单：codes/05/I18N2Test/WebContent/WEB-INF/content/loginForm.jsp

```
<%@ page language="java" contentType="text/html; charset=UTF-8"
    pageEncoding="UTF-8"%>
<%@taglib prefix= "form" uri= "http://www.springframework.org/tags/form" %>
<%@taglib prefix= "spring" uri= "http://www.springframework.org/tags" %>
<!DOCTYPE html PUBLIC "-//W3C//DTD HTML 4.01 Transitional//EN" "http://www.
w3.org/TR/html4/loose.dtd">
<html>
<head>
<meta http-equiv="Content-Type" content="text/html; charset=UTF-8">
<title>测试基于 SessionLocaleResolver 的国际化</title>
</head>
<body>
<!-- 用户可以选择切换语言环境 -->
<a href="loginForm?request_locale=zh_CN">中文</a> | <a href="loginForm?request_
locale=en_US">英文</a>
<br/>
<h3><spring:message code="title"/></h3>
<form:form modelAttribute="user" method="post" action="login" >
    <table>
        <tr>
            <td><spring:message code="loginname"/></td>
            <td><form:input path="loginname"/></td>
        </tr>
        <tr>
            <td><spring:message code="password"/></td>
            <td><form:input path="password"/></td>
        </tr>
        <tr>
            <td><input type="submit" value="<spring:message code="submit"/>"/></td>
        </tr>
    </table>
</form:form>
</body>
</html>
```

在 loginForm.jsp 页面的最上面增加了两个超链接，分别用于切换中文和英文语言环境。注意请求后面的参数 request_locale，如果请求中文环境则传递的值是“zh_CN”，请求英文环境则传递的值是“en_US”。

程序清单：codes/05/I18N2Test/WebContent/WEB-INF/springmvc-config.xml

```
<mvc:interceptors>
    <!-- 国际化操作拦截器如果采用基于（Session/Cookie）则必须配置 -->
     <bean class="org.springframework.web.servlet.i18n.LocaleChangeInterceptor" />
</mvc:interceptors>
<!-- SessionLocaleResolver 配置 -->
<bean id="localeResolver" class="org.springframework.web.servlet.i18n.
SessionLocaleResolver" />
```

注意，此处将默认配置的 localeResolver 换成了 SessionLocaleResolver。

最后修改之前的动态获取跳转页面的 loginForm()方法。

程序清单：codes/05/I18N2Test/src/org/fkit/controller/UserController

```
import java.util.Locale;
import javax.servlet.http.HttpServletRequest;
import org.fkit.domain.User;
import org.springframework.context.i18n.LocaleContextHolder;
import org.springframework.stereotype.Controller;
```

```
import org.springframework.ui.Model;
import org.springframework.web.bind.annotation.ModelAttribute;
import org.springframework.web.bind.annotation.PostMapping;
import org.springframework.web.bind.annotation.RequestMapping;
import org.springframework.web.servlet.i18n.SessionLocaleResolver;
import org.springframework.web.servlet.support.RequestContext;

@Controller
public class UserController{
@RequestMapping(value="/loginForm")
    public String loginForm(
            @PathVariable String formName,
            String request_locale,
            Model model,
            HttpServletRequest request){
        System.out.println("request_locale = " + request_locale);
        if(request_locale != null){
            // 设置中文环境
            if(request_locale.equals("zh_CN")){
                Locale locale = new Locale("zh", "CN");
            request.getSession().setAttribute(SessionLocaleResolver.LOCALE_
             SESSION_ATTRIBUTE_NAME,locale);
            }
            // 设置英文环境
            else if(request_locale.equals("en_US")){
                Locale locale = new Locale("en", "US");
                request.getSession().setAttribute(SessionLocaleResolver.LOCALE_
                 SESSION_ATTRIBUTE_NAME,locale);
            }
            // 使用之前的语言环境
            else {
                request.getSession().setAttribute(SessionLocaleResolver.LOCALE_
                 SESSION_ATTRIBUTE_NAME,LocaleContextHolder.getLocale());
            }
        }
        User user = new User();
        model.addAttribute("user",user);
        // 跳转页面
        return "/loginForm";
    }
}
```

loginForm 方法根据提交的 request_locale 参数值，获取 Session 对象，并调用 setAttribute 方法进行语言环境切换。

部署 I18N2Test 这个 Web 应用，在浏览器中输入如下 URL 来测试应用：

```
http://localhost:8080/I18N2Test/loginForm
```

会看到如图 5.6 所示的界面。

图 5.6 测试基于 Session 的国际化

页面最上方有两个超链接，用于切换中文和英文语言环境。因为当前系统默认为中文环境，所以初始显示的是中文页面。

单击“英文”超链接，页面会切换为英文语言环境，如图 5.7 所示。

图 5.7 测试基于 Session 的国际化

Eclipse 控制台输出结果如下：

```
request_locale = en_US
```

说明单击超链接时传递了 request_locale 参数，而 UserController 的 loginForm 方法根据提交的参数值，进行了语言环境切换。

输入登录名“fkit”，密码“123456”，单击“提交”按钮，请求被提交到 Controller。而后跳转到 success.jsp 页面，该页面将会根据语言环境显示欢迎语句。

5.2.3 CookieLocaleResolver

CookieLocaleResolver 不是默认的语言区域解析器，需要显式对其进行配置。使用它，Spring MVC 会从 Cookie 中获取用户所设置的语言区域，来确定使用哪个语言区域。

示例：基于 Cookie 的国际化实现

程序清单：codes/05/I18N3Test/WebContent/WEB-INF/springmvc-config.xml

```
<mvc:interceptors>
    <!-- 国际化操作拦截器如果采用基于（Session/Cookie）则必须配置 -->
     <bean class="org.springframework.web.servlet.i18n.LocaleChangeInterceptor" />
</mvc:interceptors>
<!-- CookieLocaleResolver 配置 -->
<bean id="localeResolver" class="org.springframework.web.servlet.i18n.
CookieLocaleResolver" />
```

注意，此处将默认配置的 localeResolver 换成了 CookieLocaleResolver。

修改之前的动态获取跳转页面的 loginForm()方法。

程序清单：codes/05/I18N3Test/src/org/fkit/controller/UserController

```
import java.util.Locale;
import javax.servlet.http.HttpServletRequest;
import javax.servlet.http.HttpServletResponse;
import org.apache.commons.logging.Log;
import org.apache.commons.logging.LogFactory;
import org.fkit.domain.User;
import org.springframework.context.i18n.LocaleContextHolder;
import org.springframework.stereotype.Controller;
import org.springframework.ui.Model;
import org.springframework.web.bind.annotation.ModelAttribute;
import org.springframework.web.bind.annotation.PathVariable;
import org.springframework.web.bind.annotation.PostMapping;
import org.springframework.web.bind.annotation.RequestMapping;
import org.springframework.web.servlet.i18n.CookieLocaleResolver;
import org.springframework.web.servlet.support.RequestContext;
```

```
@Controller
public class UserController{
@RequestMapping(value="/loginForm")
    public String loginForm(
            @PathVariable String formName,
            String request_locale,
            Model model,
            HttpServletRequest request,
            HttpServletResponse response){
        System.out.println("request_locale = " + request_locale);
        if(request_locale != null){
            if(request_locale.equals("zh_CN")){
                Locale locale = new Locale("zh", "CN");
                (new CookieLocaleResolver()).setLocale (request, response, locale);
            }
            else if(request_locale.equals("en_US")){
                Locale locale = new Locale("en", "US");
                (new CookieLocaleResolver()).setLocale (request, response, locale);
            }else {
                (new CookieLocaleResolver()).setLocale (request, response,
                    LocaleContextHolder.getLocale());
            }
        }
        User user = new User();
        model.addAttribute("user",user);
        // 跳转页面
        return "/loginForm";
    }
}
```

loginForm 方法根据提交的 request_locale 参数值，创建 CookieLocaleResolver 对象，并调用 setLocale 方法将语言环境设置在 Cookie 中，从而进行语言环境切换。

部署 I18N2Test 这个 Web 应用，在浏览器中输入如下 URL 来测试应用：

```
http://localhost:8080/I18N3Test/loginForm
```

会看到如图 5.6 所示的界面。

单击“英文”超链接，页面切换为英文语言环境，如图 5.7 所示。

通过 Firefox 浏览器的菜单→开发者→切换工具箱→网络可以查看发送的请求，其中的请求头和返回的响应头如图 5.8 所示。

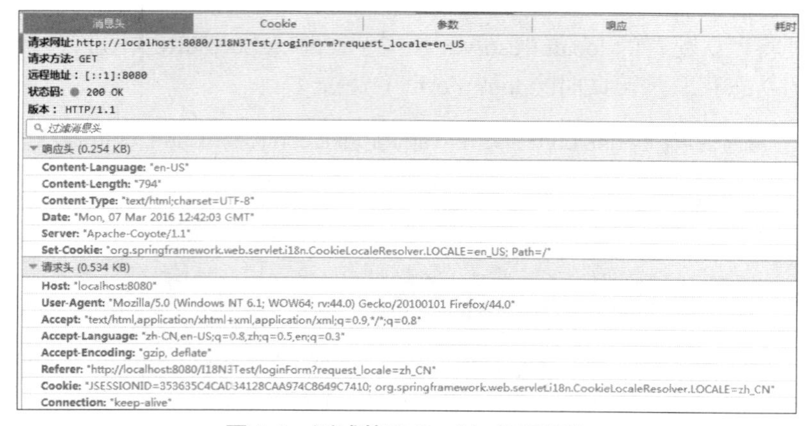

图 5.8　测试基于 Cookie 的国际化

可以看到，请求头中传递的是 CookieLocaleResolver.LOCALE=zh_CN，而响应头中则是 CookieLocaleResolver.LOCALE=en_US，这说明程序通过 Cookie 进行了语言环境切换。

输入登录名“fkit”，密码“123456”，单击“提交”按钮，请求被提交到 Controller。而后跳转到 success.jsp 页面，该页面将会根据语言环境显示欢迎语句。

 5.3 本章小结

本章介绍了 Spring MVC 的国际化知识，包括 Spring MVC 的国际化文件 messageSource、国际化语言区域解析器接口 LocaleResolver 以及接口的常用类 AcceptHeaderLocaleResolver、SessionLocaleResolver 和 CookieLocaleResolver 的使用。

第 6 章将重点介绍 Spring MVC 的数据转换、格式化和数据校验。

CHAPTER

6

第6章 Spring MVC的数据转换、格式化和数据校验

本章要点

- Spring MVC 的数据绑定流程
- 使用 ConversionService 转换数据
- 使用@IntitBinder 转换数据
- 使用 WebBindingInitializer 转换数据
- 使用 Formatter 和 FormatterRegistrar 格式化数据
- 使用 AnnotationFormatterFactory 格式化数据
- Spring 的 Validation 校验框架
- JSR 303 校验

Spring MVC 会根据请求方法签名不同，将请求消息中的信息以一定的方式转换并绑定到请求方法的参数中。在请求消息到达真正调用处理方法的这一段时间内，Spring MVC 还会完成很多其他的工作，包括请求信息转换、数据转换、数据格式化以及数据校验等。

6.1 数据绑定流程

Spring MVC 通过反射机制对目标处理方法的签名进行分析，并将请求消息绑定到处理方法的参数上。数据绑定的核心部件是 DataBinder，其运行机制如图 6.1 所示。

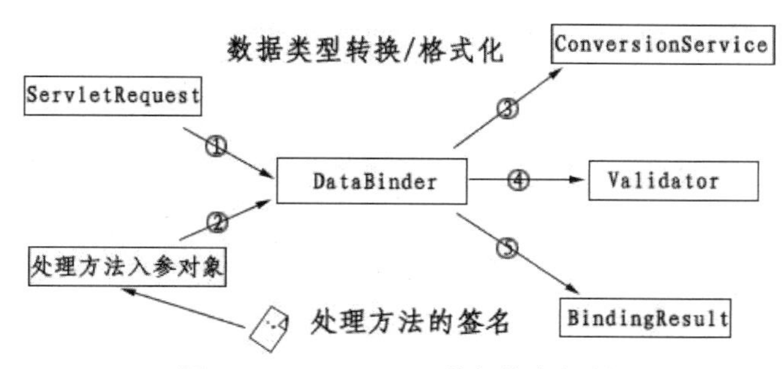

图 6.1 Spring MVC 数据绑定机制

Spring MVC 框架将 ServletRequest 对象及处理方法的参数对象实例传递给 DataBinder，DataBinder 调用装配在 Spring Web 上下文中的 ConversionService 组件进行数据类型转换、数据格式化工作，并将 ServletRequest 中的消息填充到参数对象中。然后再调用 Validator 组件对已经绑定了请求消息数据的参数对象进行数据合法性校验，并最终生成数据绑定结果 BindingResult 对象。BindingResult 包含已完成数据绑定的参数对象，还包含相应的校验错误对象，Spring MVC 抽取 BindingResult 中的参数对象及校验错误对象，将它们赋给处理方法的相应参数。

6.2 数据转换

在 Java 语言中，在 java.beans 包中提供了一个 ProperyEditor 接口来进行数据转换。ProperyEditor 的核心功能是将一个字符串转换为一个 Java 对象，以便根据界面的输入或配置文件中的配置字符串构造出一个 Java 对象。

但是 ProperyEditor 存在以下不足：

- 只能用于字符串和 Java 对象的转换，不适用于任意两个 Java 类型直接的转换。
- 对源对象及目标对象所在的上下文信息（如注解等）不敏感，在类型转换时不能利用这些上下文信息实施高级转换逻辑。

Spring 从 3.0 开始，添加了一个通用的类型转换模块，该类型转换模块位于 org.springframework.core.convert 包中。Spring 希望用这个类型转换体系替换 Java 标准的 ProperyEditor 接口。但是由于历史原因，Spring 还是同时支持两者。我们可以在 Spring MVC 处理方法的参数绑定中使用它们进行数据转换。

6.2.1 ConversionService

org.springframework.core.convert.ConversionService 是 Spring 类型转换体系的核心接口，在该接口中定义了以下 4 个方法：

- boolean canConvert(Class<?> sourceType,Class<?> targetType)。判断是否可以将一个 Java 类转换为另一个 Java 类。
- boolean canConvert(TypeDescriptor sourceType,TypeDescriptor targetType)。需要转换的类将以成员变量的方式出现，TypeDescriptor 不但描述了需要转换类的信息，还描述了类的上下文信息，例如成员变量上的注解成员变量是否以数组、集合或 Map 的方式呈现等。类型转换逻辑可以利用这些信息做出各种灵活的控制。
- <T> T convert(Object source,Class<T> targetType)。将源类型对象转换为目标类型对象。
- Object convert(Object source,TypeDescriptor sourceType,TypeDescriptor targetType)。将对象从源类型对象转换为目标类型对象，通常会利用到类中的上下文信息。

可以利用 org.springframework.context.support.ConversionServiceFactoryBean 在 Spring 的上下文中定义一个 ConversionService。Spring 将自动识别出上下文中的 ConversionService，并在 Spring MVC 处理方法的参数绑定中使用它进行数据转换。示例配置代码如下：

```
<bean id="conversionService"
   class="org.springframework.context.support.ConversionServiceFactoryBean"/>
```

在 ConversionServiceFactoryBean 中可以内置很多的类型转换器，使用它们可以完成大多数 Java 类型的转换工作，其除了包括将 Spring 对象转换为各种基础类型的对象外，还包括 String、Number、Array、Collection、Map、Properties 及 Object 之间的转换器。

可以通过 ConversionServiceFactoryBean 的 converters 属性注册自定义的类型转换器，示例配置代码如下：

```
<bean id="conversionService"
     class="org.springframework.context.support.ConversionServiceFactoryBean">
     <property name="converters">
         <list>
             <bean class="org.fkit.converter.StringToDateConverter"/>
         </list>
     </property>
 </bean>
```

6.2.2 Spring 支持的转换器

Spring 在 org.springframework.core.convert.converter 包中定义了 3 种类型的转换器接口，我们可以实现其中任意一种转换器接口，并将它作为自定义转换器注册到 ConversionServiceFactoryBean 当中。这 3 种类型转换器接口如下所示：

- Converter<S,T>。Conve0072ter<S,T>接口是 Spring 中最简单的一个转换器接口，该接口中只有一个方法：

```
T convert(S source)
```

该方法负责将 S 类型的对象转换为 T 类型的对象。

- ConverterFactory<S,R>。如果希望将一种类型的对象转换为另一种类型及其子类对象，比如将 String 转换为 Number 以及 Number 的子类 Integer、Double 等对象，就需要一系列的 Converter，如 StringToInteger、StringToDouble 等。ConverterFactory<S,R>接口的作用就是将相同系列的多个 Converter 封装在一起。该接口中也只有一个方法：

```
<T extends R> Converter<S,T> getConverter(Class<T> targetType)
```

S 为转换的源类型，R 为目标类型的基类，T 为 R 的子类。

- GenericConverter。Converter<S,T>接口只是负责将一个类型对象转换为另一个类型的对象，它并没有考虑类型对象上下文信息，因此并不能完成“复杂”类型的转换工作。

GenericConverter 接口会根据源类对象及目标类对象的上下文信息进行类型转换。该接口中定义了两个方法：

```
Set<GenericConverter.ConvertiblePair> getConvertibleTypes()
Object convert(Object source,TypeDescriptor sourceType,TypeDescriptor targetType)
```

ConvertiblePair 封装了源类型和目标类型，而 TypeDescriptor 包含了需要转换的类型对象的上下文信息，因此 GenericConverter 接口的 convert()方法可以利用这些上下文信息完成类型转换的工作。

示例：使用 ConversionService 转换数据

程序清单：codes/06/ConverterTest/WebContent/WEB-INF/content/registerForm.jsp

```
<h3>注册页面</h3>
<form action="register" method="post">
     <table>
         <tr>
             <td><label>登录名: </label></td>
             <td><input type="text" id="loginname" name="loginname" ></td>
         </tr>
          <tr>
             <td><label>生日: </label></td>
             <td><input type="text" id="birthday" name="birthday" ></td>
         </tr>
         <tr>
             <td><input id="submit" type="submit" value="登录"></td>
         </tr>
     </table>
</form>
```

registerForm.jsp 是一个简单的注册页面，其传递一个登录名和一个用户的生日信息。

程序清单：codes/06/ConverterTest/src/org/fkit/domain/User

```
import java.io.Serializable;
import java.util.Date;
public class User implements Serializable{
   private String loginname;
   private Date birthday;
   public User() {
      super();
   }
   public String getLoginname() {
      return loginname;
   }
   public void setLoginname(String loginname) {
      this.loginname = loginname;
   }
   public Date getBirthday() {
      return birthday;
   }
   public void setBirthday(Date birthday) {
      this.birthday = birthday;
   }
}
```

User 提供了 loginname 和 birthday 属性，用于接收 JSP 页面传入的数据。注意，birthday 属性的类型是一个 java.util.Date，而 JSP 页面传入的控件数据类型都是 String，这里就需要将 String 转换成 Date 对象。

程序清单：codes/06/ConverterTest/src/org/fkit/controller/UserController

```
import org.fkit.domain.User;
import org.springframework.stereotype.Controller;
import org.springframework.ui.Model;
import org.springframework.web.bind.annotation.ModelAttribute;
import org.springframework.web.bind.annotation.GetMapping;
import org.springframework.web.bind.annotation.PostMapping;

@Controller
public class UserController{

    @GetMapping(value="/registerForm")
    public String registerForm(){
        // 跳转到注册页面
        return "registerForm";
    }
    @PostMapping(value="/register")
    public String register(
            @ModelAttribute User user,
            Model model) {
        System.out.println(user);
        model.addAttribute("user", user);
        return "success";
    }
}
```

UserController 类的 register 只是简单地接收请求数据，并将其设置到 User 对象当中。

此外，还需要在 web.xml 文件中配置 Spring MVC 的前端控制器 DispatcherServlet。因为每次配置基本相同，故此处不再赘述，读者可自行配置。

部署 ConverterTest 这个 Web 应用，在浏览器中输入如下 URL 来测试应用：

```
http://localhost:8080/ConverterTest/registerForm
```

输入登录名和生日信息，如图 6.2 所示。

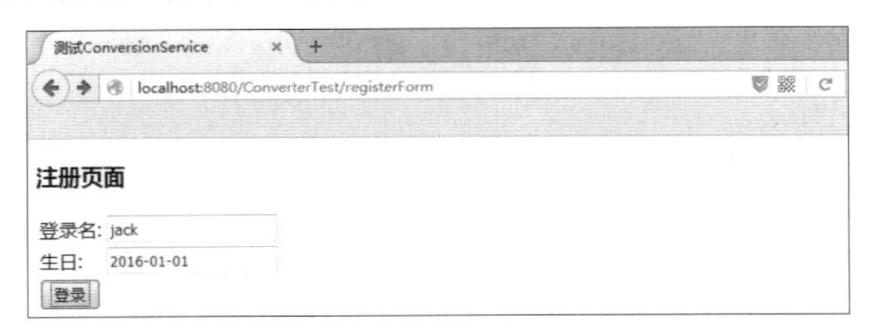

图 6.2　测试 ConversionService

由于页面传入的是字符串数据“2016-01-01”，而处理方法中的参数 User 对象的 birthday 属性的类型是 Date，因此出现数据转换异常，异常信息如下所示：

```
    Handler execution resulted in exception: org.springframework.validation.
BeanPropertyBindingResult: 1 errors
    Field error in object 'user' on field 'birthday':
    nested exception is java.lang.IllegalStateException: Cannot convert value of type
[java.lang.String] to required type [java.util.Date] for property 'birthday': no
matching editors or conversion strategy found]
```

接下来，开发自定义的转换器，将传递的字符串转换成 Date 类型。

程序清单：codes/06/ConverterTest/src/org/fkit/converter/StringToDateConverter

```
import java.text.SimpleDateFormat;
import java.util.Date;
import org.springframework.core.convert.converter.Converter;
```

```
// 实现 Converter<S,T>接口
public class StringToDateConverter implements Converter<String, Date>{
    // 日期类型模板：如 yyyy-MM-dd
    private String datePattern;
    public void setDatePattern(String datePattern) {
        this.datePattern = datePattern;
    }
    // Converter<S,T>接口的类型转换方法
    @Override
    public Date convert(String date) {
        try {
            SimpleDateFormat dateFormat = new SimpleDateFormat(this.datePattern);
            // 将日期字符串转换成 Date 类型返回
            return dateFormat.parse(date);
        } catch (Exception e) {
            e.printStackTrace();
            System.out.println("日期转换失败!");
            return null;
        }
    }
}
```

在 springmvc-config.xml 中加入自定义字符转换器。

程序清单：codes/06/ConverterTest /WebContent/WEB-INF/springmvc-config.xml

```
<!-- 装配自定义的类型转换器 -->
<mvc:annotation-driven conversion-service="conversionService"/>
<!-- 自定义的类型转换器 -->
<bean id="conversionService"
      class="org.springframework.context.support.ConversionServiceFactoryBean">
      <property name="converters">
          <list>
              <bean class="org.fkit.converter.StringToDateConverter"
              p:datePattern="yyyy-MM-dd"></bean>
          </list>
      </property>
</bean>
```

在 springmvc-config.xml 配置文件中，使用了<mvc:annotation-driven/>标签，该标签可以简化 Spring MVC 的相关配置，自动注册 RequestMappingHandlerMapping 与 RequestMappingHandlerAdapter 两个 Bean，这是 Spring MVC 为@Controllers 注解分发请求所必需的。

除此之外，<mvc:annotation-driven/>标签还会注册一个默认的 ConversionService，即 FormattingConversionServiceFactoryBean，以满足大多数类型转换的需求。现在由于需要注册一个自定义的 StringToDateConverter 转换类，因此，需要显式定义一个 ConversionService 覆盖<mvc:annotation-driven/>中的默认实现类，而这一步需要通过设置 converters 属性来完成。

在 StringToDateConverter 的 Bean 装配中，还给属性 datePattern 赋值为"yyyy-MM-dd"，即日期格式。在装配好这个 ConversionService 之后，就可以在任何控制器的处理方法中使用这个转换器了。

程序清单：codes/06/ConverterTest/WebContent/WEB-INF/content/success.jsp

```
<body>
登录名：${requestScope.user.loginname }<br>
生日：<fmt:formatDate value="${requestScope.user.birthday}"
    pattern="yyyy 年 MM 月 dd 日"/><br>
</body>
```

再次在浏览器中输入如下 URL 来测试应用：

```
http://localhost:8080/ConverterTest/registerForm
```

输入登录名和生日信息，单击“登录”按钮，转换器会自动将输入的日期字符串转换成 Date 类型，并将其设置到 User 对象的 birthday 属性当中。控制台输出如下所示：

```
User [loginname=jack, birthday=Fri Jan 01 00:00:00 CST 2016]
```

可以看到，User 对象的 birthday 属性已经获得 JSP 页面传入的日期值。

跳转到 success.jsp 页面，显示如图 6.3 所示。

图 6.3　测试 ConversionService

示例：使用@InitBinder 添加自定义编辑器转换数据

Spring MVC 在支持新的转换器框架的同时，也支持 JavaBeans 的 PropertyEditor。可以在控制器中使用@InitBinder 添加自定义的编辑器。

程序清单：codes/06/Converter2Test/src/org/fkit/converter/DateEditor

```
import java.beans.PropertyEditorSupport;
import java.text.ParseException;
import java.text.SimpleDateFormat;
import java.util.Date;
// 自定义属性编辑器
public class DateEditor extends PropertyEditorSupport {
    // 将传入的字符串数据转换成 Date 类型
    @Override
    public void setAsText(String text) throws IllegalArgumentException {
        SimpleDateFormat dateFormat = new SimpleDateFormat("yyyy-MM-dd");
        try {
            Date date = dateFormat.parse(text);
            setValue(date);
        } catch (ParseException e) {
            e.printStackTrace();
        }
    }
}
```

程序清单：codes/06/Converter2Test/src/org/fkit/controller/UserController

```
// 在控制器初始化时注册属性编辑器
 @InitBinder
public void initBinder(WebDataBinder binder){
// 注册自定义编辑器
binder.registerCustomEditor(Date.class, new DateEditor());
}
```

在 UserController()方法中增加一个 initBinder()方法，并使用@InitBinde 注解，该注解会在控制器初始化时注册属性编辑器。WebDataBinder 对象用于处理请求消息和处理方法的绑定工作。binder.registerCustomEditor()方法对传入的 Date 类型使用 DateEditor 类进行转换。

注意，使用了@InitBinder 注解，不再需要在<mvc:annotation-driven/>标签中装配自定义转换器。

部署 Converter2Test 这个 Web 应用，在浏览器中输入如下 URL 来测试应用：

```
http://localhost:8080/Converter2Test/registerForm
```

输入登录名和生日信息，测试结果如图 6.3 所示。

示例：使用 WebBindingInitializer 注册全局自定义编辑器转换数据

如果希望在全局范围内使用自定义的编辑器，则可以通过实现 WebBindingInitializer 接口并在该实现类中注册自定义编辑器完成。

程序清单：codes/06/Converter3Test/src/org/fkit/binding/DateBindingInitializer

```
import java.util.Date;
import org.fkit.converter.DateEditor;
import org.springframework.web.bind.WebDataBinder;
import org.springframework.web.bind.support.WebBindingInitializer;
// 实现 WebBindingInitializer 接口
public class DateBindingInitializer implements WebBindingInitializer {
      @Override
     public void initBinder(WebDataBinder binder) {
   // 注册自定义编辑器
   binder.registerCustomEditor(Date.class, new DateEditor());

}}
```

DateBindingInitializer 类实现 WebBindingInitializer 接口，并在 initBinder()方法中注册自定义编辑器 DateEditor 类，该类的实现和 Converter2Test 中的一致。

UserController 类中不需要再使用@InitBinder 注解注释的方法，而是在 springmvc-config.xml 配置文件中配置全局的自定义编辑器。

程序清单：codes/06/Converter3Test /WebContent/WEB-INF/springmvc-config.xml

```
<!-- 通过 AnnotationMethodHandlerAdapter 装配自定义编辑器 -->
    <bean
        class="org.springframework.web.servlet.mvc.annotation.
RequestMappingHandlerAdapter">
        <property name="webBindingInitializer">
            <bean class="org.fkjava.binding.DateBindingInitializer" />
        </property>
    </bean>
```

注意，这里使用了 RequestMappingHandlerAdapter 装配自定义编辑器，不能再使用默认配置方案<mvc:annotation-driven/>。

部署 Converter3Test 这个 Web 应用，在浏览器中输入如下 URL 来测试应用：

```
http://localhost:8080/Converter3Test/registerForm
```

输入登录名和生日信息，测试结果如图 6.3 所示。

6.2.3 多种转换器的优先顺序

对于同一个类型的对象来说，如果既在 ConversionService 中装配了自定义的转换器，又通过 WebBindingInitializer 接口装配了全局的自定义编辑器，同时还在控制器中通过@InitBinder 装配了自定义的编辑器，此时 Spring MVC 将按照以下的优先顺序查找对应的编辑器：

（1）查询通过@InitBinder 装配的自定义编辑器。

（2）查询通过 ConversionService 装配的自定义转换器。

（3）查询通过 WebBindingInitializer 接口装配的全局自定义编辑器。

6.3 数据格式化

Spring 使用 Converter 转换器进行源类型对象到目标类型对象的转换，Spring 的转换器并不承担输入以及输出信息格式化的工作。如果需要转换的源类型数据是从客户端界面中传过来的，比如日期、时间、数字、货币等数据，它们往往都拥有一定的格式。在不同的本地化环境中，同一类型的数据还会相应地呈现不同的显示格式。

如何从格式化的数据中获取真正的数据以完成数据绑定，并将处理完成的数据输出为格式化的数据是 Spring 格式化框架需要解决的问题。Spring 从 3.0 开始引入了格式化转换框架，这个框架位于 org.springframework.format 包中，其中最重要的是 Formatter<T>接口。

之前的 Converter 完成任意 Object 与 Object 之间的类型转换，而 Formatter 完成任意 Object 与 String 之间的类型转换，即格式化和解析，它和 PropertyEditor 功能类似，可以替代 PropertyEditor 来进行对象的解析和格式化，而且支持细粒度的字段级别的格式化解析。Formatter 只能将 String 转换成另一种 Java 类型。例如，将 String 转换成 Date，但它不能将 Long 转换成 Date。因此 Formatter 更适用于 Web 层的数据转换。而 Converter 则可以用在任意层中。因此，在 Spring MVC 的应用程序当中，如果想转换表单中的用户输入，则建议选择 Formatter，而不是 Converter。

Formatter 格式化转换是 Spring 通用的，定义在 org.springframework.format 包中，其不仅仅在 Spring Web MVC 场景下使用。在 org.springframework.format 包中定义的接口如下：

- **Printer<T>接口。**格式化显示接口，其将 T 类型的对象根据 Locale 信息以某种格式进行打印显示（即返回字符串形式）。该接口中定义了一个 print 方法，其根据本地化信息将数据输出为不同格式的 T 类型字符串。

```
String print(T object,Locale locale)
```

- **Parser<T>接口。**解析接口，其根据 Locale 信息解析字符串到 T 类型的对象。该接口中定义了一个 parse 方法，其参考本地化信息将一个格式化的字符串转换为 T 类型的对象。

```
T parse(String text,Locale locale) throws ParseException
```

- **Formatter<T>接口。**格式化接口，继承自 Printer<T>和 Parser<T>接口，它完成 T 类型对象的格式化和解析功能。
- **FormatterRegistrar 接口。**注册格式化转换器。该接口中定义了一个 registerFormatters 方法，其参数就是 FormatterRegistry 对象，用于注册多个格式化转换器。

```
void registerFormatters(FormatterRegistry registry)
```

- **AnnotationFormatterFactory<A extends Annotation> 接口。**注解驱动的字段格式化工厂，用于创建带注解的对象字段的 Printer 和 Parser，即用于格式化和解析带注解的对象字段。该接口中定义了以下几个方法：

  Set<Class<?>> getFieldTypes()。注解 A 的应用范围，即哪些属性类可以标注 A 注解。

  Printer<?> getPrinter(A annotation, Class<?> fieldType)。根据注解 A 获取特定属性类型 Printer。

  Parser<?> getParser(A annotation,Class<?> fieldType)。根据注解 A 获取特定属性类型的 parser。

示例：使用 Formatter 格式化数据

程序清单：codes/06/FormatterTest/src/org/fkit/formatter/DateFormatter

```
import java.text.ParseException;
import java.text.SimpleDateFormat;
import java.util.Date;
import java.util.Locale;
import org.springframework.format.Formatter;
// 实现 Formatter<T>接口
public class DateFormatter implements Formatter<Date>{
    // 日期格式化对象
    private SimpleDateFormat dateFormat;
    // 构造器，通过依赖注入的日期类型创建日期格式化对象
    public DateFormatter(String datePattern) {
        this.dateFormat = new SimpleDateFormat(datePattern);
    }
    // 显示 Formatter<T>的 T 类型对象
    @Override
    public String print(Date date, Locale locale) {
        return dateFormat.format(date);
    }
    // 解析文本字符串，返回一个 Formatter<T>的 T 类型对象。
    @Override
    public Date parse(String source, Locale locale) throws ParseException {
        try {
            return dateFormat.parse(source);
        } catch (Exception e) {
            throw new IllegalArgumentException();
        }
    }
}
```

DateFormatter 类实现了 org.springframework.format.Formatter 接口。实现了接口中的两个方法：parse 方法，使用指定的 Locale 将一个 String 解析成目标 T 类型；print 方法，用于返回 T 类型的字符串表示形式。在 DateFormatter 类中使用 SimpleDateFormat 对象将 String 转换成 Date 类型，日期类型模板 yyyy-MM-dd 会通过配置文件的依赖注入设置。

接下来在 springmvc-config.xml 中加入自定义格式化转换器。

程序清单：codes/06/FormatterTest/WebContent/WEB-INF/springmvc-config.xml

```
<!-- 装配自定义格式化转换器 -->
    <mvc:annotation-driven conversion-service="conversionService"/>
    <!-- 格式化转换器 -->
     <bean id="conversionService"
     class="org.springframework.format.support.FormattingConversionServiceFactoryBean">
        <property name="formatters">
            <list>
                <bean class="org.fkit.formatter.DateFormatter" c:_0="yyyy-MM-dd"/>
            </list>
        </property>
    </bean>
```

Spring 在格式化模块中定义了一个实现 ConversionService 接口的 FormattingConversionService 实现类，该类既具有类型转换功能，又具有格式化的功能。而 FormattingConversionService-FactoryBean 工厂类用于在 Spring 上下文中构造一个 FormattingConversionService 对象，通过这个工厂类可以注册自定义的格式化转换器。

以上配置使用 FormattingConversionServiceFactoryBean 对自定义的格式化转换器 DateFormatter

进行了注册。FormattingConversionServiceFactoryBean 类有一个属性 converters，可以用它注册 Converter；有一个属性 formatters，可以用它注册 Formatter。

提示：

值得注意的是，在<mvc:annotation-driven/>标签内部默认创建的 ConversionService 实例就是一个 FormattingConversionServiceFactoryBean，有了 FormattingConversionServiceFactoryBean 之后，Spring MVC 对处理方法的参数就绑定格式化功能了。

FormatterTest 项目的其他资源和类和之前讲述的 ConverterTest 项目一致，此处不再赘述。

部署 FormatterTest 这个 Web 应用，在浏览器中输入如下 URL 来测试应用：

```
http://localhost:8080/FormatterTest/registerForm
```

测试结果和 ConverterTest 项目一致，页面传递的日期字符串被 DateFormatter 转换成 Date 类型。

以上使用实现 Formatter<T>接口的方式完成数据转换，而 Spring 本身提供了很多常用的 Formatter 实现。在 org.springframework.format.datetime 包中提供了一个用于时间对象格式化的 DateFormatter 实现类。

在 org.springframework.format.number 包中提供了 3 个用于数字对象格式化的实现类：

- NumberFormatter。用于数字类型对象的格式化。
- CurrencyFormatter。用于货币类型对象的格式化。
- PercentFormatter。用于百分数数字类型对象的格式化。

例如，如果要使用 org.springframework.format.datetime 包中提供的 DateFormatter 实现类完成字符串到日期对象的转换，则只需要在配置文件中配置就可以了：

```
<!-- 装配自定义格式化 -->
    <mvc:annotation-driven conversion-service="conversionService"/>
     <!-- 格式化 -->
     <bean id="conversionService"
        class="org.springframework.format.support.FormattingConversionServiceFactoryBean">
        <property name="formatters">
            <list>
                <bean class="org.springframework.format.datetime.DateFormatter"
                p:pattern="yyyy-MM-dd"/>
            </list>
        </property>
    </bean>
```

示例：使用 FormatterRegistrar 注册 Formatter

注册 Formatter 的另一种方法是使用 FormatterRegistrar。

程序清单：codes/06/FormatterRegistrarTest/src/org/fkit/formatter/MyFormatterRegistrar

```
import org.springframework.format.FormatterRegistrar;
import org.springframework.format.FormatterRegistry;
public class MyFormatterRegistrar implements FormatterRegistrar{
    private DateFormatter dateFormatter;
    public void setDateFormatter(DateFormatter dateFormatter) {
        this.dateFormatter = dateFormatter;
    }
    @Override
    public void registerFormatters(FormatterRegistry registry) {
```

```
            registry.addFormatter(dateFormatter);
        }
    }
```

实现 FormatterRegistrar 只需要实现一个方法，就是 registerFormatters，在该方法中添加需要注册的 Formatter。

程序清单：codes/06/FormatterRegistrarTest/WebContent/WEB-INF/springmvc-config.xml

```
    <!-- 装配自定义格式化 -->
        <mvc:annotation-driven conversion-service="conversionService"/>
    <!-- DateFormatter bean -->
        <bean id="dateFormatter" class="org.fkit.formatter.DateFormatter" c:_0="yyyy-MM-dd"/>
        <!-- 格式化 -->
         <bean id="conversionService"
            class="org.springframework.format.support.FormattingConversionServiceFactoryBean">
            <property name="formatterRegistrars">
                <set>
                    <bean class="org.fkit.formatter.MyFormatterRegistrar"
                    p:dateFormatter-ref="dateFormatter"/>
                </set>
            </property>
        </bean>
```

配置文件中不需要再注册任何 Formatter 了，而是注册 Registrar。

FormatterRegistrarTest 项目的其他资源和类和之前讲述的 FormatterTest 项目一致，此处不再赘述。

部署 FormatterRegistrarTest 这个 Web 应用，在浏览器中输入如下 URL 来测试应用：

```
    http://localhost:8080/FormatterRegistrarTest/registerForm
```

测试结果和 FormatterTest 项目一致，页面传递的日期字符串被 DateFormatter 转换成 Date 类型。

示例：使用 AnnotationFormatterFactory<A extends Annotation>格式化数据

之前的例子使用手工代码实现 Formatter 接口或在 xml 配置文件中对 Spring 提供的 Formatter 接口的实现类进行对象数据输入输出的格式化，但是现在这种硬编码的格式化方式显然已经过时了。Spring 为开发者提供了注解驱动的属性对象格式化功能：在 Bean 属性中设置，Spring MVC 处理方法参数绑定数据，模型数据输出时自动通过注解应用格式化的功能。

在 org.springframework.format.annotation 包下面定义了两个格式化的注解类型。

1. DateTimeFormat

@DateTimeFormat 注解可以对 java.util.Date、java.util.Calendar 等时间类型的属性进行标注。它支持以下几个互斥的属性。具体说明如下：

- iso。类型为 DateTimeFormat.ISO。以下是几个常用的可选值。

  DateTimeFormat.ISO.DATE：格式为 yyyy-MM-dd。

  DateTimeFormat.ISO.DATE_TIME：格式为 yyyy-MM-dd hh:mm:ss .SSSZ。

  DateTimeFormat.ISO.TIME：格式为 hh:mm:ss .SSSZ。

  DateTimeFormat.ISO.NONE：表示不使用 ISO 格式的时间。
- pattern。类型为 String，使用自定义的时间格式化字符串，如“yyyy-MM-dd hh:mm:ss”。
- style。类型为 String，通过样式指定日期时间的格式，由两位字符组成，第 1 位表示日期的样式，第 2 位表示时间的格式，以下是几个常用的可选值。

S：短日期/时间的样式。

M：中日期/时间的样式。

L：长日期/时间的样式。

F：完整日期/时间的样式。

-：忽略日期/时间的样式。

2. NumberFormat

@ NumberFormat 可对类似数字类型的属性进行标注，它拥有两个互斥的属性，具体说明如下：

- pattern。类型为 String，使用自定义的数字格式化串，如“##，###。##”。
- style。类型为 NumberFormat.Style，以下是几个常用的可选值：

  NumberFormat. CURRENCY：货币类型。

  NumberFormat. NUMBER：正常数字类型。

  NumberFormat. PERCENT：百分数类型。

程序清单：codes/06/AnnotationFormatterTest/WebContent/WEB-INF/content/testForm.jsp

```
<h3>测试表单数据格式化</h3>
<form action="test" method="post">
    <table>
        <tr>
            <td><label>日期类型: </label></td>
            <td><input type="text" id="birthday" name="birthday" ></td>
        </tr>
        <tr>
            <td><label>整数类型: </label></td>
            <td><input type="text" id="total" name="total" ></td>
        </tr>
        <tr>
            <td><label>百分数类型: </label></td>
            <td><input type="text" id="discount" name="discount" ></td>
        </tr>
        <tr>
            <td><label>货币类型: </label></td>
            <td><input type="text" id="money" name="money" ></td>
        </tr>
        <tr>
            <td><input id="submit" type="submit" value="提交"></td>
        </tr>
    </table>
</form>
```

程序清单：codes/06/AnnotationFormatterTest/src/org/fkit/domain/User

```
import java.io.Serializable;
import java.util.Date;
import org.springframework.format.annotation.DateTimeFormat;
import org.springframework.format.annotation.NumberFormat;
import org.springframework.format.annotation.NumberFormat.Style;
public class User implements Serializable{
    // 日期类型
    @DateTimeFormat(pattern="yyyy-MM-dd")
    private Date birthday;
    // 正常数字类型
   @NumberFormat(style=Style.NUMBER, pattern="#,###")
    private int total;
```

```
    // 百分数类型
    @NumberFormat(style=Style.PERCENT)
    private double discount;
    // 货币类型
    @NumberFormat(style=Style.CURRENCY)
    private double money;
    // 省略 set/get 方法......
}
```

User 类的多个属性使用了@DateTimeFormat 和@NumberFormat 注解，用于将页面传递的 String 转换成对应的格式化数据。

程序清单：codes/06/AnnotationFormatterTest/src/org/fkit/controller/FormatterController

```
import org.fkit.domain.User;
import org.springframework.stereotype.Controller;
import org.springframework.ui.Model;
import org.springframework.web.bind.annotation.GetMapping;
import org.springframework.web.bind.annotation.ModelAttribute;
import org.springframework.web.bind.annotation.PostMapping;
@Controller
public class FormatterController{

   @GetMapping(value="/testForm")
 public String registerForm(){
    // 跳转到测试表单数据格式化
    return "testForm";
}
@PostMapping(value="/test")
public String test(
           @ModelAttribute User user,
           Model model) {
        System.out.println(user);
        model.addAttribute("user", user);
        return "success";
     }
}
```

程序清单：codes/06/AnnotationFormatterTest/WebContent/WEB-INF/content/success.jsp

```
<%@ taglib prefix="form" uri="http://www.springframework.org/tags/form" %>
<h3>测试表单数据格式化</h3>
<form:form modelAttribute="user" method="post" action="" >
<table>
   <tr>
      <td>日期类型:</td>
      <td><form:input path="birthday"/></td>
   </tr>
   <tr>
      <td>整数类型:</td>
      <td><form:input path="total"/></td>
   </tr>
   <tr>
      <td>百分数类型:</td>
      <td><form:input path="discount"/></td>
   </tr>
   <tr>
      <td>货币类型:</td>
      <td><form:input path="money"/></td>
   </tr>
</table>
```

如果希望在视图页面中将模型属性数据以格式化的方式进行渲染，则需要使用 Spring 的页面标签显示模型数据。所以 success.jsp 中使用了<form:form modelAttribute="user">标签，并且绑定了 User 对象。

程序清单：codes/06/AnnotationFormatterTest/WebContent/WEB-INF/springmvc-config.xml

```
<!-- 默认装配方案 -->
<mvc:annotation-driven/>
```

在配置文件中只是使用了默认的<mvc:annotation-driven/>标签，而该标签内部默认创建的 ConversionService 实例就是一个 FormattingConversionServiceFactoryBean，这样就可以支持注解驱动的格式化功能了。

部署 AnnotationFormatterTest 这个 Web 应用，在浏览器中输入如下 URL 来测试应用：

```
http://localhost:8080/AnnotationFormatterTest/testForm
```

输入要提交的数据，如图 6.4 所示。

单击“提交”按钮，将表单数据提交到 FormatterController 控制器，并在 test 方法的入参过程中完成数据格式化。而后跳转到 success.jsp 页面，如果 6.5 所示。

图 6.4　测试 Formatter

图 6.5　测试 Formatter 结果

可以看到，数据已经被格式化并输出在视图页面当中。

6.4　数据校验

数据校验也是所有 Web 应用必须处理的问题。因为 Web 应用的开放性，网络上所有的浏览者都可以自由使用该应用，因此该应用通过输入页面收集的数据是非常复杂的，不仅会包含正常用户的误输入，还可能包含恶意用户的非法输入。一个健壮的应用系统必须将这些非法输入阻止在应用之外，防止这些非法输入的数据进入系统，这样才可以保证系统不受影响。

非法的输入，轻则会导致系统非正常中断，重则会导致系统崩溃。应用程序必须能正常处理表现层接收的各种数据，通常的做法是遇到非法输入时应用程序直接返回，提示用户必须重新输入，也就是将那些非法输入过滤掉。这种对非法输入的过滤，就是输入校验，也称为“数据校验”。

输入校验分为客户端校验和服务器端校验，客户端校验主要是过滤正常用户的误操作，通常通过 JavaScript 代码完成；服务器端校验是整个应用阻止非法数据的最后防线，主要通过在应用中编程实现。

客户端校验的主要作用是防止正常用户的误输入，这仅能对输入进行初步过滤；对于恶意用户的恶意行为，客户端校验将无能为力。因此，客户端校验绝不可代替服务器端校验。当然，客户端校验也绝不可少，因为 Web 应用的大部分用户都是正常用户，他们的输入可能包含了大量的误输入，客户端校验能把这些误输入阻止在客户端外，从而降低服务器的负载。

Spring MVC 提供了强大的数据校验功能，其中有两种方法可以验证输入：一种是利用 Spring 自带的 Validation 校验框架；另一种是利用 JSR 303（Java 验证规范）实现校验功能。

6.4.1 Spring 的 Validation 校验框架

Spring 拥有自己独立的数据校验框架。Spring 在进行数据绑定时，可同时调用校验框架来完成数据校验工作。

Spring 的校验框架在 org.springframework.validation 包中，其中重要的接口和类如下：

- Validator。最重要的接口。该接口有两个方法：

  boolean supports(Class<?> clazz)。该校验器能够对 clazz 类型的对象进行校验。

  void validate(Object target,Errors errors)。对目标类 target 进行校验，并将校验错误记录在 errors 当中。
- Errors。Spring 用来存放错误信息的接口。Spring MVC 框架在将请求数据绑定到入参对象后，就会调用校验框架实施校验，而校验结果保存在处理方法的入参对象之后的参数对象当中。这个保存校验结果的参数对象必须是 Errors 或者 BindingResult 类型。一个 Errors 对象中包含了一系列的 FieldError 和 ObjectError 对象。FieldError 表示与被校验的对象中的某个属性相关的一个错误。BindingResult 扩展了 Errors 接口，同时可以获取数据绑定结果对象的信息。
- ValidationUtils。Spring 提供的一个关于校验的工具类。它提供了多个为 Errors 对象保存错误的方法。
- LocalValidatorFactoryBean。位于 org.springframework.validation.beanvalidation 包中，该类既实现了 Spring 的 Validator 接口，也实现了 JSR 303 的 Validator 接口。只要在 Spring 容器中定义一个 LocalValidatorFactoryBean，即可将其注入到需要数据校验的 Bean 中。定义一个 LocalValidatorFactoryBean 的 Bean 非常简单：

```
<bean id="validator"
class="org.springframework.validation.beanvalidation.LocalValidatorFactoryBean"/>
```

<mvc:annotation-driven/>会默认装配好一个 LocalValidatorFactoryBean，所以在实际开发中不需要手动配置 LocalValidatorFactoryBean。需要注意的是，Spring 本身没有提供 JSR 303 的实现，如果要使用 JSR 303 完成验证，则必须将 JSR 303 的实现（注入 Hibernate Validator）jar 文件加入到应用程序的类路径下，这样 Spring 会自动加载并装配好 JSR 303 的实现。

示例：测试 Spring 的 Validation 校验

程序清单：codes/06/ValidatorTest/WebContent/WEB-INF/content/loginForm.jsp

```
<%@taglib prefix= "form" uri= "http://www.springframework.org/tags/form" %>
<html>
<head>
<meta http-equiv="Content-Type" content="text/html; charset=UTF-8">
<title>测试 Validator 接口验证</title>
</head>
<body>
<h3>登录页面</h3>
<!-- 绑定 user -->
<form:form modelAttribute="user" method="post" action="login" >
    <table>
        <tr>
            <td>登录名:</td>
            <td><form:input path="loginname"/></td>
            <!-- 显示 loginname 属性的错误信息 -->
            <td><form:errors path="loginname" cssStyle= "color:red"/></td>
        </tr>
        <tr>
```

```
            <td>密码:</td>
            <td><form:input path="password"/></td>
            <!-- 显示 password 属性的错误信息 -->
            <td><form:errors path="password" cssStyle= "color:red"/></td>
        </tr>
        <tr>
            <td><input type="submit" value="提交"/></td>
        </tr>
    </table>
</form:form>
```

页面使用<form:errors>标签显示属性的错误信息。

程序清单：codes/06/ValidatorTest/src/org/fkit/domain/User

```
import java.io.Serializable;
public class User implements Serializable{
    private String loginname;
    private String password;
    // 省略 set/get 方法
}
```

程序清单：codes/06/ValidatorTest/src/org/fkit/validator/UserValidator

```
import org.fkit.domain.User;
import org.springframework.stereotype.Repository;
import org.springframework.validation.Errors;
import org.springframework.validation.ValidationUtils;
import org.springframework.validation.Validator;

// 实现 Spring 的 Validator 接口
@Repository("userValidator")
public class UserValidator implements Validator {
    // 该校验器能够对 clazz 类型的对象进行校验。
    @Override
    public boolean supports(Class<?> clazz) {
        // User 指定的 Class 参数所表示的类或接口是否相同，或是否是其超类或超接口。
        return User.class.isAssignableFrom(clazz);
    }
    // 对目标类 target 进行校验，并将校验错误记录在 errors 当中
    @Override
    public void validate(Object target, Errors errors) {
        /**
        使用 ValidationUtils 中的一个静态方法 rejectIfEmpty()来对 loginname 属性进行校验，
        若'loginname'属性是 null 或者空字符串，就拒绝验证通过。
        */
        ValidationUtils.rejectIfEmpty(errors, "loginname", null, "登录名不能为空");
        ValidationUtils.rejectIfEmpty(errors, "password", null, "密码不能为空");
        User user = (User)target;
        if(user.getLoginname().length() > 10){
            // 使用 Errors 的 rejectValue 方法验证
            errors.rejectValue("loginname", null, "用户名不能超过 10 个字符");
        }
        if(user.getPassword() != null
              && !user.getPassword().equals("")
              && user.getPassword().length() < 6){
            errors.rejectValue("password", null, "密码不能小于 6 位");
        }
    }
}
```

UserValidato 实现了 Spring 的 Validator 接口，其可以对 User 对象进行数据校验，并分别使用 ValidationUtils 的 rejectIfEmpty 方法和 Errors 的 rejectValue 方法对 User 进行数据校验。

@Repository("userValidator")注解将该对象声明为 Spring 容器中的一个 Bean，名字为“userValidator”。

程序清单：codes/06/ValidatorTest/src/org/fkit/controller/UserController

```
import org.fkit.domain.User;
import org.fkit.validator.UserValidator;
import org.springframework.beans.factory.annotation.Autowired;
import org.springframework.beans.factory.annotation.Qualifier;
import org.springframework.stereotype.Controller;
import org.springframework.ui.Model;
import org.springframework.validation.Errors;
import org.springframework.web.bind.annotation.GetMapping;
import org.springframework.web.bind.annotation.ModelAttribute;
import org.springframework.web.bind.annotation.PathVariable;
import org.springframework.web.bind.annotation.PostMapping;

@Controller
public class UserController{

    // 注入 UserValidator 对象
    @Autowired
    @Qualifier("userValidator")
    private UserValidator userValidator;

    @GetMapping(value="/loginForm")
     public String loginForm(Model model){
        User user = new User();
        model.addAttribute("user",user);
        // 跳转到登录页面
        return "loginForm";
    }

     @PostMapping(value="/login")
     public String login(
             @ModelAttribute User user,
             Model model,
             Errors errors) {
         System.out.println(user);
         model.addAttribute("user", user);
         // 调用 userValidator 的验证方法
         userValidator.validate(user, errors);
         // 如果验证不通过跳转到 loginForm 视图
         if(errors.hasErrors()){
             return "loginForm";
         }
         return "success";
     }
}
```

login 方法对传进来的参数进行校验，注意方法的最后一个参数 errors，该参数是一个 Spring 校验框架的 Errors 对象。在该方法中调用了之前写的 userValidator 类进行数据校验，如果校验失败，则跳转到“loginForm”视图。

部署 ValidatorTest 这个 Web 应用，在浏览器中输入如下 URL 来测试应用：

```
http://localhost:8080/ValidatorTest/loginForm
```

不输入登录名和密码，直接提交，如图 6.6 所示。

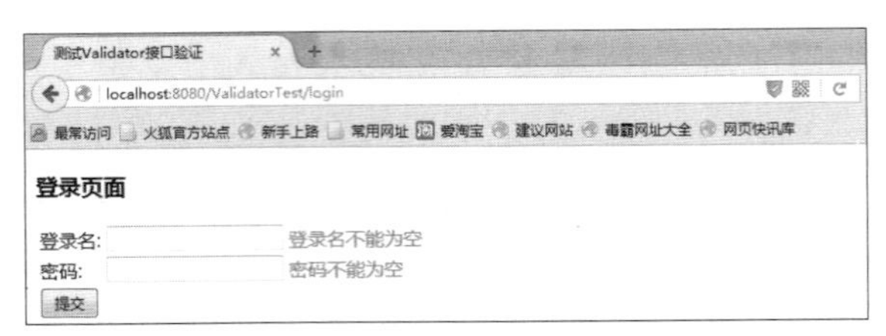

图 6.6　测试 Validation 框架（不输入登录名、密码）

可以看到，校验框架校验后返回了错误信息并将其显示在页面上。

输入不合法的登录名和密码，提交，结果如图 6.7 所示。

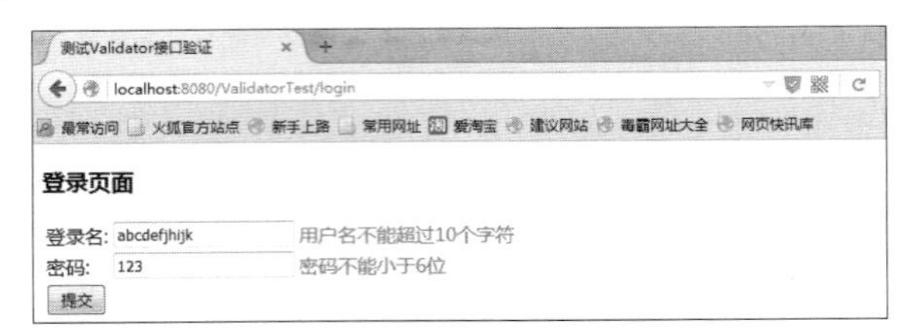

图 6.7　测试 Validation 框架（输入不合法的登录名、密码）

由于早期 Spring 就提供了 Validation 框架，所以之前的很多应用都使用 Validation 框架进行数据校验。由于 Validation 框架通过硬编码完成数据校验，在实际开发中会显得比较麻烦，因此现代开发更加推荐使用 JSR 303 完成数据校验。

6.4.2　JSR 303 校验

JSR 303 是 Java 为 Bean 数据合法性校验所提供的一个标准规范，叫作 Bean Validation。2009 年 12 月 Java EE 6 发布，Bean Validation 作为一个重要特性被包含其中，用于对 JavaBean 中的字段值进行验证。官方参考实现是 Hibernate Validator。

Bean Validation 为 JavaBean 验证定义了相应的元数据类型和 API。在应用程序中，通过在 Bean 属性上标注类似于@NotNull、@Max 等标准的注解指定校验规则，并通过标注的验证接口对 Bean 进行验证。Bean Validation 是一个运行时的数据验证框架，在验证之后验证的错误信息会被马上返回。

读者可以通过 http://jcp.org/en/jsr/detail?id=303 了解 JSR 303 的详细内容。

JSR 303 是一个规范，它的核心接口是 javax.validation.Validator，该接口根据目标对象类中所标注的校验注解进行数据校验，并得到校验结果。JSR 303 目前有两个实现，第一个实现是 Hibernate Validator，可以从以下网站下载：

```
https://sourceforge.net/projects/hibernate/files/hibernate-validator/
```

第二个实现是 Apache bval，可以从以下网站下载：

```
http://bval.apache.org/downloads.html
```

JSR 303 中定义了一套可标注在成员变量、属性方法上的校验注解，如表 6.1 所示。

表 6.1　JSR 303 注解

注解	功能	范例
@Null	验证对象是否为 null	@Null String desc;
@NotNull	验证对象是否不为 null，无法检查长度为 0 的字符串，用于验证基本数据类型	@NotNull String name;

续表

注解	功能	范例
@NotBlank	检查约束字符串是不是 Null，被 Trim 的长度是否大于 0。只作用于字符串，且会去掉前后空格	@NotBlank String name;
@AssertTrue	验证 Boolean 对象是否为 true	@AssertTrue boolean isEmpty;
@AssertFalse	验证 Boolean 对象是否为 false	@AssertFalse boolean isEmpty;
@Max(value)	验证 Number 和 String 对象是否小于等于指定的值	@Max(18) Int age;
@Min(value)	验证 Number 和 String 对象是否大于等于指定的值	@Max(60) Int age;
@DecimalMax(value)	被标注的值必须不大于约束中指定的最大值.。这个约束的参数是一个通过 BigDecimal 定义的最大值的字符串表示，小数存在精度	@DecimalMax(1.1) BigDecimal price;
@DecimalMin(value)	被标注的值必须不小于约束中指定的最小值.。这个约束的参数是一个通过 BigDecimal 定义的最小值的字符串表示，小数存在精度	@DecimalMax(0.5) BigDecimal price;
@Digits(integer,fraction)	验证字符串是否是符合指定格式的数字，integer 指定整数精度，fraction 指定小数精度	@Digits(integer=5,fraction=2) BigDecimal price;
@Size(min, max)	验证对象（Array、Collection、Map、String）长度是否在给定的范围之内	@Size(min=15, max=60) Int age;
@Email	验证是否是合法的邮件地址	@Email String email;
@Past	验证 Date 和 Calendar 对象是否在当前时间之前	@Past Date birthDate;
@Future	验证 Date 和 Calendar 对象是否在当前时间之后	@Future Date shippingDate;
@Pattern	验证 String 对象是否符合正则表达式的规则	@Pattern(regexp="[1][3,8][3,6,9][0-9]{8}") String phone;

Hibernate Validator 是 JSR 303 的一个参考实现，除了支持所有标准的校验注解之外，它还扩展了如表 6.2 所示的注解。

表 6.2 Hibernate Validator 扩展的注解

注解	功能	范例
@URL	验证是否是合法的 URL	@URL String url;
@CreditCardNumber	验证是否是合法的信用卡号码	@CreditCardNumber String creditCard;
@Length(min, max)	验证字符串的长度必须在指定的范围内	@Length(min=6, max=8) String password;
@NotEmpty	检查元素是否为 Null 或者 Empty。用于 Array、Collection、Map、String	@NotEmpty String name;
@Range(min,max,message)	验证属性值必须在合适的范围内	@Range(min=18, max=60,message=" 学生的年龄必须在 18 岁到 60 岁之间") Int age;

示例：测试 JSR 303 校验

本例使用 Hibernate Validator 实现，本书成书时最高版本是 6.0.5。下载它之后将 hibernate-validator-6.0.5.Final.jar、hibernate-validator-annotation-processor-6.0.5.Final.jar、hibernate-validator-cdi-6.0.5.Final.jar 和 lib/required 下的 validation-api-2.0.0.Final.jar、javax.el-3.0.1-b08.jar、classmate-1.3.1.jar 和 jboss-logging-3.3.0. Final.jar 加入到项目当中。

程序清单：codes/06/JSR303Test/WebContent/WEB-INF/content/registerForm.jsp

```
<%@taglib prefix= "form" uri= "http://www.springframework.org/tags/form" %>
<html>
<head>
<meta http-equiv="Content-Type" content="text/html; charset=UTF-8">
<title>测试 JSR 303</title>
</head>
<body>
<h3>注册页面</h3>
<form:form modelAttribute="user" method="post" action="login" >
    <table>
        <tr>
            <td>登录名:</td>
            <td><form:input path="loginname"/></td>
            <td><form:errors path="loginname" cssStyle= "color:red"/></td>
        </tr>
        <tr>
            <td>密码:</td>
            <td><form:input path="password"/></td>
            <td><form:errors path="password" cssStyle= "color:red"/></td>
        </tr>
        <tr>
            <td>用户名:</td>
            <td><form:input path="username"/></td>
            <td><form:errors path="username" cssStyle= "color:red"/></td>
        </tr>
        <tr>
            <td>年龄:</td>
            <td><form:input path="age"/></td>
            <td><form:errors path="age" cssStyle= "color:red"/></td>
        </tr>
        <tr>
            <td>邮箱:</td>
            <td><form:input path="email"/></td>
            <td><form:errors path="email" cssStyle= "color:red"/></td>
        </tr>
        <tr>
            <td>生日:</td>
            <td><form:input path="birthDay"/></td>
            <td><form:errors path="birthDay" cssStyle= "color:red"/></td>
        </tr>
        <tr>
            <td>电话:</td>
            <td><form:input path="phone"/></td>
            <td><form:errors path="phone" cssStyle= "color:red"/></td>
        </tr>
        <tr>
            <td><input type="submit" value="提交"/></td>
        </tr>
    </table>
</form:form>
</body>
</html>
```

registerForm.jsp 是一个注册页面，用于提交用户注册信息，注册信息包括用户名、密码、邮箱、电话等。之后将在后台使用 JSR 303 进行验证。

程序清单：codes/06/JSR303Test/src/org/fkit/domain/User

```
import java.io.Serializable;
import java.util.Date;
import javax.validation.constraints.Past;
import javax.validation.constraints.Pattern;
import javax.validation.constraints.Email;
import org.hibernate.validator.constraints.Length;
import javax.validation.constraints.NotBlank;
import org.hibernate.validator.constraints.Range;
import org.springframework.format.annotation.DateTimeFormat;
public class User implements Serializable{
    @NotBlank(message="登录名不能为空")
    private String loginname;
    @NotBlank(message="密码不能为空")
    @Length(min=6,max=8,message="密码长度必须在 6 位到 8 位之间")
    private String password;
    @NotBlank(message="用户名不能为空")
    private String username;
    @Range(min=15, max=60,message="年龄必须在 15 岁到 60 岁之间")
    private int age;
    @Email(message="必须是合法的邮箱地址")
    private String email;
    @DateTimeFormat(pattern="yyyy-MM-dd")
    @Past(message="生日必须是一个过去的日期")
    private Date birthDate;
    @Pattern(regexp="[1][3,8][3,6,9][0-9]{8}",message="无效的电话号码")
    private String phone;
    // 省略 set/get 方法
}
```

User 类使用了 Hibernate Validator 的注解对前台提交的数据进行验证。

程序清单：codes/06/JSR303Test/src/org/fkit/controller/UserController

```
import javax.validation.Valid;
import org.fkit.domain.User;
import org.springframework.stereotype.Controller;
import org.springframework.ui.Model;
import org.springframework.validation.Errors;
import org.springframework.web.bind.annotation.GetMapping;
import org.springframework.web.bind.annotation.ModelAttribute;
import org.springframework.web.bind.annotation.PostMapping;
@Controller
public class UserController{
   @GetMapping(value="/registerForm")
 public String registerForm(Model model){
    User user = new User();
    model.addAttribute("user",user);
    // 跳转到注册页面
    return "registerForm";
}

// 数据校验使用@Valid，后面跟着 Errors 对象保存校验信息
 @PostMapping(value="/login")
     public String login(
             @Valid @ModelAttribute  User user,
             Errors  errors,
             Model model)  {
         System.out.println(user);
         if(errors.hasErrors()){
             return "registerForm";
         }
         model.addAttribute("user", user);
```

```
        return "success";
    }
}
```

在 UserController 中使用@Valid 注解对提交的数据进行校验，后面跟着 Errors 对象保存校验信息。如果 errors 中有错误信息，则返回 registerForm 页面，验证通过则跳转到 success 页面。

程序清单：codes/06/ JSR303Test/WebContent/WEB-INF/content/success.jsp

```
<%@ taglib uri="http://java.sun.com/jsp/jstl/fmt" prefix="fmt"%>
<html>
<head>
<meta http-equiv="Content-Type" content="text/html; charset=UTF-8">
<title>测试 JSR 303</title>
</head>
<body>
<h3>测试 JSR 303</h3>
登录名：${requestScope.user.loginname }<br>
密码：${requestScope.user.password }<br>
用户名：${requestScope.user.username }<br>
年龄：${requestScope.user.age }<br>
邮箱：${requestScope.user.email }<br>
生日：<fmt:formatDate value="${requestScope.user.birthday}"
    pattern="yyyy 年 MM 月 dd 日"/><br>
电话：${requestScope.user.phone }<br>
</body>
</html>
```

由于<mvc:annotation-driven/>会默认装配好一个 LocalValidatorFactoryBean，因此 springmvc-config.xml 配置文件中只是基本配置，不需要增加其他的配置。

部署 JSR303Test 这个 Web 应用，在浏览器中输入如下 URL 来测试应用：

```
http://localhost:8080/JSR303Test/registerForm
```

结果如图 6.8 所示。

输入错误的注册信息，直接单击“提交”按钮，然后后台验证不通过，显示如图 6.9 所示。

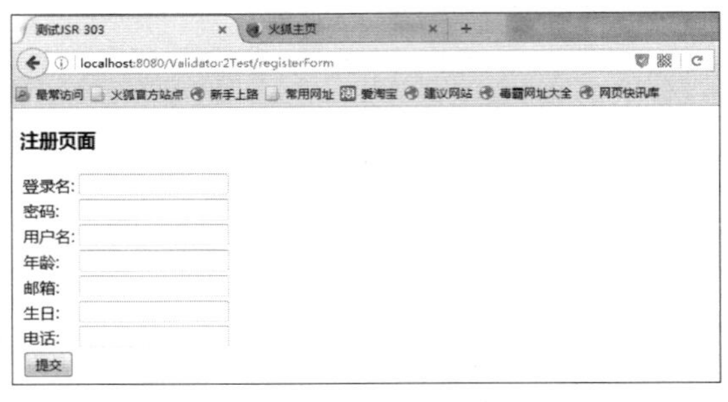

图 6.8　测试 JSR 303 规范

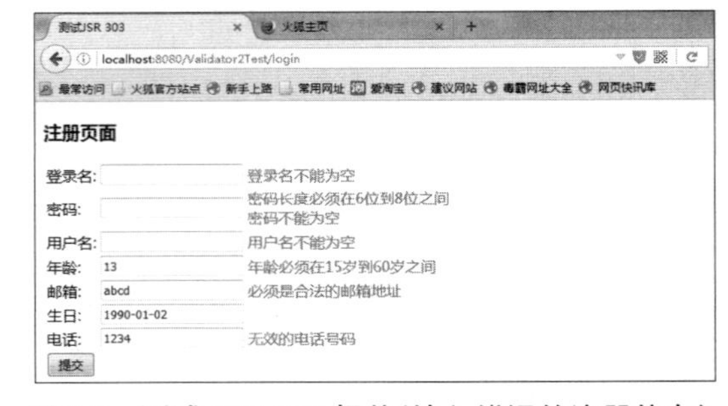

图 6.9　测试 JSR 303 规范（输入错误的注册信息）

输入符合校验规则的注册信息，通过验证后会跳转到成功页面，如图 6.10 所示。

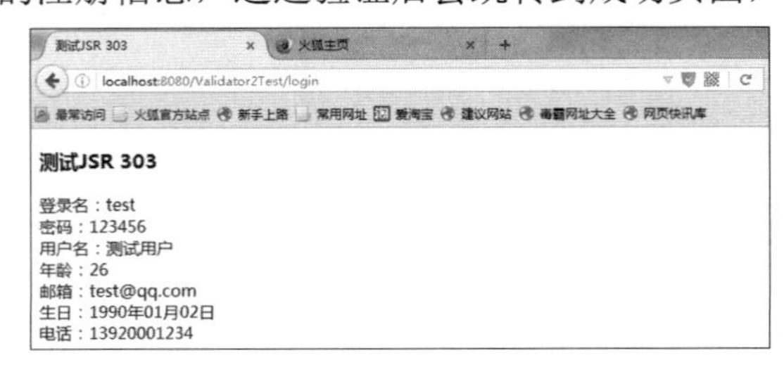

图 6.10　测试 JSR 303 规范（输入符合校验规则的注册信息）

上面是使用注解的 message 属性输出错误信息，而在实际项目中，我们希望错误信息更加人性化、更具可读性，同时还希望显示国际化的错误信息。接下来我们就为项目加入国际化的错误信息。

Spring MVC 支持国际化显示数据校验的错误信息。每个属性在数据绑定和数据校验发生错误时，都会生成一个对应的 FieldError 对象，FieldError 对象实现了 org.springframework.context.MessageSourceResolvable 接口，顾名思义，MessageSourceResolvable 是可用国际化资源进行解析的对象。MessageSourceResolvable 接口有如下 3 个方法：

- Object[] getArguments()。返回一组参数对象。
- String[] getCodes()。返回一组消息代码，每一个代码对应一个属性资源，可以使用 getArguments()返回的参数对资源属性值进行参数替换。
- String getDefaultMessage()。默认的消息，如果没有装配相应的国际化资源，那么显示的所有错误信息都是默认的。

当一个属性校验失败后，校验框架会为该属性生成 4 个消息代码，这些代码以校验注解类名为前缀，结合类名、属性名以及属性类型名生成多个对应的消息代码。

例如之前的 User 类的 loginname 属性上标注了一个@NotBlank 注解，当该属性的值不满足@NotBlank 所定义的限制规则时，就会产生以下 4 种错误代码：

- NotBlank.user.loginname。根据类名、属性名产生的错误代码。
- NotBlank.loginname。根据属性名产生的错误代码。
- NotBlank.java.lang.String。根据属性类型产生的错误代码。
- NotBlank。根据验证注解名产生的错误代码。

当使用 Spring MVC 标签显示错误信息时，Spring MVC 会查看 Web 上下文是否装配了对应的国际化消息，如果没有，则显示默认的错误消息，否则使用国际化消息对错误代码进行显示。

知道错误对象的错误码是对应国际化消息的键名称后，接下来就非常简单了，定义两个国际化资源文件，在国际化资源文件中为错误代码定义相应的本地化消息内容。

程序清单：codes/06/ JSR303Test /src/message_en_US.properties

```
NotBlank.user.loginname= Loginname is not null
NotBlank.user.password= Password is not null
Length.user.password=Password length must be between 6 and 8
NotBlank.user.username= Username is not null
Range.user.age=Age must be between the ages of 15 to 60
Email.user.email=Must be a legitimate email address
Past.user.birthday=Birthday must be a date in the past
Pattern.user.phone=Invalid phone number
```

程序清单：codes/06/ JSR303Test /src/message_zh_CN.properties

```
NotBlank.user.loginname=登录名不能为空
NotBlank.user.password=密码不能为空
Length.user.password=密码长度必须在 6 位到 8 位之间
NotBlank.user.username=用户名不能为空
Range.user.age=年龄必须在 15 岁到 60 岁之间
Email.user.email=必须是合法的邮箱地址
Past.user.birthday=生日必须是一个过去的日期
Pattern.user.phone=无效的电话号码
```

接下来还需要在 springmvc-config.xml 配置文件中增加国际化的配置。

程序清单：codes/06/ JSR303Test /WebContent/WEB-INF/springmvc-config.xml

```
<bean id="messageSource"
      class="org.springframework.context.support.ResourceBundleMessageSource">
       <!-- 国际化资源文件名 -->
       <property name="basenames" value="message"/>
   </bean>
```

重新部署运行项目，读者可以自行切换语言环境来测试国际化错误信息的显示。

6.5 本章小结

本章介绍了 Spring MVC 的数据转换、格式化和数据校验。数据转换 Converter 可以将任意类型转换成另一种类型，格式化 Formatter 则只能将 String 转换成另一种 Java 类型，Formatter 更适用于 Web 层。

对于数据校验，现阶段更多的是使用 JSR 303 验证规范，需要重点掌握。

第 7 章将重点介绍 Spring MVC 的文件上传下载和拦截器机制。

CHAPTER

7

第 7 章 Spring MVC 的文件上传下载和拦截器机制

本章要点

- Spring MVC 的文件上传
- Spring MVC 的文件下载
- Spring MVC 拦截器机制

在实际企业项目开发中，文件上传下载和拦截器是非常常用的企业级功能，本章重点介绍这两个功能。

 7.1 文件上传

文件上传是项目开发当中最常用的功能。为了能上传文件，必须将表单的 method 设置为 POST，并将 enctype 设置为 multipart/form-data。只有在这种情况下，浏览器才会把用户选择的文件二进制数据发送给服务器。

一旦将 enctype 设置为 multipart/form-data，浏览器即会采用二进制流的方式来处理表单数据，而对于文件上传的处理则涉及在服务器端解析原始的 HTTP 响应。2003 年，Apache Software Foundation 发布了开源的 Commons FileUpload 组件，其很快成为 Servlet/JSP 程序员上传文件的最佳选择。

Servlet 3.0 规范的 HttpServletRequest 已经提供了方法来处理文件上传，但这种上传需要在 Servlet 中完成。而 Spring MVC 则提供了更简单的封装。

Spring MVC 为文件上传提供了直接的支持，这种支持是用即插即用的 MultipartResolver 实现的。Spring MVC 使用 Apache Commons FileUpload 技术实现了一个 MultipartResolver 实现类：CommonsMultipartResolver。因此，Spring MVC 的文件上传还需要依赖 Apache Commons FileUpload 的组件。

Apache Commons FileUpload 的组件共有两个，最新版本分别是 commons-fileupload-1.3.3.jar 和 commons-io-2.6.jar。

commons-fileupload-1.3.3.jar 的官网下载地址为：http://commons.apache.org/proper/commons-fileupload/download_fileupload.cgi。

commons-io-2.6.jar 的官网下载地址为：http://commons.apache.org/proper/commons-io/download_io.cgi

示例：Spring MVC 的文件上传

程序清单：codes/07/FileUploadTest/WebContent/WEB-INF/content/uploadForm.jsp

```
<html>
<head>
<title>文件上传</title>
</head>
<body>
    <h2>文件上传</h2>
    <form action="upload" enctype="multipart/form-data" method="post">
        <table>
            <tr>
                <td>文件描述:</td>
                <td><input type="text" name="description"></td>
            </tr>
            <tr>
                <td>请选择文件:</td>
                <td><input type="file" name="file"></td>
            </tr>
            <tr>
                <td><input type="submit" value="上传"></td>
            </tr>
        </table>
```

```
    </form>
</body>
</html>
```

负责上传文件的表单和一般表单有一些区别，负责上传文件的表单的编码类型必须是“multipart/form-data”。

程序清单：codes/07/FileUploadTest/src/org/fkit/controller/FileUploadController

```
import java.io.File;
import java.net.URLEncoder;
import javax.servlet.http.HttpServletRequest;
import org.apache.commons.io.FileUtils;
import org.fkit.domain.User;
import org.springframework.http.MediaType;
import org.springframework.http.ResponseEntity;
import org.springframework.http.ResponseEntity.BodyBuilder;
import org.springframework.stereotype.Controller;
import org.springframework.ui.Model;
import org.springframework.web.bind.annotation.GetMapping;
import org.springframework.web.bind.annotation.ModelAttribute;
import org.springframework.web.bind.annotation.PostMapping;
import org.springframework.web.bind.annotation.RequestHeader;
import org.springframework.web.bind.annotation.RequestParam;
import org.springframework.web.multipart.MultipartFile;

@Controller
public class FileUploadController{
// 上传文件会自动绑定到MultipartFile
 @PostMapping(value="/upload")
 public String upload(HttpServletRequest request,
        @RequestParam("description") String description,
        @RequestParam("file") MultipartFile file) throws Exception{
    System.out.println(description);
    // 如果文件不为空，写入上传路径
    if(!file.isEmpty()){
        // 上传文件路径
        String path = request.getServletContext().getRealPath(
        "/images");
        // 上传文件名
        String filename = file.getOriginalFilename();
        File filepath = new File(path,filename);
        // 判断路径是否存在，如果不存在就创建一个
    if (!filepath.getParentFile().exists()) {
        filepath.getParentFile().mkdirs();
    }
    // 将上传文件保存到一个目标文件当中
        file.transferTo(new File(path+File.separator+ filename));
     System.out.println("上传文件路径: " + (path+File.separator+ filename));
        return "success";
    }else{
        return "error";
    }
 }
}
```

Spring MVC 会将上传文件绑定到 MultipartFile 对象。MultipartFile 提供了获取上传文件内容、文件名等方法。通过 transferTo()方法可以将文件上传。MultipartFile 对象中的常用方法如下：

➢ byte[] getBytes()。获取文件数据。

- String getContentType()。获取文件 MIME 类型，如 image/jpeg 等。
- InputStream getInputStream()。获取文件流。
- String getName()。获取表单中文件组件的名称。
- String getOriginalFilename()。获取上传文件的原名。
- long getSize()。获取文件的字节大小，单位为 byte。
- boolean isEmpty()。是否有上传的文件。
- void transferTo(File dest)。将上传文件保存到一个目标文件中。

Spring MVC 上下文中默认没有装配 MultipartResolver，因此默认情况下其不能处理文件上传工作。如果想使用 Spring 的文件上传功能，则需要在上下文中配置 MultipartResolver。

程序清单：codes/07/FileUploadTest/WebContent/WEB-INF/springmvc-config.xml

```
<?xml version="1.0" encoding="UTF-8"?>
<beans xmlns="http://www.springframework.org/schema/beans"
   xmlns:xsi="http://www.w3.org/2001/XMLSchema-instance"
   xmlns:p="http://www.springframework.org/schema/p"
   xmlns:c="http://www.springframework.org/schema/c"
   xmlns:mvc="http://www.springframework.org/schema/mvc"
   xmlns:context="http://www.springframework.org/schema/context"
   xsi:schemaLocation="
      http://www.springframework.org/schema/beans
      http://www.springframework.org/schema/beans/spring-beans.xsd
      http://www.springframework.org/schema/mvc
      http://www.springframework.org/schema/mvc/spring-mvc.xsd
      http://www.springframework.org/schema/context
      http://www.springframework.org/schema/context/spring-context.xsd">

   <!-- Spring 可以自动扫描 base-pack 下面的包或者子包下面的 java 文件，
      如果扫描到有 Spring 的相关注解的类，则把这些类注册为 Spring 的 Bean -->
   <context:component-scan base-package="org.fkit.controller"/>
   <!-- 默认装配方案 -->
   <mvc:annotation-driven/>
   <!-- 静态资源处理 -->
   <mvc:default-servlet-handler/>

   <!-- 视图解析器  p:prefix 属性表示前缀  p:suffix 表示后缀  -->
    <bean id="viewResolver"
        class="org.springframework.web.servlet.view.InternalResourceViewResolver"
        p:prefix="/WEB-INF/content/" p:suffix=".jsp"/>

   <!-- 文件上传配置 -->
    <bean id="multipartResolver"
      class="org.springframework.web.multipart.commons.CommonsMultipartResolver">
      <!-- 上传文件大小上限，单位为字节 (10MB)  -->
      <property name="maxUploadSize">
         <value>10485760</value>
      </property>
      <!-- 请求的编码格式，必须和 jSP 的 pageEncoding 属性一致，以便正确读取表单的内容，默认为
ISO-8859-1 -->
      <property name="defaultEncoding">
         <value>UTF-8</value>
      </property>
   </bean>

</beans>
```

需要注意的是，CommonsMultipartResolver 必须依赖于 Apache Commons FileUpload 的组件，所以需要将 Apache 的 Commons FileUpload 的 jar 包放到项目的类路径下。

部署 FileUploadTest 这个 Web 应用，在浏览器中输入如下 URL 来测试应用：

```
http://localhost:8080/FileUploadTest/uploadForm
```

输入文件描述信息并选择上传文件，如图 7.1 所示。

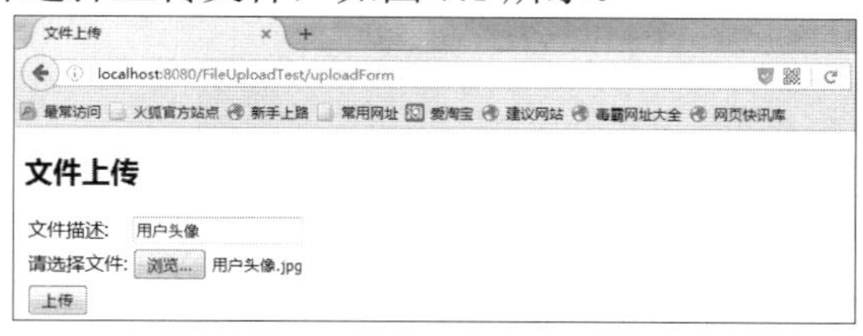

图 7.1 测试文件上传页面

单击“上传”按钮，文件会被上传并保存到服务器部署项目的 images 文件夹下面。

示例：使用对象接收上传文件

在实际的项目开发中，很多时候上传的文件会作为对象的属性被保存。Spring MVC 的处理也非常简单。

程序清单：codes/07/FileUploadTest/WebContent/WEB-INF/content/registerForm.jsp

```
<html>
<head>
<title>用户注册</title>
</head>
<body>
    <h2>用户注册</h2>
    <form action="register" enctype="multipart/form-data" method="post">
        <table>
            <tr>
                <td>用户名:</td>
                <td><input type="text" name="username"></td>
            </tr>
            <tr>
                <td>请上传头像:</td>
                <td><input type="file" name="image"></td>
            </tr>
            <tr>
                <td><input type="submit" value="注册"></td>
            </tr>
        </table>
    </form>
</body>
</html>
```

程序清单：codes/07/FileUploadTest/src/org/fkit/domain/User

```
public class User implements Serializable{
    private String username;
    // 对应上传的 file，类型为 MultipartFile，上传文件会自动绑定到 image 属性
    private MultipartFile image;
    // 省略 set/get 方法
}
```

程序清单：codes/07/FileUploadTest/src/org/fkit/controller/FileUploadController

```
@PostMapping(value="/register")
 public String register(HttpServletRequest request,
        @ModelAttribute User user,
        Model model)throws Exception{
    System.out.println(user.getUsername());
```

```
        // 如果文件不为空，写入上传路径
        if(!user.getImage().isEmpty()){
            // 上传文件路径
            String path = request.getServletContext().getRealPath(
            "/images");
            // 上传文件名
            String filename = user.getImage().getOriginalFilename();
            File filepath = new File(path,filename);
            // 判断路径是否存在，如果不存在就创建一个
            if (!filepath.getParentFile().exists()) {
                filepath.getParentFile().mkdirs();
            }
            // 将上传文件保存到一个目标文件当中
            user.getImage().transferTo(new File(path+File.separator+ filename));
            // 将文件名添加到 Model 中
            model.addAttribute("filename", user.getImage().getOriginalFilename());
         System.out.println("上传文件路径: " + (path+File.separator+ filename));
            // 跳转到下载页面
            return "userInfo";
        }else{
            return "error";
        }
    }
```

程序清单：codes/07/FileUploadTest/WebContent/WEB-INF/content/userInfo.jsp

```
    <h3>文件下载</h3>
    <a href="javascript:window.location.href='download?filename='+
encodeURIComponent( '${requestScope.filename }' )">
    ${requestScope.filename }
    </a>
```

因为传递的文件名有可能是中文字符，所以超链接中使用了 JavaScript 脚本的 encodeURIComponent 函数进行编码。

在浏览器中输入如下 URL 来测试应用：

```
    http://localhost:8080/FileUploadTest/registerForm
```

输入用户名并选择上传用户头像，单击“注册”按钮上传用户头像，然后跳转到下载页面，如图 7.2 所示。

图 7.2　文件下载页面

7.2　文件下载

示例：Spring MVC 的文件下载

可能很多读者会觉得，文件下载太简单，直接在页面给出一个超链接，该链接的 href 属性等于要下载文件的文件名，不就可以实现文件下载了吗？这样做大部分时候的确可以实现文件下载，但如果该文件的文件名为中文文件名，在某些早期的浏览器上就会导致下载失败（使用最新的 Firefox、Opera、Chrome、Safari 都可以正常下载文件名为中文的文件）。

Spring MVC 提供了一个 ResponseEntity 类型，使用它可以很方便地定义返回的 BodyBuilder、HttpHeaders 和 HttpStatus。

程序清单：codes/07/FileUploadTest/src/org/fkit/controller/FileUploadController

```
@GetMapping(value="/download")
    public ResponseEntity<byte[]> download(HttpServletRequest request,
           @RequestParam("filename") String filename,
           @RequestHeader("User-Agent") String userAgent
           )throws Exception{
       // 下载文件路径
       String path = request.getServletContext().getRealPath(
            "/images");
       // 构建 File
       File file = new File(path+File.separator+ filename);
       // ok 表示 Http 协议中的状态 200
       BodyBuilder builder = ResponseEntity.ok();
       // 内容长度
       builder.contentLength(file.length());
       // application/octet-stream ： 二进制流数据（最常见的文件下载）。
       builder.contentType(MediaType.APPLICATION_OCTET_STREAM);
       // 使用 URLDecoder.decode 对文件名进行解码
       filename = URLEncoder.encode(filename, "UTF-8");
       // 设置实际的响应文件名，告诉浏览器文件要以附件形式打开
       // 不同的浏览器，处理方式不同，要根据浏览器版本进行区别判断
       if (userAgent.indexOf("MSIE") > 0) {
            // 如果是 IE，只需要用 UTF-8 字符集进行 URL 编码即可
            builder.header("Content-Disposition", "attachment; filename=" + filename);
       } else {
            // 而 FireFox、Chrome 等浏览器，则需要说明编码的字符集
            // 注意 filename 后面有个*号，在 UTF-8 后面有两个单引号！
            builder.header("Content-Disposition", "attachment; filename*=UTF-8''" +
filename);
       }
       return builder.body(FileUtils.readFileToByteArray(file));
}
```

download 处理方法接收到页面传递的文件名 filename 后，使用 Apache Commons FileUpload 组件的 FileUtils 读取项目的 images 文件夹下的该文件，并将其构建成 ResponseEntity 对象返回客户端下载。

使用 ResponseEntity 对象，可以很方便地定义返回的 BodyBuilder、HttpHeaders 和 HttpStatus。BodyBuilder 对象用来构建返回的 Body；HttpHeaders 类型代表的是 Http 协议中的头信息；HttpStatus 类型代表的是 Http 协议中的状态。上面代码中的 MediaType，代表的是 Internet Media Type，即互联网媒体类型，也叫作 MIME 类型。在 Http 协议消息头中，使用 Content-Type 来表示具体请求中的媒体类型信息。有关 BodyBuilder、MediaType 和 HttpStatus 类的详细信息参考 Spring MVC 的 API 文档。

单击下载页面的超链接，显示文件正在下载。结果如图 7.3 所示。

单击“浏览”按钮，选择下载文件保存的路径，然后单击“确定”按钮，文件即会被顺利下载并保存。

图 7.3 文件下载

提示

文件下载示例在最新的 Windows 系统、Linux 系统、Mac OS 系统经过实测，最新的 Firefox、Opera、Chrome、Safar 等浏览器都可以正常下载文件。

7.3 拦截器

Interceptor 拦截器是 Spring MVC 中相当重要的功能，它的主要作用是拦截用户的请求并进行相应的处理。比如通过拦截器来进行用户权限验证，或者用来判断用户是否已经登录等。

Spring MVC 拦截器是可插拔式的设计。如果需要使用某个拦截器，只需要在配置文件中应用该拦截器即可；如果不需要使用该拦截器，只需要在配置文件中取消应用该拦截器。不管是否应用某个拦截器，对 Spring MVC 框架不会有任何影响。

7.3.1 HandlerInterceptor 接口

Spring MVC 中的 Interceptor 拦截器拦截请求是通过实现 HandlerInterceptor 接口来完成的。在 Spring MVC 中定义一个 Interceptor 拦截器非常简单，通常在要定义的 Interceptor 拦截器类中实现 Spring 的 HandlerInterceptor 接口，或者继承抽象类 HandlerInterceptorAdapter。

HandlerInterceptor 接口中定义了三个方法，Spring MVC 就是通过这三个方法来对用户的请求进行拦截处理的。

- boolean preHandle (HttpServletRequest request, HttpServletResponse response, Object handle)。顾名思义，该方法将在请求处理之前被调用。Spring MVC 中的 Interceptor 实行的是链式调用，即在一个应用中或者说在一个请求中可以同时存在多个 Interceptor。每个 Interceptor 的调用会依据它的声明顺序依次执行，而且最先执行的是 Interceptor 中的 preHandle 方法，所以可以在这个方法中进行一些前置的初始化操作或者是对当前请求的一个预处理，也可以在这个方法中进行一些判断来决定请求是否要继续进行下去。该方法的返回值是 Boolean 类型的，当返回值为 false 时，表示请求结束，后续的 Interceptor 和 Controller 都不会再执行；当返回值为 true 时就会继续调用下一个 Interceptor 的 preHandle 方法；如果已经是最后一个 Interceptor，就会调用当前请求的 Controller 方法。
- void postHandle(HttpServletRequest request, HttpServletResponse response,Object handler, ModelAndView mv)。该方法和之后的 afterCompletion 方法都只能在当前所属的 Interceptor 的 preHandle 方法的返回值为 true 时才能被调用。postHandle 方法，顾名思义，就是在当前请求被处理之后，也就是 Controller 方法被调用之后执行，但是它会在 DispatcherServlet 进行视图返回渲染之前被调用，所以我们可以在这个方法中对 Controller 处理之后的 ModelAndView 对象进行操作。postHandle 方法被调用的方向跟 preHandle 是相反的，也就是说先声明的 Interceptor 的 postHandle 方法反而会后执行，这和 Struts2 里面的 Interceptor 的执行过程类似。
- void afterCompletion(HttpServletRequest request,HttpServletResponse response, Object handler, Exception exception)。该方法也是在当前所属的 Interceptor 的 preHandle 方法的返回值为 true 时才会执行。顾名思义，该方法将在整个请求结束之后，也就是在 DispatcherServlet 渲染了对应的视图之后执行。这个方法的主要作用是进行资源清理。

示例：拦截器实现用户权限验证

本小节通过拦截器完成一个用户权限验证的功能。即用户必须登录之后才可以访问网站首页，如果没有登录就直接访问网站首页，则拦截器会拦截请求，并将请求重新转发到登录页面，同时提示用户需要先登录再访问网站。

程序清单：codes/07/InterceptorTest/WebContent/WEB-INF/content/loginForm.jsp

```
<h3>登录页面</h3>
<form action="login" method="post">
    <!-- 提示信息 -->
    <font color="red">${requestScope.message }</font>
    <table>
        <tr>
            <td><label>登录名：</label></td>
            <td><input type="text" id="loginname" name="loginname" ></td>
        </tr>
        <tr>
            <td><label>密码：</label></td>
            <td><input type="password" id="password" name="password" ></td>
        </tr>
        <tr>
            <td><input type="submit" value="登录"></td>
        </tr>
    </table>
</form>
```

程序清单：codes/07/InterceptorTest/src/org/fkit/controller/UserController

```
import org.fkit.domain.User;
import org.springframework.stereotype.Controller;
import org.springframework.web.bind.annotation.PostMapping;
import org.springframework.web.servlet.ModelAndView;
@Controller
public class UserController {

    /**
     * 处理/login 请求
     * */
    @PostMapping(value="/login")
     public ModelAndView login(
            String loginname,String password,
            ModelAndView mv,
            HttpSession session){
        // 模拟数据库根据登录名和密码查找用户，判断用户登录
        if(loginname != null && loginname.equals("fkit")
                && password!= null && password.equals("123456")){
            // 模拟创建用户
            User user = new User();
            user.setLoginname(loginname);
            user.setPassword(password);
            user.setUsername("管理员");
            // 登录成功，将 user 对象设置到 HttpSession 作用域
            session.setAttribute("user", user);
            // 转发到 main 请求
            mv.setViewName("redirect:main");
        }else{
            // 登录失败，设置失败提示信息，并跳转到登录页面
            mv.addObject("message", "登录名或密码错误，请重新输入!");
            mv.setViewName("loginForm");
        }
```

```
        return mv;
    }
}
```

UserController 类的 login 方法用来处理登录请求，本示例没有使用数据库存储数据，只是简单地模拟了用户登录，只要用户输入的登录名是“jack”，密码是“123456”，则验证通过，并创建一个 User 对象保存到 HttpSession 当中，同时将请求使用客户端跳转到 main 请求；如果登录失败，则设置失败提示信息到 ModelAndView 对象，同时将请求使用客户端跳转到 loginForm 请求，即登录页面。

程序清单：codes/07/InterceptorTest/src/org/fkit/controller/BookController

```
import java.util.ArrayList;
import java.util.List;
import org.fkit.domain.Book;
import org.springframework.stereotype.Controller;
import org.springframework.ui.Model;
import org.springframework.web.bind.annotation.RequestMapping;
@Controller
public class BookController {
    /**
     * 处理“/main”请求
    * */
    @RequestMapping(value="/main")
     public String main(Model model){
        // 模拟数据库获得所有图书集合
        List<Book> book_list = new ArrayList<Book>();
        book_list.add(new Book("java.jpg","疯狂 Java 讲义(附光盘)","李刚 编著",74.2));
        book_list.add(new Book("ee.jpg","轻量级 Java EE 企业应用实战","李刚 编著",59.2));
        book_list.add(new Book("android.jpg","疯狂 Android 讲义(附光盘)","李刚 编著",60.6));
        book_list.add(new Book("ajax.jpg","疯狂 Ajax 讲义(附光盘)","李刚 编著",66.6));
        // 将图书集合添加到 model 当中
        model.addAttribute("book_list", book_list);
        // 跳转到 main 页面
        return "main";
    }
}
```

BookController 类的 main 方法用来处理网站首页的请求，该方法获得所有图书信息，并将它们设置到 Model 当中，然后传递到 main 页面。本示例没有使用数据库存储数据，只是简单地创建了一个集合模拟从数据库获取图书信息。

程序清单：codes/07/InterceptorTest/WebContent/WEB-INF/content/main.jsp

```
<h3>欢迎[${sessionScope.user.username }]访问</h3>
<br>
<table border="1">
    <tr>
        <th>封面</th><th>书名</th><th>作者</th><th>价格</th>
    </tr>
    <c:forEach items="${requestScope.book_list }" var="book">
        <tr>
            <td><img src="images/${book.image }" height="60"></td>
            <td>${book.name }</td>
            <td>${book.author }</td>
            <td>${book.price }</td>
        </tr>
    </c:forEach>
</table>
```

接下来，设计一个拦截器验证用户是否登录，如果用户没有登录，不可以访问除登录页面和登录请求之外的所有Controller。

程序清单：codes/07/InterceptorTest/src/org/fkit/interceptor/AuthorizationInterceptor

```
import javax.servlet.http.HttpServletRequest;
import javax.servlet.http.HttpServletResponse;
import org.fkit.domain.User;
import org.springframework.web.servlet.HandlerInterceptor;
import org.springframework.web.servlet.ModelAndView;
/**
 * 拦截器必须实现HandlerInterceptor接口
 * */
public class AuthorizationInterceptor  implements HandlerInterceptor {

    // 不拦截"/loginForm"和"/login"请求
    private static final String[] IGNORE_URI = {"/loginForm", "/login"};

     /**
     * 该方法将在整个请求完成之后执行， 主要作用是清理资源，
     * 该方法也只能在当前 Interceptor 的 preHandle 方法的返回值为 true 时才会执行。
     */
    @Override
    public void afterCompletion(HttpServletRequest request,
            HttpServletResponse response, Object handler, Exception exception)
            throws Exception {
        System.out.println("AuthorizationInterceptor afterCompletion --> ");

    }
    /**
     * 该方法将在 Controller 的方法调用之后执行， 方法中可以对 ModelAndView 进行操作 ，
     * 该方法也只能在当前 Interceptor 的 preHandle 方法的返回值为 true 时才会执行。
     */
    @Override
    public void postHandle(HttpServletRequest request, HttpServletResponse response,
            Object handler, ModelAndView mv) throws Exception {
        System.out.println("AuthorizationInterceptor postHandle --> ");

    }
     /**
     * preHandle 方法是进行处理器拦截用的，该方法将在 Controller 处理之前被调用，
     * 该方法的返回值为true时拦截器才会继续往下执行，该方法的返回值为false的时候整个请求就结束了。
     */
    @Override
    public boolean preHandle(HttpServletRequest request, HttpServletResponse response,
            Object handler) throws Exception {
        System.out.println("AuthorizationInterceptor preHandle --> ");
        // flag 变量用于判断用户是否登录，默认为 false
        boolean flag = false;
        //获取请求的路径进行判断
        String servletPath = request.getServletPath();
        // 判断请求是否需要拦截
        for (String s : IGNORE_URI) {
            if (servletPath.contains(s)) {
                flag = true;
                break;
            }
        }
        // 拦截请求
        if (!flag){
            // 1.获取 session 中的用户
            User user = (User) request.getSession().getAttribute("user");
            // 2.判断用户是否已经登录
            if(user == null){
```

```
            // 如果用户没有登录，则设置提示信息，跳转到登录页面
             System.out.println("AuthorizationInterceptor 拦截请求：");
             request.setAttribute("message", "请先登录再访问网站");
             request.getRequestDispatcher("loginForm").forward(request, response);
        }else{
            // 如果用户已经登录，则验证通过，放行
             System.out.println("AuthorizationInterceptor 放行请求：");
             flag = true;
        }
      }
      return flag;
   }
}
```

在 springmvc-config.xml 文件中配置拦截器。

程序清单：codes/07/InterceptorTest/WebContent/WEB-INF/springmvc-config.xml

```
<!-- Spring MVC 拦截器定义 -->
   <mvc:interceptors>
       <mvc:interceptor>
          <!-- 拦截所有的请求 -->
          <mvc:mapping path="/*"/>
          <!-- 使用 bean 定义一个 Interceptor -->
           <bean class="org.fkit.interceptor.AuthorizationInterceptor"/>
       </mvc:interceptor>
    </mvc:interceptors>
```

部署 InterceptorTest 这个 Web 应用，在浏览器中输入如下 URL 来测试应用：

```
http://localhost:8080/InterceptorTest/main
```

如果没有登录，直接访问 main 请求，拦截器会拦截请求，验证用户是否登录，此时用户若没有登录，则跳转到登录页面，如图 7.4 所示。

输入登录名“fkit”和密码“123456”后，显示用户登录成功，而后跳转到网站首页，如图 7.5 所示。

图 7.4　拦截器验证用户权限

7.4　本章小结

本章介绍了 Spring MVC 的文件上传和文件下载，Spring MVC 对文件上传提供了完美的封装，使得用户能以非常简便的方式同时上传多个文件；Spring MVC 专门提供了 ResponseEntity 类型，用于实现文件下载。

接下来介绍了 Spring MVC 拦截器，通过引用拦截器机制，Spring MVC 框架可以使用可插拔方式管理各种功能。

第 8 章将重点介绍 MyBatis 持久化框架知识。

图 7.5　网站首页

CHAPTER

8

第8章 MyBatis简介

本章要点

- 什么是 ORM
- 流行的 ORM 框架
- MyBatis 概述
- MyBatis 使用入门

8.1 ORM和MyBatis

目前流行的编程语言，例如 Java、C#等，都是面向对象的编程语言；而目前主流的数据库产品，例如 Oracle、DB2 等，依然是关系数据库。编程语言和底层数据库的发展不协调，催生出了 ORM 框架，ORM 框架可作为面向对象编程语言和数据库之间的桥梁。

确切地说，MyBatis 并不完全是一种 ORM 框架，它的设计思想和 ORM 相似，只是它允许开发人员直接编写 SQL 语句，使得访问数据库更加灵活，更准确地说，它应该是一种“SQL Mapping”框架。

8.1.1 对象/关系数据库映射（ORM）

ORM 的全称是 Object/Relation Mapping，即对象/关系数据库映射。可以将 ORM 理解成一种规范，它概述了这类框架的基本特征，完成面向对象的编程语言到关系数据库的映射。当 ORM 框架完成映射后，程序员既可以利用面向对象程序设计语言的简单易用性，又可以利用关系数据库的技术优势。因此可以把 ORM 当成应用程序和数据库的桥梁。

当使用一种面向对象的编程语言来进行应用开发时，从项目一开始就采用的是面向对象分析、面向对象设计、面向对象编程，但到了持久层数据库访问时，又必须重返关系数据库的访问方式，这是一种非常糟糕的感觉。于是人们需要一种工具，它可以把关系数据库包装成面向对象的模型，这个工具就是 ORM。

ORM 框架是面向对象程序设计语言与关系数据库发展不同步时的中间解决方案。随着面向对象数据库的发展，其理论逐步完善，最终面向对象数据库会取代关系数据库。只是这个过程不可一蹴而就，ORM 框架在此期间会蓬勃发展。但随着面向对象数据库的广泛使用，ORM 工具会逐渐消亡。

对于时下所有流行的编程语言而言，面向对象的程序设计语言代表了目前程序设计语言的主流和趋势，其具备非常多的优势。比如：

- 面向对象的建模、操作。
- 多态、继承。
- 摒弃难以理解的过程。
- 简单易用，易理解。

但数据库的发展并未能与程序设计语言同步，而且关系数据库系统的某些优势也是面向对象语言目前无法比拟的。比如：

- 大量数据查找、排序。
- 集合数据连接操作、映射。
- 数据库访问的并发、事务。
- 数据库的约束、隔离。

面对这种面向对象语言与关系数据库系统并存的局面，采用 ORM 就变成一种必然。只要依然采用面向对象程序设计语言，底层依然采用关系数据库，中间就少不了 ORM 工具。采用 ORM 框架之后，应用程序不再直接访问底层数据库，而是以面向对象的方式来操作持久化对象（例如创建、修改、删除等），而 ORM 框架则将这些面向对象的操作转换成底层的 SQL 操作。

图 8.1 显示了 ORM 工具工作的示意图。

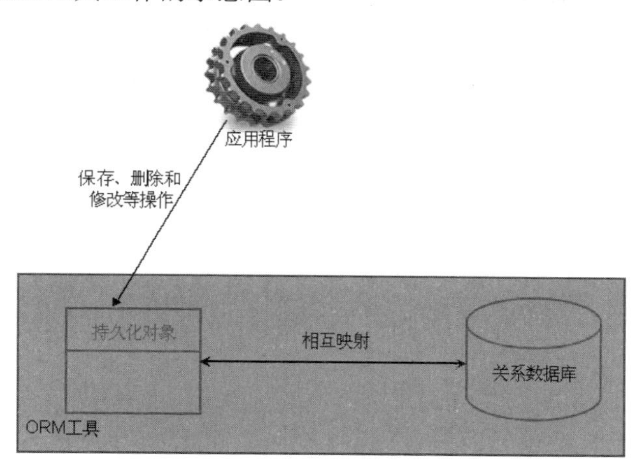

图 8.1 ORM 工具工作的示意图

正如图 8.1 所示，ORM 工具的唯一作用就是：把对持久化对象的保存、修改、删除等操作，转换成对数据库的操作。由此，程序员可以以面向对象的方式操作持久化对象，而 ORM 框架则负责将相关操作转换成对应的 SQL（结构化查询语言）操作。

8.1.2 基本映射方式

ORM 工具提供了持久化类和数据表之间的映射关系，通过这种映射关系的过渡，程序员可以很方便地通过持久化类实现对数据表的操作。实际上，所有的 ORM 工具大致都遵循相同的映射思路。ORM 有如下几条基本映射关系：

- **数据表映射类。**持久化类被映射到一个数据表。程序使用这个持久化类来创建实例、修改属性、删除实例时，系统自动会转换为对这个表进行 CRUD 操作。图 8.2 显示了这种映射关系。

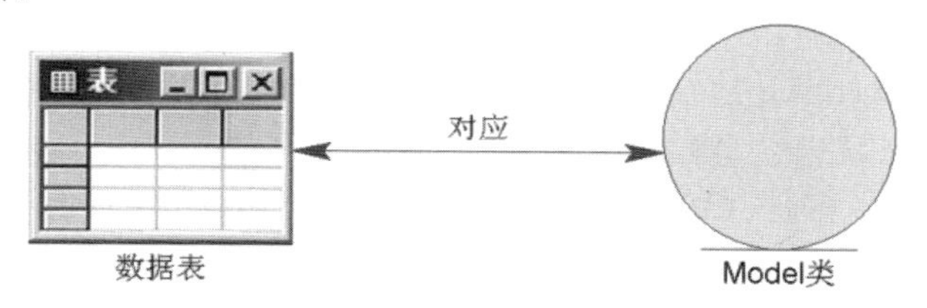

图 8.2 数据表对应 Model 类

正如图 8.2 所示，受 ORM 管理的持久化类（就是一个普通 Java 类）对应一个数据表，只要程序对这个持久化类进行操作，系统就可以将其转换成对对应数据库表的操作。

- **数据表的行映射对象（即实例）。**持久化类会生成很多实例，每个实例就对应数据表中的一行记录。当程序在应用中修改持久化类的某个实例时，ORM 工具会将其转换成对对应数据表中特定行的操作。每个持久化对象对应数据表的一行记录的示意图如图 8.3 所示。
- **数据表的列（字段）映射对象的属性。**当程序修改某个持久化对象的指定属性时（持久化实例映射到数据行），ORM 会将其转换成对对应数据表中指定数据行、指定列的操作。数据表的列被映射到对象属性的示意图如图 8.4 所示。

基于这种基本的映射方式，ORM 工具可完成对象模型和关系模型之间的相互映射。由此可见，在 ORM 框架中，持久化对象是一种媒介，应用程序只需操作持久化对象，ORM 框架则负责将这种操作转换为底层数据库操作。这种转换对开发者透明，开发者无须关心内部细节，从而将开发者从关系模型中解放出来，使得开发者能以面向对象的思维操作关系数据库。

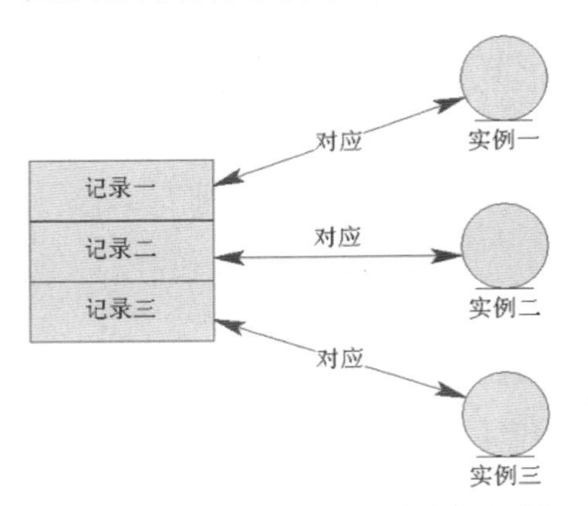

图 8.3 数据表中的记录行对应持久化对象

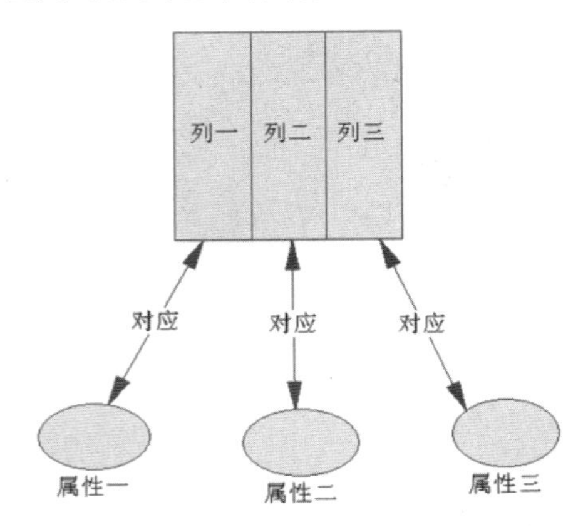

图 8.4 数据表中的列对应对象的属性

8.1.3 流行的 ORM 框架简介

目前 ORM 框架产品非常多，除了各大著名公司、组织的产品外，甚至其他一些小团队也都推出自己的 ORM 框架。目前流行的 ORM 框架有如下这些产品：

- JPA。JPA 本身只是一种 ORM 规范，并不是 ORM 产品。它是 Java EE 规范制定者向开源世界学习的结果。相对于其他开源 ORM 框架，JPA 的最大优势在于它是官方标准，因此具有通用性。如果应用程序是面向 JPA 编程，那么应用程序就可以在各种 ORM 框架之间自由切换。
- Hibernate。目前最流行的开源 ORM 框架，已经被选为 JBoss 的持久层解决方案。整个 Hibernate 项目也一并投入了 JBoss 的怀抱，而 JBoss 又加入了 Red Hat 组织。因此，Hibernate 属于 Red Hat 组织的一部分。Hibernate 灵巧的设计、优秀的性能，以及丰富的文档，都是其风靡全球的重要因素。
- MyBatis（早期名称是 iBATIS）。Apache 软件基金组织的子项目。与其称它是一种 ORM 框架，不如称它是一种“SQL Mapper”框架，它是一种“半自动化”的 ORM 的实现，曾经在 Java EE 开发中扮演非常重要的角色，但是其并不支持纯粹的面向对象操作，它允许开发人员直接编写 SQL 语句，更加灵活。

8.1.4 MyBatis 概述

MyBatis 是一个支持普通 SQL 查询、存储过程和高级映射的优秀持久层框架。MyBatis 去掉了几乎所有的 JDBC 代码和参数的手工设置以及对结果集的检索封装。MyBatis 可以使用简单的 XML 或注解进行配置和原始映射，将 Java 的 POJO（Plain Old Java Objects，普通的 Java 对象）映射成数据库中的记录。

MyBatis 作为持久层框架，其主要思想是将程序中的大量 SQL 语句剥离出来，配置在配置文件中，以实现 SQL 的灵活配置。这样做的好处是将 SQL 与程序代码分离，做到可以在不修改程序代码的情况下，直接在配置文件中修改 SQL。

纵观目前主流的 ORM，无论 Hibernate 还是 JPA，都对数据库结构提供了较为完整的封装，提供了从 POJO（Plain Old Java Objects）到数据库表的全套映射机制。程序员往往只需定义好 POJO 到数据库表的映射关系，即可通过 Hibernate 或者 JPA 提供的方法完成持久层操作。程序员甚至不需要熟练掌握 SQL，Hibernate/JPA 会根据制定的存储逻辑，自动生成对应的 SQL 并调用 JDBC 接口加以执行。

大多数情况下（特别是对新项目、新系统的开发而言），这样的机制无往不利，大有一统天下的势头。但是，在一些特定的环境下，Hibernate 这种一站式的解决方案却未必适合。例如：

- 系统的部分或全部数据来自现有数据库，出于安全考虑，只对开发团队提供几条 Select SQL（或存储过程）以获取所需数据，具体的表结构不予公开。
- 开发规范中要求，所有牵涉业务逻辑部分的数据库操作，必须在数据库层由存储过程实现（就金融行业而言，工商银行、中国银行、交通银行等商业银行都曾在开发规范中严格指定）。
- 系统数据处理量巨大，性能要求极为苛刻，这往往意味着我们必须通过经过高度优化的 SQL 语句（或存储过程）才能达到系统性能设计指标。

面对这样的需求，Hibernate 不再适合解决上述问题，甚至无法使用它。此时，直接使用 JDBC 进行数据库操作实际上也是不错的选择，只是拖沓的数据库访问代码、乏味的字段读取操作令人厌烦，而“半自动化”的 MyBatis，却正好解决了这个问题。

这里的“半自动化”是相对 Hibernate 等提供了全面的数据库封装机制的“全自动化”ORM 实现而言的，“全自动”ORM 实现了 POJO 和数据库表之间的映射，以及 SQL 的自动生成和执行。而 MyBatis 的着力点，则在于 POJO 与 SQL 之间的映射关系。也就是说，使用 MyBatis 提供的 ORM 机制，对业务逻辑实现人员而言，面对的是纯粹的 Java 对象，这一点与通过 Hibernate 实现 ORM 而言基本一致。而对于具体的数据操作，Hibernate 会自动生成 SQL 语句，而 MyBatis 则并不会为程序员在运行期间自动生成 SQL。具体的 SQL 需要程序员编写，然后通过映射配置文件，将 SQL 所需的参数及返回的结果字段映射到指定 POJO。

相对 Hibernate 等“全自动”ORM 机制而言，MyBatis 以 SQL 开发的工作量和数据库移植性上的让步，为系统设计提供了更大的自由空间。作为“全自动”ORM 实现的一种有益补充，MyBatis 的存在具有特别的意义。

8.2 MyBatis 入门

MyBatis 的用法非常简单，我们只要在 Java 项目中引入 MyBatis 框架，就能以面向对象的方式操作关系数据库。

8.2.1 MyBatis 下载和安装

本书成书之前，MyBatis 的最新稳定版本是 mybatis-3.4.5，本书所用的代码也是基于该版本测试通过的。下载和安装 MyBatis 请按如下步骤进行：

① 登录 http://www.mybatis.org 站点，即可在页面上看到一个“MyBatis-3.4.5”超链接，单击该超链接即可开始下载 MyBatis 的压缩包。

② 解压缩刚下载的压缩包，得到一个名为 mybatis-3.4.5 的文件夹，该文件夹下包含如下文件结构：

- mybatis-3.4.5.jar。MyBatis 的核心类库。
- mybatis-3.4.5.pdf。MyBatis 的参考文档。
- mybatis-3.4.5-javadoc.jar。MyBatis 的 API 文档。
- mybatis-3.4.5-sources.jar。MyBatis 的源代码。
- lib。MyBatis 所依赖的第三方 jar 包。

③ 将解压缩路径中的 mybatis-3.4.5.jar 包添加到应用程序的类加载路径中，既可以通过添加环境变量的方式来添加，也可以使用 Ant 或 IDE 工具来管理应用程序的类加载路径。

④ 如果直接在控制台编译使用了 MyBatis API 的类，则需要将 mybatis-3.4.5.jar 包位置添加到 CLASSPATH 里。如果使用 Ant 或者 Eclipse 等 IDE 工具，则无须修改环境变量。

经过上面的步骤，就可以在应用程序中使用 MyBatis 框架的功能了。

提示：

由于 MyBatis 的底层依然是基于 JDBC 的，因此在应用程序中使用 MyBatis 执行持久化时同样少不了 JDBC 驱动。本示例程序底层采用了 MySQL 数据库，因此还需要将 MySQL 数据库驱动添加到应用程序的类加载路径中。

示例：MyBatis 的数据库操作入门

新建一个项目 MyBatisQs，为了让 Web 应用具有 MyBatis 支持的功能，将 mybatis-3.4.5 解压文件夹下 mybatis-3.4.5.jar 和 lib 文件夹下所有 MyBatis 框架所依赖的第三方 jar 包复制到 Web 应用的 lib 文件夹下，也就是 MyBatisQs\WebContent\WEB-INF\ lib 路径下。

首先，在 MySQL 数据库中创建一个新的数据库 mybatis 和 tb_user 表。SQL 脚本如下：

```
# 创建一个名称为 mybatis 的数据库
create database mybatis;
# 使用名称为 mybatis 的数据库
use mybatis;
# 如果存在 tb_user 表则删除
DROP TABLE IF EXISTS tb_user;
# 创建一个 tb_user 表，有 id、name、sex、age 列
CREATE TABLE tb_user (
  ID INT(11) PRIMARY KEY AUTO_INCREMENT,
  NAME VARCHAR(18) DEFAULT NULL,
  SEX CHAR(2) DEFAULT NULL,
  AGE INT(11) DEFAULT NULL
 );
```

在 MySQL 数据库中执行 SQL 脚本，完成创建数据库和表的操作。

前面已经介绍过，在所有的 ORM 框架中都有一个非常重要的媒介：PO（持久化对象）。持久化对象的作用就是完成持久化操作，简单地说，就是通过该对象对数据库执行增、删、改的操作，以面向对象的方式操作数据库。

应用程序无须直接访问数据库，甚至无须理会底层数据库采用何种数据库，这一切对应用程序完全透明，应用程序只需创建、修改、删除持久化对象即可；与此同时，MyBatis 则负责把这种操作转换为对指定数据库表的操作。

MyBatis 中的 PO 是非常简单的，其是非侵入式的设计，完全采用普通的 Java 对象作为持久化对象使用。下面即是一个 POJO（普通的、传统的 Java 对象）类。

程序清单：codes/08/MyBatisQs/src/org/fkit/domain/User.java

```
import java.io.Serializable;
public class User implements Serializable{
    // 用户标识
    private Integer id;
    // 用户姓名
    private String name;
    // 用户性别
    private String sex;
    // 用户年龄
    private Integer age;
    // 无参数构造器
    public User() {
        super();
    }
    // 有参数构造器
    public User( String name, String sex, Integer age) {
        super();
        this.name = name;
        this.sex = sex;
        this.age = age;
    }
     // setter 和 getter 方法
    public Integer getId() {
        return id;
    }
    public void setId(Integer id) {
        this.id = id;
    }
    public String getName() {
        return name;
    }
    public void setName(String name) {
        this.name = name;
    }
    public String getSex() {
        return sex;
    }
    public void setSex(String sex) {
        this.sex = sex;
    }
    public Integer getAge() {
        return age;
    }
    public void setAge(Integer age) {
        this.age = age;
    }
}
```

仔细看上面这个类的代码，会发现这个类与普通的 JavaBean 没有任何区别。实际上，MyBatis 直接采用了 POJO（普通的、传统的 Java 对象）作为持久化类，这就是 MyBatis 被称非低侵入式设计的原因。MyBatis 不要求持久化类继承任何父类，或者实现任何接口，这样可保证代码不被污染。

对于 MyBatis 来说，现在还不理解持久化类 User 和数据库表之间的对应关系，也不理解持久化类的属性与数据表的各个列之间的对应关系。MyBatis 是通过 XML 文件去完成持久化类和数据库表之间的映射关系的。

程序清单：codes/08/MyBatisQs/src/org/fkit/mapper/UserMapper.xml

```
<?xml version="1.0" encoding="UTF-8"?>
<!DOCTYPE mapper PUBLIC "-//mybatis.org//DTD Mapper 3.0//EN"
"http://mybatis.org/dtd/mybatis-3-mapper.dtd">
<mapper namespace="org.fkit.mapper.UserMapper">
  <insert id="save" parameterType="org.fkit.domain.User" useGeneratedKeys="true">
     INSERT INTO TB_USER(name,sex,age) VALUES(#{name},#{sex},#{age})
  </insert>
</mapper>
```

上面的 XML 配置中定义了一条 insert 语句，详细解释如下：

（1）<mapper namespace="org.fkit.mapper.UserMapper">，为这个 mapper 指定一个唯一的 namespace，namespace 的值习惯上设置成包名+SQL 映射文件名，这样就能够保证 namespace 的值是唯一的，例如 namespace="org.fkit.mapper.UserMapper"就是 org.fkit.mapper（包名）+UserMapper（UserMapper.xml 文件去除后缀）。

（2）在 insert 标签中编写了 SQL 插入语句，设置 insert 标签的 id 属性值为 save。id 属性值必须是唯一的，不能够重复。

（3）使用 parameterType 属性指明插入时使用的参数类型。

（4）使用 useGeneratedKeys="true"表示使用数据库的自动增长策略，这需要底层数据库的支持。

（5）insert 标签中只有一条标准的 insert 语句，用来向 TB_USER 表插入一条数据，#{name}表示取参数中的对象的 name 属性值。

接下来，通过这个持久化类来完成对数据库的操作：插入一条数据。

对于 MyBatis 来说，现在还不知道需要连接哪个数据库，以及连接数据库时所用的连接池、用户名和密码等详细信息。这些信息对于所有的持久化类都是通用的，MyBatis 把这些通用信息称为根配置信息，根配置信息需要使用配置文件指定。

程序清单：codes/08/MyBatisQs/src/mybatis-config.xml

```
<?xml version="1.0" encoding="UTF-8" ?>
<!DOCTYPE configuration
  PUBLIC "-//mybatis.org//DTD Config 3.0//EN"
  "http://mybatis.org/dtd/mybatis-3-config.dtd">
  <!-- XML 配置文件包含对 MyBatis 系统的核心设置 -->
<configuration>
    <!-- 指定 MyBatis 所用日志的具体实现 -->
    <settings>
        <setting name="logImpl" value="LOG4J"/>
    </settings>
    <!-- 环境配置，即连接的数据库。 -->
    <environments default="mysql">
    <environment id="mysql">
    <!-- 指定事务管理类型，type="JDBC"指直接简单使用了 JDBC 的提交和回滚设置 -->
      <transactionManager type="JDBC"/>
      <!-- dataSource 指数据源配置，POOLED 是 JDBC 连接对象的数据源连接池的实现。 -->
```

```
        <dataSource type="POOLED">
          <property name="driver" value="com.mysql.jdbc.Driver"/>
          <property name="url" value="jdbc:mysql://127.0.0.1:3306/mybatis"/>
          <property name="username" value="root"/>
          <property name="password" value="root"/>
        </dataSource>
      </environment>
    </environments>
    <!-- mappers 告诉了MyBatis 去哪里找持久化类的映射文件 -->
    <mappers>
      <mapper resource="org/fkit/mapper/UserMapper.xml"/>
    </mappers>
  </configuration>
```

MyBatis 根配置文件默认被命名为 mybatis-config.xml，应用程序运行时需要先加载该文件。

MyBatis 配置文件是一个 XML 文件，该文件第一行是 XML 文件声明，指定该 XML 文件的版本和存储该文件所用的字符集。

MyBatis 配置文件的根元素是<configuration.../>，根元素中有<settings.../>子元素，该元素有很多子元素，本示例只是配置了日志信息，之后可以在控制台看到输出的 SQL 语句，其在调试中非常有用。根元素中还有< environments.../>子元素，用来配置 MyBatis 的环境，即连接的数据库，该子元素用于将 SQL 映射应用于多种数据库中。每个数据库对应一个 SqlSessionFactory，可以配置多种环境，但只能为 SqlSessionFactory 实例选择一个环境，default 属性表示选择的环境。<environment.../>子元素用于配置一个环境，<transactionManager.../>子元素用来配置 MyBatis 当中的事务管理，JDBC 属性表示直接简单使用 JDBC 的提交和回滚设置。<dataSource.../>子元素用来配置数据源，MyBatis 并不推荐采用 DriverManager 来连接数据库，而是推荐使用数据源来管理数据库连接，这样能保证最好的性能。该元素依次有很多<property.../>子元素，这些<property.../>子元素用于配置 MyBatis 连接数据库的必要信息，如连接数据库的驱动、URL、用户名、密码等信息。

提示

数据源是一种用来提高数据库连接性能的常规手段，数据源会负责维持一个数据库连接池，当程序创建数据源实例时，系统会一次性地创建多个数据库连接，并把这些数据库连接保存在连接池中。当程序需要进行数据库访问时，无须重新获得数据库连接，而是从连接池中取出一个空闲的数据库连接，当程序使用数据库连接访问数据库结束后，无须关闭数据库连接，而是将数据库连接归还给连接池即可。通过这种方式，就可避免频繁地获取数据库连接、关闭数据库连接所导致的性能下降。

根元素中还有< mappers.../>子元素，它可以支持多个< mapper.../>子元素，每个< mapper.../>子元素用于指定一个持久化配置文件。

下面是完成数据插入的代码。

程序清单：codes/08/MyBatisQs/src/org/fkit/tst/MyBatisTest.java

```
import java.io.InputStream;
import org.apache.ibatis.io.Resources;
import org.apache.ibatis.session.SqlSession;
import org.apache.ibatis.session.SqlSessionFactory;
import org.apache.ibatis.session.SqlSessionFactoryBuilder;
```

```
import org.fkit.domain.User;

public class MyBatisTest {

    public static void main(String[] args) {
        // 创建 Session 实例
        SqlSession sqlSession = null;
        try (// 读取 mybatis-config.xml 文件
                InputStream is = Resources.getResourceAsStream("mybatis-config.xml");
            ){
            // 初始化 mybatis，创建 SqlSessionFactory 类的实例
            SqlSessionFactory sqlSessionFactory = new SqlSessionFactoryBuilder()
                    .build(is);
            sqlSession = sqlSessionFactory.openSession();
            // 创建 User 对象
            User user = new User("admin", "男", 26);
            // 插入数据
            sqlSession.insert("org.fkit.mapper.UserMapper.save", user);
            // 提交事务
            sqlSession.commit();
        } catch (Exception e) {
            // 回滚事务
            sqlSession.rollback();
            e.printStackTrace();
        }finally{
            try {
                // 关闭 sqlSession
                if(sqlSession != null) sqlSession.close();
            } catch (Exception e) {
                e.printStackTrace();
            }
        }

    }
}
```

上面持久化操作的代码非常简单。程序先创建一个 User 对象，再使用 SqlSession 的 insert() 方法来保存 User 对象即可，这是完全对象化的操作方式，可以说非常简单明了。当 Java 程序以面向对象的方式来操作持久化对象时，MyBatis 负责将这种操作转换为底层 SQL 操作。

执行持久化操作之前，为了查看控制台输出的 SQL 语句，需要加入日志框架 LOG4J 的相关 jar 包，该 jar 包在 mybatis-3.4.5 解压文件夹下的 lib 文件夹中可以找到。并在 CLASSPATH 下增加一个 log4j.properties 文件。

程序清单：codes/08/MyBatisQs/src/log4j.properties

```
# 全局的日志配置
log4j.rootLogger=ERROR, stdout
# MyBatis 的日志配置
log4j.logger.org.fkit.mapper.UserMapper=DEBUG
# 控制台输出
log4j.appender.stdout=org.apache.log4j.ConsoleAppender
log4j.appender.stdout.layout=org.apache.log4j.PatternLayout
log4j.appender.stdout.layout.ConversionPattern=%5p [%t] - %m%n
```

配置文件 log4j.propertie 的余下内容是针对日志格式的，这一内容已经超出本书的范围。

关于 Log4j 的更多内容，读者请自行参考 Log4j 官方文档。

运行 MyBatisTest 类的 main()方法，运行完成后，可以看到 mybatis 数据库中的 TB_USER 表中包含了 User 实例对应的记录，如图 8.5 所示。

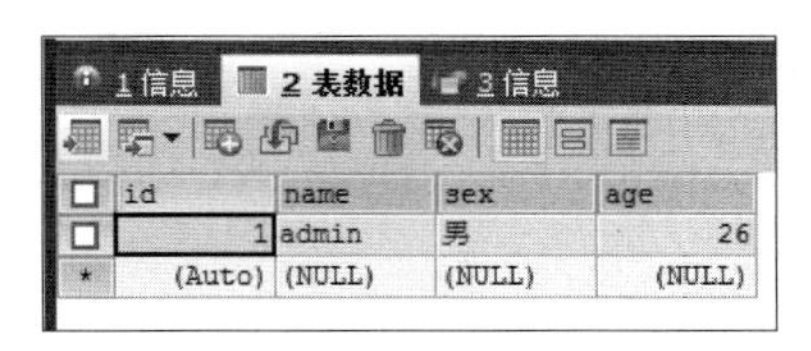

图 8.5 使用 MyBatis 成功插入记录

同时，在控制台可以观察到执行时的日志信息，其中包括了 MyBatis 所执行的 SQL 语句。

```
DEBUG [main] ==>  Preparing: INSERT INTO TB_USER(name,sex,age) VALUES(?,?,?)
DEBUG [main] ==> Parameters: admin(String), 男(String), 26(Integer)
DEBUG [main] <==    Updates: 1
```

有开发经验的读者可以通过日志信息了解 MyBatis 的执行过程。

正如上面程序中粗体代码所示，在执行 session.insert("org.fkit.mapper.UserMapper.save", user)之前，先要获取 SqlSession 对象。PO 只有在 SqlSession 的管理下才可完成数据库访问。为了使用 MyBatis 进行持久化操作，通常需要执行如下操作步骤：

1. 开发持久化类 PO 和编写持久化操作的 Mapper.xml，在其中定义要执行的 SQL 语句。
2. 获取 SqlSessionFactory。
3. 获取 SqlSession。
4. 用面向对象的方式操作数据库。
5. 关闭事务，关闭 SqlSession。

对 PO 的操作必须在 SqlSession 的管理下才能同步到数据库。SqlSession 由 SqlSessionFactory 工厂产生，SqlSessionFactory 是数据库编译后的内存镜像，通常一个应用对应一个 SqlSessionFactory 对象。SqlSessionFactory 对象通过加载 mybatis-config.xml 配置文件生成。

上面使用 MyBatis 添加了一条记录，对比 MyBatis 和 JDBC 两种操作方式，不难发现 MyBatis 的两个显著优点：

- 只需要在 Mapper.xml 配置文件中编写 SQL 语句，在应用程序中就可以采用 OO 方式来访问数据库。
- 在 JDBC 访问过程中大量的 checked 异常被包装成 MyBatis 的 Runtime 异常，从而不再要求程序必须处理所有异常。

8.3 本章小结

本章介绍了流行的 ORM 模式，包括现阶段流行的 ORM 框架。还介绍了 MyBatis 的概念和入门操作。

第 9 章将重点介绍 MyBatis 的基本用法。

CHAPTER

9

第9章 MyBatis的基本用法

本章要点

- SqlSessionFactory 对象
- SqlSession 对象
- MyBatis 根配置文件详解
- MyBatis 日志信息配置详解
- Mapper 映射文件详解

9.1　MyBatis 体系结构

MyBatis 的持久化解决方案将用户从原始的 JDBC 访问中解放出来，用户只需要定义需要操作的 SQL 语句，无须关注底层的 JDBC 操作，就可以以面向对象的方式进行持久层操作。底层数据库连接的获取、数据访问的实现、事务控制和数据库连接的关闭等都无须用户关心，从而将应用层从底层的 JDBC/JTA API 抽取出来。通过配置文件管理 JDBC 连接，让 MyBatis 完成持久化访问的实现。

MyBatis 中的常用对象有 SqlSessionFactory 和 SqlSession。

9.1.1　SqlSessionFactory

SqlSessionFactory 是 MyBatis 的关键对象，它是单个数据库映射关系经过编译后的内存镜像。SqlSessionFactory 对象的实例可以通过 SqlSessionFactoryBuilder 对象来获得，而 SqlSessionFactoryBuilder 则可以从 XML 配置文件或一个预先定制的 Configuration 的实例构建出 SqlSessionFactory 的实例。每一个 MyBatis 的应用程序都以一个 SqlSessionFactory 对象的实例为核心。其也是线程安全的，SqlSessionFactory 一旦被创建，应该在应用执行期间都存在。在应用运行期间不需要重复创建多次，建议使用单例模式。SqlSessionFactory 是创建 SqlSession 的工厂。

SqlSessionFactory 的常用方法如下：

- SqlSession openSession()。创建 SqlSession 对象。

9.1.2　SqlSession

SqlSession 是 MyBatis 的关键对象，它是执行持久化操作的对象，类似于 JDBC 中的 Connection。它是应用程序与持久存储层之间执行交互操作的一个单线程对象，也是 MyBatis 执行持久化操作的关键对象。SqlSession 对象完全包含以数据库为背景的所有执行 SQL 操作的方法，它的底层封装了 JDBC 连接，可以用 SqlSession 实例来直接执行已映射的 SQL 语句。每个线程都应该有它自己的 SqlSession 实例。SqlSession 的实例不能被共享，也是线程不安全的，绝对不能将 SqlSession 实例的引用放在一个类的静态字段甚至是实例字段中。也绝不能将 SqlSession 实例的引用放在任何类型的管理范围中，比如 Serlvet 当中的 HttpSession 对象中。使用完 SqlSession 之后关闭 Session 很重要，应该确保使用 finally 块来关闭它。

SqlSession 的常用方法如下：

- int insert(String statement)。插入方法，参数 statement 是在配置文件中定义的<insert.../>元素的 id，返回执行 SQL 语句所影响的行数。
- int insert(String statement,Object parameter)。插入方法，参数 statement 是在配置文件中定义的<insert.../>元素的 id，parameter 是插入所需的参数，通常是对象或者 Map，返回执行 SQL 语句所影响的行数。
- int update(String statement)。更新方法，参数 statement 是在配置文件中定义的<update.../>元素的 id，返回执行 SQL 语句所影响的行数。
- int update(String statement,Object parameter)。更新方法，参数 statement 是在配置文件中定义的<update.../>元素的 id，parameter 是插入所需的参数，通常是对象或者 Map，返回执行 SQL 语句所影响的行数。
- int delete(String statement)。删除方法，参数 statement 是在配置文件中定义的<delete.../>

元素的 id。返回执行 SQL 语句所影响的行数。

- int delete(String statement,Object parameter)。删除方法，参数 statement 是在配置文件中定义的<delete.../>元素的 id，parameter 是插入所需的参数，通常是对象或者 Map，返回执行 SQL 语句所影响的行数。
- <T> T selectOne(String statement)。查询方法，参数 statement 是在配置文件中定义的<select.../>元素的 id。返回执行 SQL 语句查询结果的泛型对象，通常查询结果只有一条数据时才使用。
- <T> T selectOne(String statement,Object parameter)。查询方法，参数 statement 是在配置文件中定义的<select.../>元素的 id，parameter 是查询所需的参数，通常是对象或者 Map，返回执行 SQL 语句查询结果的泛型对象，通常查询结果只有一条数据时才使用。
- <E> List<E> selectList(String statement)。查询方法，参数是在配置文件中定义的<select.../>元素的 id，返回执行 SQL 语句查询结果的泛型对象的集合。
- <E> List<E> selectList(String statement,Object parameter)。查询方法，参数 statement 是在配置文件中定义的<select.../>元素的 id，parameter 是查询所需的参数，通常是对象或者 Map，返回执行 SQL 语句查询结果的泛型对象的集合。
- <E> List<E> selectList(String statement,Object parameter,RowBounds rowBounds)。查询方法，参数 statement 是在配置文件中定义的<select.../>元素的 id，parameter 是查询所需的参数，通常是对象或者 Map。RowBounds 对象用于分页，它的两个属性：offset 指查询的当前页数；limit 指当前页显示多少条数据。返回执行 SQL 语句查询结果的泛型对象的集合。
- <K,V> Map<K,V> selectMap(String statement,String mapKey)。查询方法，参数 statement 是在配置文件中定义的<select.../>元素的 id，mapKey 是返回数据的其中一个列名，执行 SQL 语句查询的结果将会被封装成一个 Map 集合返回，key 就是参数 mapKey 传入的列名，value 是封装的对象。
- <K,V> Map<K,V> selectMap(String statement,Object parameter,String mapKey)。查询方法，参数 statement 是在配置文件中定义的<select.../>元素的 id，parameter 是查询所需的参数，通常是对象或者 Map，mapKey 是返回数据的其中一个列名，执行 SQL 语句查询的结果将会被封装成一个 Map 集合返回，key 就是参数 mapKey 传入的列名，value 是封装的对象。
- <K,V> Map<K,V>selectMap(String statement,Object parameter,String mapKey, RowBounds rowBounds)。查询方法，参数 statement 是在配置文件中定义的<select.../>元素的 id，parameter 是查询所需的参数，通常是对象或者 Map，mapKey 是返回数据的其中一个列名，RowBounds 对象用于分页。执行 SQL 语句查询的结果将会被封装成一个 Map 集合返回，key 就是参数 mapKey 传入的列名，value 是封装的对象。
- void select(String statement,ResultHandler handler)。查询方法，参数 statement 是在配置文件中定义的<select.../>元素的 id，ResultHandler 对象用来处理查询返回的复杂结果集，通常用于多表查询。
- void select(String statement,Object parameter,ResultHandler handler)。查询方法，参数 statement 是在配置文件中定义的<select.../>元素的 id，parameter 是查询所需的参数，通常是对象或者 Map，ResultHandler 对象用来处理查询返回的复杂结果集，通常

用于多表查询。

- void select(String statement,Object parameter,RowBounds rowBounds,ResultHandler handler)。查询方法，参数 statement 是在配置文件中定义的<select.../>元素的 id，parameter 是查询所需的参数，通常是对象或者 Map，RowBounds 对象用于分页，ResultHandler 对象用来处理查询返回的复杂结果集，通常用于多表查询。
- void commit()。提交事务。
- void rollback()。回滚事务。
- void close()。关闭 SqlSession 对象。
- Connection getConnection()。获得 JDBC 的数据库连接对象。
- <T> T getMapper(Class<T> type)。返回 mapper 接口的代理对象，该对象关联了 SqlSession 对象，开发者可以通过该对象直接调用方法操作数据库，参数 type 是 Mapper 的接口类型。MyBatis 官方手册建议通过 mapper 对象访问 MyBatis。

提示

在实际应用中很少会直接使用 DriverManager 来获取数据库连接，通常都会使用 DataSource 来获取数据库连接，SqlSessionFactory 底层封装了 DataSource。

9.2 深入 MyBatis 的配置文件

通过上面的介绍，我们可以知道 MyBatis 的持久化操作离不开 SqlSessionFactory 对象，这个对象是整个数据库映射关系经过编译后的内存镜像，该对象的 openSession()方法可以打开 SqlSession 对象。该对象由 SqlSessionFactoryBuilder 加载 MyBatis 的配置文件产生。

再来回顾一下之前的代码：

```
    // 读取 mybatis-config.xml 文件
    InputStream inputStream = Resources.getResourceAsStream("mybatis-config.xml");
    // 初始化 MyBatis，创建 SqlSessionFactory 类的实例
    SqlSessionFactory sqlSessionFactory = new SqlSessionFactoryBuilder().build
(inputStream);
    // 创建 Session 实例
    SqlSession session = sqlSessionFactory.openSession();
```

上述代码的功能是根据配置文件 mybatis-config.xml ，创建 SqlSessionFactory 对象，然后产生 SqlSession，执行 SQL 语句。而 MyBatis 的初始化就发生在第三句：

```
    SqlSessionFactory sqlSessionFactory = new SqlSessionFactoryBuilder().build
(inputStream);
```

现在就让我们看看第三句到底发生了什么。

MyBatis 初始化基本过程：

SqlSessionFactoryBuilder 根据传入的输入流生成 Configuration 对象，然后根据 Configuration 对象创建默认的 SqlSessionFactory 实例。

初始化的基本过程如图 9.1 所示。

由图 9.1 可知，MyBatis 初始化要经过以下几步：

1. 调用 SqlSessionFactoryBuilder 对象的 build(inputStream)方法。
2. SqlSessionFactoryBuilder 会根据输入流 inputStream 等信息创建 XMLConfigBuilder 对象。
3. SqlSessionFactoryBuilder 调用 XMLConfigBuilder 对象的 parse()方法。

④ XMLConfigBuilder 对象解析 XML 配置文件返回 Configuration 对象。

⑤ SqlSessionFactoryBuilder 根据 Configuration 对象创建一个 DefaultSessionFactory 对象。

⑥ SqlSessionFactoryBuilder 返回 DefaultSessionFactory 对象给客户端，供客户端使用。

由此可见，SqlSessionFactory 是根据 MyBatis 的配置文件 mybatis-config.xml 创建的。下面我们就来重点介绍 MyBatis 的配置文件中常用属性的意义。

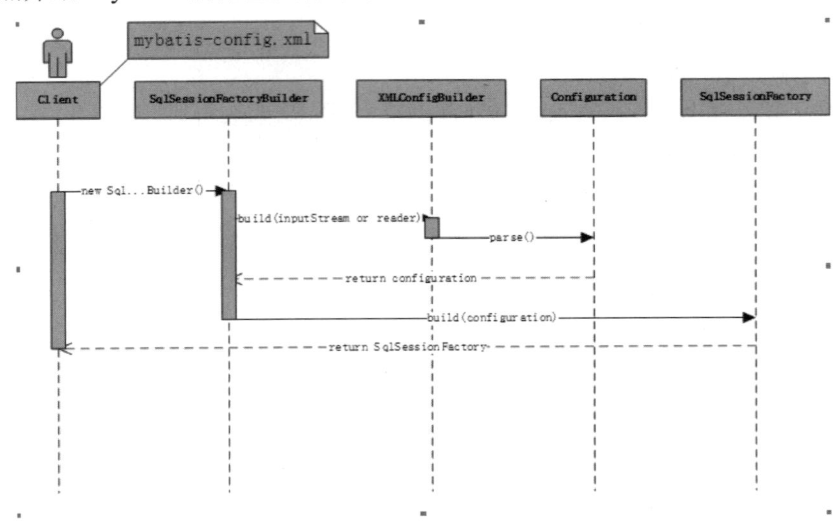

图 9.1 MyBatis 初始化基本流程

9.2.1 MyBatis 的配置文件结构

MyBatis 的配置文件包含了影响 MyBatis 行为的信息。文档的结构如下：

顶层 configuration 配置

- properties 属性
- settings 设置
- typeAliases 类型命名
- typeHandlers 类型处理器
- objectFactory 对象工厂
- plugins 插件
- environments 环境
  - environment 环境变量
  - transactionManager 事务管理器
  - dataSource 数据源
- databaseIdProvider 数据库厂商标识
- mappers 映射器

9.2.2 properties 属性

这些属性都是可外部配置且可动态替换的，既可以在典型的 Java 属性文件中配置，亦可通过 properties 元素的子元素来传递。

可以在 CLASSPATH 中增加一个 db.properties 的 Java 属性文件。

```
driver=com.mysql.jdbc.Driver
url=jdbc:mysql://127.0.0.1:3306/mybatis
```

```
username=root
password=root
```

在配置文件中配置<properties.../>属性：

```
<properties resource="db.properties"/>
```

其中的属性就可以在整个配置文件中使用来替换需要动态配置的属性值。

```
<dataSource type="POOLED">
<property name="driver" value="${driver}"/>
<property name="url" value="${url}"/>
<property name="username" value="${username}"/>
<property name="password" value="${password}"/>
</dataSource>
```

driver、url、username 和 password 属性将会由 db.properties 文件中对应的值来替换。这样就为配置提供了诸多灵活选择。

9.2.3 settings 设置

这是 MyBatis 中极为重要的调整设置，它们会改变 MyBatis 的运行时行为。表 9.1 描述了设置中各项的参数、默认值等。

表 9.1 settings 设置的详细说明

设置参数	描述	有效值	默认值
cacheEnabled	该配置影响所有映射器中配置的缓存的全局开关	true | false	true
lazyLoadingEnabled	延迟加载的全局开关。当开启时，所有关联对象都会延迟加载。特定关联关系中可通过设置 fetchType 属性来覆盖该项的开关状态	true | false	false
aggressiveLazyLoading	当启用时，对任意延迟属性的调用会使带有延迟加载属性的对象完整加载；反之，每种属性将会按需加载	true | false	true
multipleResultSetsEnabled	是否允许单一语句返回多结果集（需要兼容驱动）	true | false	true
useColumnLabel	使用列标签代替列名。不同的驱动在这方面会有不同的表现，具体可参考相关驱动文档或通过测试这两种不同的模式来观察所用驱动的结果	true | false	true
useGeneratedKeys	允许 JDBC 支持自动生成主键，需要驱动兼容。如果设置为 true 则这个设置强制使用自动生成主键，尽管一些驱动不能兼容但仍可正常工作（比如 Derby）	true | false	false
autoMappingBehavior	指定 MyBatis 应如何自动映射列到字段或属性。NONE 表示取消自动映射；PARTIAL 只会自动映射没有定义嵌套结果集映射的结果集。FULL 会自动映射任意复杂的结果集（无论是否嵌套）	NONE|PARTIAL|FULL	PARTIAL
autoMappingUnknownColumnBehavior	NONE:什么都不做；WARNING：输出警告；FAILING：映射失败，抛出 SqlSessionException；	NONE|WARNING|FAILING	NONE

续表

设置参数	描述	有效值	默认值
defaultExecutorType	配置默认的执行器。SIMPLE 就是普通的执行器；REUSE 执行器会重用预处理语句（prepared statements）；BATCH 执行器将重用语句并执行批量更新	SIMPLE| REUSE| BATCH	SIMPLE
defaultStatementTimeout	设置超时时间，它决定驱动等待数据库响应的秒数	integer	没有设置(null)
defaultFetchSize	默认返回的结果集的大小	integer	没有设置(null)
safeRowBoundsEnabled	允许在嵌套语句中使用分页（RowBounds）	true | false	false
mapUnderscoreToCamelCase	是否开启自动驼峰命名规则（camel case）映射	true | false	false
localCacheScope	MyBatis 利用本地缓存机制（Local Cache）防止循环引用（circular references）和加速重复嵌套查询。默认值为 SESSION，这种情况下会缓存一个会话中执行的所有查询。若设置值为 STATEMENT，则本地会话仅用在语句执行上，对相同 SqlSession 的不同调用将不会共享数据	SESSION | STATEMENT	SESSION
jdbcTypeForNull	当没有为参数提供特定的 JDBC 类型时，为空值指定 JDBC 类型。某些驱动需要指定列的 JDBC 类型，多数情况直接用一般类型即可，比如 NULL、VARCHAR 或 OTHER	NULL| VARCHAR| OTHER	OTHER
lazyLoadTriggerMethods	指定哪个对象的方法触发一次延迟加载	方法名的 list 集合	equals,clone,hashCode,toString
defaultScriptingLanguage	指定动态 SQL 生成的默认语言	一个类型别名或完全限定类名	org.apache.ibatis.scripting.xmltags.XMLDynamicLanguageDriver
callSettersOnNulls	指定当结果集中值为null时是否调用映射对象的 setter（map 对象时为 put）方法，这对于有 Map.keySet() 依赖或 null 值初始化时是有用的。注意基本类型（int、boolean 等）是不能设置成 null 的	true | false	false
logPrefix	指定 MyBatis 增加到日志名称的前缀	String	没有设置(null)
logImpl	指定 MyBatis 所用日志的具体实现，未指定时将自动查找	SLF4J|LOG4J|LOG4J2|JDK_LOGGING|COMMONS_LOGGING | STDOUT_LOGGING | NO_LOGGING	没有设置(null)
proxyFactory	指定 MyBatis 创建具有延迟加载能力的对象所用到的代理工具	CGLIB | JAVASSIST	JAVASSIST (MyBatis 3.3 or above)

一个配置完整的 settings 元素的示例如下：

```
<settings>
  <setting name="cacheEnabled" value="true"/>
  <setting name="lazyLoadingEnabled" value="true"/>
  <setting name="multipleResultSetsEnabled" value="true"/>
```

```
    <setting name="useColumnLabel" value="true"/>
    <setting name="useGeneratedKeys" value="false"/>
    <setting name="autoMappingBehavior" value="PARTIAL"/>
    <setting name="autoMappingUnknownColumnBehavior" value="WARNING"/>
    <setting name="defaultExecutorType" value="SIMPLE"/>
    <setting name="defaultStatementTimeout" value="25"/>
    <setting name="defaultFetchSize" value="100"/>
    <setting name="safeRowBoundsEnabled" value="false"/>
    <setting name="mapUnderscoreToCamelCase" value="false"/>
    <setting name="localCacheScope" value="SESSION"/>
    <setting name="jdbcTypeForNull" value="OTHER"/>
    <setting name="lazyLoadTriggerMethods" value="equals,clone,hashCode,toString"/>
</settings>
```

9.2.4 typeAliases 类型命名

类型别名是为 Java 类型设置的一个短的名字。它只和 XML 配置有关，存在的意义仅在于用来减少类完全限定名的冗余。

```
<typeAliases>
    <typeAlias  alias="user" type="org.fkit.domain.User"/>
</typeAliases>
```

当这样配置时，user 可以用在任何使用 org.fkit.domain.User 的地方。

也可以指定一个包名，MyBatis 会在包名下面搜索需要的 JavaBean。

```
<typeAliases>
  <package name="org.fkit.domain"/>
</typeAliases>
```

每一个在包 org.fkit.domain 中的 JavaBean，在没有注解的情况下，会使用 Bean 的首字母小写的非限定类名来作为它的别名。比如 org.fkit.domain.User 的别名为 user；若有注解，则别名为其注解值。

```
@Alias("user")
public class User {
    ...
}
```

MyBatis 已经为许多常见的 Java 类型内建了相应的类型别名（见表 9.2）。它们都是大小写不敏感的，需要注意的是由基本类型名称重复导致的特殊处理。

表 9.2 MyBatis 默认别名

别名	映射的类型
_byte	byte
_long	long
_short	short
_int	int
_integer	int
_double	double
_float	float
_boolean	boolean
string	String
byte	Byte
long	Long
short	Short
int	Integer
integer	Integer
double	Double

续表

别名	映射的类型
float	Float
boolean	Boolean
date	Date
bigdecimal	BigDecimal
object	Object
map	Map
hashmap	HashMap
list	List
arraylist	ArrayList
collection	Collection
iterator	Iterator

9.2.5 typeHandlers 类型处理器

无论是 MyBatis 在预处理语句（PreparedStatement）中设置一个参数时，还是从结果集中取出一个值时，都会用类型处理器将获取的值以合适的方式转换成 Java 类型。表 9.3 描述了一些默认的类型处理器。

表 9.3 MyBatis 默认的类型处理器

类型处理器	Java 类型	JDBC 类型
BooleanTypeHandler	java.lang.Boolean, boolean	数据库兼容的 BOOLEAN
ByteTypeHandler	java.lang.Byte, byte	数据库兼容的 NUMERIC 或 BYTE
ShortTypeHandler	java.lang.Short, short	数据库兼容的 NUMERIC 或 SHORT INTEGER
IntegerTypeHandler	java.lang.Integer, int	数据库兼容的 NUMERIC 或 INTEGER
LongTypeHandler	java.lang.Long, long	数据库兼容的 NUMERIC 或 LONG INTEGER
FloatTypeHandler	java.lang.Float, float	数据库兼容的 NUMERIC 或 FLOAT
DoubleTypeHandler	java.lang.Double, double	数据库兼容的 NUMERIC 或 DOUBLE
BigDecimalTypeHandler	java.math.BigDecimal	数据库兼容的 NUMERIC 或 DECIMAL
StringTypeHandler	java.lang.String	CHAR、VARCHAR
ClobReaderTypeHandler	java.io.Reader	无
ClobTypeHandler	java.lang.String	CLOB、LONGVARCHAR
NStringTypeHandler	java.lang.String	NVARCHAR、NCHAR
NClobTypeHandler	java.lang.String	NCLOB
BlobInputStreamTypeHandler	java.io.InputStream	无
ByteArrayTypeHandler	byte[]	数据库兼容的字节流类型
BlobTypeHandler	byte[]	BLOB, LONGVARBINARY
DateTypeHandler	java.util.Date	TIMESTAMP
DateOnlyTypeHandler	java.util.Date	DATE
TimeOnlyTypeHandler	java.util.Date	TIME
SqlTimestampTypeHandler	java.sql.Timestamp	TIMESTAMP
SqlDateTypeHandler	java.sql.Date	DATE
SqlTimeTypeHandler	java.sql.Time	TIME
ObjectTypeHandler	Any	OTHER 或未指定类型
EnumTypeHandler	Enumeration Type	VARCHAR，任何兼容的字符串类型，存储枚举的名称（而不是索引）
EnumOrdinalTypeHandler	Enumeration Type	任何兼容的 NUMERIC 或 DOUBLE 类型，存储枚举的索引（而不是名称）

9.2.6 objectFactory 对象工厂

MyBatis 每次创建结果对象的新实例时，它都会使用一个对象工厂（ObjectFactory）实例来完成。默认的对象工厂需要做的仅仅是实例化目标类，要么通过默认构造方法，要么在参数映射存在的时候通过参数构造方法来实例化。如果想覆盖对象工厂的默认行为，则可以通过创建自己的对象工厂来实现。

```
// 自定义对象工厂
public class ExampleObjectFactory extends DefaultObjectFactory {
  public Object create(Class type) {
    return super.create(type);
  }
  public Object create(Class type, List<Class> constructorArgTypes, List<Object>
constructorArgs) {
    return super.create(type, constructorArgTypes, constructorArgs);
  }
  public void setProperties(Properties properties) {
    super.setProperties(properties);
  }
  public <T> boolean isCollection(Class<T> type) {
    return Collection.class.isAssignableFrom(type);
  }
}
```

在 MyBatis 配置文件中配置自定义对象工厂。

```
<objectFactory type="org.fkit.factory.ExampleObjectFactory">
  <property name="someProperty" value="100"/>
</objectFactory>
```

ObjectFactory 接口很简单，它包含两个创建方法：一个是处理默认构造方法的；另外一个是处理带参数的构造方法的。最后，setProperties 方法可以被用来配置 ObjectFactory，在初始化 ObjectFactory 实例后，objectFactory 元素体中定义的属性会被传递给 setProperties 方法。

9.2.7 environments 配置环境

MyBatis 的环境配置实际就是数据源的配置。MyBatis 可以配置多种环境，这种机制使得 MyBatis 可以将 SQL 映射应用于多种数据库中。例如，开发、测试和生产环境需要有不同的配置；多个生产数据库想使用相同的 SQL 映射等等。

提示

尽管可以配置多个环境，但是每个 SqlSessionFactory 实例只能选择一个环境，即每个数据库对应一个 SqlSessionFactory 实例。所以，如果你想连接两个数据库，就需要创建两个 SqlSessionFactory 实例，每个数据库对应一个。而如果是三个数据库，就需要三个实例，依此类推。

环境示例配置如下：

```
<environments default="development">
  <environment id="development">
    <transactionManager type="JDBC">
      <property name="..." value="..."/>
    </transactionManager>
    <dataSource type="POOLED">
      <property name="driver" value="${driver}"/>
      <property name="url" value="${url}"/>
      <property name="username" value="${username}"/>
      <property name="password" value="${password}"/>
```

```
    </dataSource>
  </environment>
</environments>
```

注意这里的关键点：

默认的环境 id（比如，default="development"）。

每个 environment 元素定义的环境 id（比如，id="development"）。

事务管理器的配置（比如，type="JDBC"）。

数据源的配置（比如，type="POOLED"）。

环境 id 可以随意命名，建议简洁有意义，而默认环境一定要匹配定义的其中一个环境 id。

<transactionManager.../>表示事务管理器配置，在 MyBatis 中有 JDBC 和 MANAGED 两种类型的事务管理器：

- JDBC。这个配置直接使用了 JDBC 的提交和回滚设置，它依赖于从数据源得到的连接来管理事务范围。
- MANAGED。这个配置几乎没做什么。它从来不提交或回滚一个连接，而是让容器来管理事务的整个生命周期（比如 Java EE 应用服务器的上下文）。默认情况下它会关闭连接，然而一些容器并不希望这样，可以将 closeConnection 属性设置为 false 来阻止它默认的关闭行为。

transactionManager 的 MANAGED 配置示例如下：

```
<transactionManager type="MANAGED">
  <property name="closeConnection" value="false"/>
</transactionManager>
```

如果开发者使用 Spring + MyBatis，则没有必要配置事务管理器，因为 Spring 模块会使用自带的管理器来覆盖 MyBatis 的事务管理器配置。

<dataSource.../>表示数据源配置，在 MyBatis 中有 UNPOOLED、POOLED 和 JNDI 三种数据源类型：

- UNPOOLED。这个数据源的实现只是每次被请求时打开和关闭连接。它对没有性能要求的简单应用程序是一个很好的选择。不同的数据库在这方面表现也是不一样的。UNPOOLED 类型的数据源仅仅需要配置以下 5 种属性：
  - driver。这是 JDBC 驱动的 Java 类的完全限定名（并不是 JDBC 驱动中可能包含的数据源类）。
  - url。这是数据库的 JDBC URL 地址。
  - username。登录数据库的用户名。
  - password。登录数据库的密码。
  - defaultTransactionIsolationLevel。默认的连接事务隔离级别。
- POOLED：这种数据源的实现利用“池”的概念将 JDBC 连接对象组织起来，避免了创建新的连接实例时所必需的初始化和认证时间。这是一种使得并发 Web 应用快速响应请求的流行处理方式。除了上述提到的 UNPOOLED 的 5 种属性外，还可以使用更多属性来配置 POOLED 的数据源：
  - poolMaximumActiveConnections。在任意时间可以存在的活动（也就是正在使用）连接数量，默认值是 10。
  - poolMaximumIdleConnections。任意时间可能存在的空闲连接数。
  - poolMaximumCheckoutTime。在被强制返回之前，池中连接被检出（checked out）

时间，默认值为 20000 ms（即 20 s）。

- poolTimeToWait。这是一个底层设置，如果获取连接花费相当长的时间，它会给连接池打印状态日志并重新尝试获取一个连接（避免在误配置的情况下一直安静地失败），默认值为 20000 ms（即 20 s）。
- poolPingQuery。发送到数据库的侦测查询，用来检验连接是否处在正常工作秩序中并准备接受请求。默认是“NO PING QUERY SET”，这会导致多数数据库驱动失败时带有一个恰当的错误消息。
- poolPingEnabled。是否启用侦测查询。若开启，也必须使用一个可执行的 SQL 语句设置 poolPingQuery 属性（最好是一个非常快的 SQL），默认值为 false。
- poolPingConnectionsNotUsedFor。配置 poolPingQuery 的使用频度。其可以被设置成匹配具体的数据库连接超时时间，来避免不必要的侦测，默认值为 0（即所有连接每一时刻都被侦测，当然仅当 poolPingEnabled 为 true 时适用）。
- JNDI。这个数据源的实现是为了能在如 EJB 或应用服务器这类容器中使用，容器可以集中或在外部配置数据源，然后放置一个 JNDI 上下文的引用。这种数据源配置只需要两个属性：

  initial_context。这个属性用来在 InitialContext 中寻找上下文（即 initialContext.lookup(initial_context)）。这是个可选属性，如果忽略，那么 data_source 属性将会直接从 InitialContext 中寻找。

  data_source。这是引用数据源实例位置的上下文路径。若提供了 initial_context 配置则会在其返回的上下文中进行查找，没有提供则直接在 InitialContext 中查找。

dataSource 的 JDNI 配置示例如下：

```
<dataSource type="JNDI">
    <property name="initial_context" value="java:/comp/env"/>
    <property name="data_source" value="fkjavads"/>
</dataSource>
```

其中“java:/comp/env”是 Tomcat 服务器的前缀，每个 Web 服务器的前缀都不一样，具体请查看 Web 服务器相关文档。

9.2.8 mapper 映射器

MyBatis 需要开发者自己写 SQL 语句，mapper 映射器告诉 MyBatis 到哪里去找映射文件，进而找到这些 SQL 语句。在实际开发中可以使用相对于类路径的资源引用或完全限定资源定位符（包括 file:/// 的 URL），以及类名和包名等。例如：

```
<!-- 使用类路径查找资源文件 -->
<mappers>
   <mapper resource="org/fkit/mapper/UserMapper.xml"/>
</mappers>
<!--使用本地文件 -->
<mappers>
   <mapper url="file:///C:/mapper/UserMapper.xml"/>
</mappers>
<!-- 使用接口类 -->
<mappers>
   <mapper class="org.fkit.mapper.UserMapper"/>
</mappers>
<!-- 使用包名 -->
<mappers>
```

```
    <package name="org.fkit.mapper"/>
</mappers>
```

mapper映射器会告诉MyBatis去哪里找映射文件，剩下的细节就是每个SQL映射文件了，也就是接下来我们要重点讨论的。

9.3　MyBatis 日志信息配置

使用 MyBatis 的时候，经常需要输出 SQL 语句、参数信息、查询结果等日志信息，为此 MyBatis 也提供了非常简单有效的解决方案。

MyBatis 内置的日志工厂提供日志功能，具体的日志实现有以下几种工具：

- SLF4J
- Apache Commons Logging
- Log4j 2
- Log4j
- JDK logging

具体选择哪个日志实现工具由 MyBatis 的内置日志工厂决定。它会使用最先找到的（按上文列举的顺序查找）。如果一个都未找到，日志功能就会被禁用。

不少应用服务器的 classpath 中已经包含 Apache Commons Logging，如 Tomcat 和 WebShpere，所以 MyBatis 会把它作为具体的日志实现。记住这点非常重要。这将意味着，在诸如 WebSphere 的环境中——WebSphere 提供了 Apache Commons Logging 的私有实现，你的 Log4j 配置将被忽略。不过，如果你的应用部署在一个包含 Apache Commons Logging 的环境里，而你又想用其他的日志框架比如 Log4j，可以通过在 MyBatis 的配置文件 mybatis-config.xml 里面添加一项 setting（配置）来选择一个不同的日志实现。这也是 MyBatis 推荐的做法。

```
<configuration>
  <settings>
    ...
    <setting name="logImpl" value="LOG4J"/>
    ...
  </settings>
</configuration>
```

这样就是告诉 MyBatis 当前项目的日志实现使用 Log4j，Log4j 的配置信息就会起作用。

logImpl 可选的值有：SLF4J、LOG4J、LOG4J2、JDK_LOGGING、COMMONS_LOGGING、STDOUT_LOGGING、NO_LOGGING 或者是实现了接口 org.apache.ibatis.logging.Log 的类的完全限定类名，并且这个类的构造函数需要以一个字符串（String 类型）为参数。具体可以参考 org.apache.ibatis.logging.slf4j.Slf4jImpl.java 的实现。

MyBatis 可以对包、类、命名空间和全限定的语句记录日志。

具体怎么做，视使用的日志框架而定，这里以 Log4j 为例。配置日志功能非常简单：首先增加依赖的 jar 包，如 log4j.jar，log4j.jar 可以直接在 mybatis-3.4.5 解压文件夹下的 lib 文件夹中找到，也可以自己去官网下载。

再添加配置文件，有 log4j.properties 或 log4j.xml 两种，下面分别讲解两种配置文件。

9.3.1　log4j.properties 配置日志

比如需要记录这个 mapper 接口的日志：

```
package org.fkit.mapper;
```

```
public interface UserMapper {
  @Select("SELECT * FROM tb_user WHERE id = #{id}")
  User selectUser(int id);
}
```

在应用的 CLASSPATH 中增加一个名称为 log4j.properties 的文件，文件的具体内容如下：

```
# 全局日志环境配置
log4j.rootLogger=ERROR, stdout
# MyBatis 日志环境配置...
log4j.logger.org.fkit.mapper.UserMapper=TRACE
# 控制台输出...
log4j.appender.stdout=org.apache.log4j.ConsoleAppender
log4j.appender.stdout.layout=org.apache.log4j.PatternLayout
log4j.appender.stdout.layout.ConversionPattern= %5p [%t] - %m%n
```

添加以上配置后，Log4j 就会把 org.fkit.mapper.UserMapper 的 TRACE（详细执行）日志记录下来，对于应用中的其他类则仅仅记录 ERROR（错误信息）。

也可以将日志从整个 mapper 接口级别调整到语句级别，从而实现更细粒度的控制。如下配置只记录 selectUser 语句的日志：

```
log4j.logger.org.fkit.mapper.UserMapper.selectUser=TRACE
```

也可以对一组 mapper 接口记录日志，只要对 mapper 接口所在的包开启日志功能即可：

```
log4j.logger.org.fkit.mapper =TRACE
```

某些查询可能会返回大量的数据，如果只想记录其执行的 SQL 语句该怎么办？为此，MyBatis 中 SQL 语句的日志级别被设为 DEBUG（JDK Logging 中为 FINE），结果日志的级别为 TRACE（JDK Logging 中为 FINER）。所以，只要将日志级别调整为 DEBUG 即可：

```
log4j.logger.org.fkit.mapper =DEBUG
```

如果要记录日志的是类似下面的 mapper 文件而不是 mapper 接口又该怎么办呢？

```
<?xml version="1.0" encoding="UTF-8" ?>
<!DOCTYPE mapper
PUBLIC "-//mybatis.org//DTD Mapper 3.0//EN"
"http://mybatis.org/dtd/mybatis-3-mapper.dtd">
<mapper namespace="org.fkit.mapper.UserMapper">
   <select id="selectUser" resultType="Blog">
      select * from tb_user where id = #{id}
   </select>
</mapper>
```

只要对命名空间增加日志记录功能即可：

```
log4j.logger.org.fkit.mapper =TRACE
```

9.3.2 log4j.xml 配置日志

之前的 Java 项目都是采取 properties 文件作为配置文件，而最新的项目大多采用 XML 文件作为配置文件。

在应用的 CLASSPATH 中增加一个名称为 log4j.xml 的文件，文件的具体内容如下：

```
<?xml version="1.0" encoding="UTF-8"?>
<!DOCTYPE log4j:configuration SYSTEM "log4j.dtd">
<log4j:configuration xmlns:log4j="http://jakarta.apache.org/log4j/">
   <appender name="STDOUT" class="org.apache.log4j.ConsoleAppender">
      <layout class="org.apache.log4j.PatternLayout">
         <param name="ConversionPattern" value=" %5p [%t] - %m%n" />
      </layout>
   </appender>
    <logger name="org.fkit.mapper.UserMapper">
```

```
        <level value="TRACE" />
    </logger>
    <root>
        <level value="ERROR" />
        <appender-ref ref="STDOUT" />
    </root>
</log4j:configuration>
```

添加以上配置后，Log4j 就会把 org.fkit.mapper.UserMapper 的 TRACE（详细执行）日志记录下来，对于应用中的其他类则仅仅记录 ERROR（错误信息）。

也可以将日志从整个 mapper 接口级别调整到语句级别，从而实现更细粒度的控制。如下配置只记录 selectUser 语句的日志：

```
<logger name="org.fkit.mapper.UserMapper.selectUser ">
    <level value="TRACE" />
</logger>
```

也可以对一组 mapper 接口记录日志，只要对 mapper 接口所在的包开启日志功能即可：

```
<logger name="org.fkit.mapper">
    <level value="TRACE" />
</logger>
```

某些查询可能会返回大量的数据，如果只想记录其执行的 SQL 语句该怎么办？为此，MyBatis 中 SQL 语句的日志级别被设为 DEBUG（JDK Logging 中为 FINE），结果日志的级别为 TRACE（JDK Logging 中为 FINER)。所以，只要将日志级别调整为 DEBUG 即可：

```
<logger name="org.fkit.mapper">
    <level value="DEBUG" />
</logger>
```

如果要记录日志的是类似下面的 mapper 文件而不是 mapper 接口又该怎么办呢？

```
<?xml version="1.0" encoding="UTF-8" ?>
<!DOCTYPE mapper
PUBLIC "-//mybatis.org//DTD Mapper 3.0//EN"
"http://mybatis.org/dtd/mybatis-3-mapper.dtd">
<mapper namespace="org.fkit.mapper.UserMapper">
    <select id="selectUser" resultType="Blog">
        select * from tb_user where id = #{id}
    </select>
</mapper>
```

只要对命名空间增加日志记录功能即可：

```
<logger name="org.fkit.mapper.UserMapper">
        <level value="TRACE" />
    </logger>
```

9.4　深入 Mapper XML 映射文件

MyBatis 的真正强大之处在于它的映射语句，这也是它的魔力所在。由于它的功能异常强大，映射器的 XML 文件就显得相对简单。如果拿它跟具有相同功能的 JDBC 代码进行对比，你会立即发现省掉了将近 95%的代码。MyBatis 就是针对 SQL 构建的，并且比普通的方法做得更好。

SQL 映射文件常用的元素如下：

- select。映射查询语句。
- insert。映射插入语句。

- update。映射更新语句。
- delete。映射删除语句。
- sql。可被其他语句引用的可重用语句块。
- cache。给定命名空间的缓存配置。
- cache-ref。其他命名空间缓存配置的引用。
- resultMap。最复杂也是最强大的元素，用来描述如何从数据库结果集中加载对象。

提示

parameterMap 已废弃！它是老式风格的参数映射。

9.4.1 select

select 元素用来映射查询语句，它是 MyBatis 中最常用的元素之一。

执行简单查询的 select 元素是非常简单的。例如：

```
<select id="selectUser" parameterType="int" resultType="hashmap">
  SELECT * FROM TB_USER WHERE ID = #{id}
</select>
```

这个语句被称作 selectUser，其接受一个 int（或 Integer）类型的参数，并返回一个 HashMap 类型的对象，HashMap 中的键是列名，值便是结果行中的对应值。

注意参数符号#{id}，这是告诉 MyBatis 创建一个预处理语句参数。通过 JDBC，这样的一个参数在 SQL 中会由一个“?”来标识，并被传递到一个新的预处理语句中。以上 MyBatis 配置文件执行时会生成如下 JDBC 代码：

```
String selectUser = "SELECT * FROM TB_USER WHERE ID=?";
PreparedStatement ps = conn.prepareStatement(selectUser);
ps.setInt(1,id);
```

select 元素有很多属性可以配置，它们用来决定每条语句的作用细节。例如：

```
<select
  id=" selectUser"
  parameterType="int"
  resultType="hashmap"
  resultMap="userResultMap"
  flushCache="false"
  useCache="true"
  timeout="10000"
  fetchSize="256"
  statementType="PREPARED"
  resultSetType="FORWARD_ONLY">
```

select 元素的属性描述如下：

- id。在命名空间中唯一的标识符，可以被用来引用这条语句。
- parameterType。将会传入这条语句的参数类的完全限定名或别名。这个属性是可选的，因为 MyBatis 可以通过 TypeHandler 推断出具体传入语句的参数，默认值为 unset。
- resultType。从这条语句中返回的期望类型的类的完全限定名或别名。注意如果是集合情形，那应该是集合可以包含的类型，而不能是集合本身。返回时可以使用 resultType 或 resultMap，但不能同时使用。
- resultMap。外部 resultMap 的命名引用。结果集的映射是 MyBatis 最强大的特性，对其有一个很好的理解的话，许多复杂映射的情形都能迎刃而解。返回时可以使用 resultMap 或 resultType，但不能同时使用。

- flushCache。如果设置为 true，则任何时候只要语句被调用，都会导致本地缓存和二级缓存都被清空，默认值为 false。
- useCache。如果设置为 true，将会导致本条语句的结果被二级缓存，在 select 元素当中默认值为 true。
- timeout。这个设置是在抛出异常之前，驱动程序等待数据库返回请求结果的秒数。默认值为 unset（依赖驱动）。
- fetchSize。其尝试使得驱动程序每次批量返回的结果行数和这个设置值相等。默认值为 unset（依赖驱动）。
- statementType。值为 STATEMENT、PREPARED 或 CALLABLE。这会让 MyBatis 分别使用 JDBC 中的 Statement、PreparedStatement 或 CallableStatement，默认值为 PREPARED。
- resultSetType。结果集的类型，值为 FORWARD_ONLY、SCROLL_SENSITIVE 或 SCROLL_INSENSITIVE，默认值为 unset（依赖驱动）。
- databaseId。如果配置了 databaseIdProvider，MyBatis 会加载所有的不带 databaseId 或匹配当前 databaseId 的语句；如果带或者不带的语句都有，则不带的会被忽略。
- resultOrdered。这个设置仅针对嵌套结果 select 语句适用：如果为 true，就是假设包含了嵌套结果集或分组，这样的话当返回一个主结果行的时候，就不会发生对前面结果集引用的情况。这就使得在获取嵌套的结果集时不至于导致内存不够用。默认值为 false。
- resultSets。这个设置仅对多结果集的情况适用，它将列出语句执行后返回的结果集并给每个结果集起一个名称，名称是逗号分隔的。

9.4.2 insert、update 和 delete

insert、update 和 delete 元素用来映射 DML 语句，是 MyBatis 中最常用的元素之一。insert、update 和 delete 元素配置和 select 非常接近。例如：

```
<insert
  id="insertUser"
  parameterType="org.fkit.domain.User"
  flushCache="true"
  statementType="PREPARED"
  keyProperty=""
  keyColumn=""
  useGeneratedKeys=""
  timeout="20">

<update
  id="updateUser"
  parameterType="org.fkit.domain.User "
  flushCache="true"
  statementType="PREPARED"
  timeout="20">

<delete
  id="deleteUser"
  parameterType="org.fkit.domain.User "
  flushCache="true"
  statementType="PREPARED"
```

```
  timeout="20">
```

insert、update 和 delete 元素的属性大多和 select 的一致，它们特有的属性描述如下：

- useGeneratedKeys。（仅对 insert 和 update 有用）这会令 MyBatis 使用 JDBC 的 getGeneratedKeys 方法来获取由数据库内部生成的主键（比如，像 MySQL 和 SQL Server 这样的关系数据库管理系统的自动递增字段），默认值为 false。
- keyProperty。（仅对 insert 和 update 有用）唯一标记一个属性，MyBatis 会通过 getGeneratedKeys 的返回值或者通过 insert 语句的 selectKey 子元素设置它的键值，默认为 unset。如果希望得到多个生成的列，也可以是逗号分隔的属性名称列表。
- keyColumn。（仅对 insert 和 update 有用）通过生成的键值设置表中的列名，这个设置仅对某些数据库（像 PostgreSQL）是必须的，当主键列不是表中的第一列时需要设置。如果希望得到多个生成的列，也可以是逗号分隔的属性名称列表。

下面是 insert、update 和 delete 语句的示例：

```
<insert id="insertUser">
  insert into TB_USER (id,username,password,email,address)
  values (#{id},#{username},#{password},#{email},#{address})
</insert>

<update id="updateUser">
  update TB_USER set
    username = #{username},
    password = #{password},
    email = #{email},
    address = #{address}
  where id = #{id}
</update>

<delete id="deleteUser">
  delete from TB_USER where id = #{id}
</delete>
```

而插入语句的配置规则更加丰富，因为在插入语句执行时很多时候是需要返回插入成功的数据生成的主键值的，所以<insert.../>元素里面有一些额外的属性和子元素用来处理主键的生成，而且根据数据库的主键生成策略不同，配置也有多种方式。

首先，如果数据库支持自动生成主键的字段（比如 MySQL 和 SQL Server），那么可以设置 useGeneratedKeys="true"，然后再把 keyProperty 设置到目标属性上就可以了（一般会设置到 id 属性上）。例如，如果上面的 TB_USER 表已经对 id 使用了自动生成的列类型，那么语句可以修改为：

```
<insert id="insertUser" useGeneratedKeys="true" keyProperty="id">
  insert into TB_USER (username,password,email,address)
  values (#{username},#{password},#{email},#{address})
</insert>
```

对于不支持自动生成类型的数据库（比如 Oracle）或可能不支持自动生成主键的 JDBC 驱动来说，MyBatis 有另外一种方法来生成主键。

```
<insert id="insertUser">
  <selectKey keyProperty="id" resultType="int" order="BEFORE">
    select SEQUENCE_TB_USER.nextval as id from dual
  </selectKey>
  insert into TB_USER
    (id, username, password, email,address)
  values
    (#{id}, #{username}, #{password}, #{email}, #{address})
</insert>
```

在上面的示例中，selectKey 元素将会首先运行，其通过查询 SEQUENCE 序列，TB_USER 的 id 会被设置，然后插入语句会被调用。

selectKey 元素描述如下：

```
<selectKey
  keyProperty="id"
  resultType="int"
  order="BEFORE"
  statementType="PREPARED">
```

- keyProperty。selectKey 语句结果应该被设置到目标属性（一般会设置到 id 属性）上。如果希望得到多个生成的列，也可以是逗号分隔的属性名称列表。
- keyColumn。匹配属性的返回结果集中的列名称。如果希望得到多个生成的列，也可以是逗号分隔的属性名称列表。
- resultType。结果的类型。MyBatis 通常可以推算出来，但是为了更加确定，建议明确写出。MyBatis 允许任何简单类型用作主键的类型，包括字符串。如果希望作用于多个生成的列，则可以使用一个包含期望属性的 Object 或一个 Map。
- order。可以被设置为 BEFORE 或 AFTER。如果设置为 BEFORE，那么它会首先选择主键，设置 keyProperty 然后执行插入语句。如果设置为 AFTER，那么先执行插入语句，然后是 selectKey 元素。
- statementType。与前面相同，MyBatis 支持 STATEMENT、PREPARED 和 CALLABLE 语句的映射类型，分别代表 Statement、PreparedStatement 和 CallableStatement 类型。

9.4.3 sql

sql 元素可以被用来定义可重用的 SQL 代码段，可以包含在其他语句中。它可以被静态地（在加载参数时）参数化。不同的属性值通过包含的实例发生变化。例如：

```
<sql id="userColumns"> ${alias}.id,${alias}.username,${alias}.password </sql>
```

这个 SQL 片段可以被包含在其他语句中，例如：

```
<select id="selectUsers" resultType="map">
  select
    <include refid="userColumns"><property name="alias" value="t1"/></include>
  from some_table t1
</select>
```

属性值可以用于包含的 refid 属性或者包含的字句里面的属性，例如：

```
<sql id="sometable">
  ${prefix}Table
</sql>

<sql id="someinclude">
  from
    <include refid="${include_target}"/>
</sql>

<select id="select" resultType="map">
  select
    field1, field2, field3
  <include refid="someinclude">
    <property name="prefix" value="Some"/>
    <property name="include_target" value="sometable"/>
  </include>
</select>
```

9.4.4 参数（Parameters）

前面的所有语句中所见到的都是简单参数的例子，实际上参数是 MyBatis 非常强大的元素。对于简单参数的使用，大多数情况下参数都很少，例如：

```
<select id="selectUsers" parameterType="int" resultType="User">
  select id, username, password
  from users
  where id = #{id}
</select>
```

上面的这个示例说明了一个非常简单的命名参数映射。参数类型被设置为 int，这样这个参数就可以被设置成任何内容。原生的类型或简单数据类型（比如整型和字符串），因为没有相关属性，会完全用参数值来替代。

但是，如果传入一个复杂的对象（比如 User），行为就会有一点不同了。例如：

```
<insert id="insertUser" parameterType="User">
  insert into users (id, username, password)
  values (#{id}, #{username}, #{password})
</insert>
```

如果 User 类型的参数对象被传递到了语句中，如#{id}语句则会查找参数对象 User 的 id 属性，#{username}和#{password}也是一样，然后将它们的值传入预处理语句的参数中。

示例：测试 select、insert、update 和 delete 操作

本示例直接使用第 8 章创建的 TB_USER 表、数据库脚本、User.java 和 log4j.xml，具体请参考第 8 章内容，此处不再赘述。

在实际项目开发中，连接数据库的参数信息不会直接写在 mybatis-config.xml 中，而是通过一个 properties 文件定义连接数据库的参数信息，并在 mybatis-config.xml 中引用。

程序清单：codes/09/DMLTest/src/db.properties

```
driver=com.mysql.jdbc.Driver
url=jdbc:mysql://127.0.0.1:3306/mybatis
username=root
password=root
```

程序清单：codes/09/DMLTest/src/mybatis-config.xml

```
<!-- 引入 properties 资源文件 -->
<properties resource="db.properties"/>
<!-- 环境配置，即连接的数据库。 -->

<environments default="mysql">
    <environment id="mysql">
    <!--  指定事务管理类型，type="JDBC"指直接简单使用了 JDBC 的提交和回滚设置 -->
      <transactionManager type="JDBC"/>
      <!--  dataSource 指数据源配置，POOLED 是 JDBC 连接对象的数据源连接池的实现。 -->
      <dataSource type="POOLED">
        <property name="driver" value="${driver}"/>
        <property name="url" value="${url}"/>
        <property name="username" value="${username}"/>
        <property name="password" value="${password}"/>
      </dataSource>
    </environment>
  </environments>
```

mybatis-config.xml 中的<properties resource="db.properties"/>配置表示引入 db.properties 资源配置文件，<property name="driver" value="${driver}"/>表示 driver 的值引用 db.properties 文

件中的名称为 driver 的值 com.mysql.jdbc.Driver。${url}、${username}和${password}引用 db.properties 文件中对应的 url、username 和 password 的值。

程序清单：codes/09/DMLTest/src/org/fkit/mapper/UserMapper.xml

```
<?xml version="1.0" encoding="UTF-8"?>
<!DOCTYPE mapper PUBLIC "-//mybatis.org//DTD Mapper 3.0//EN"
"http://mybatis.org/dtd/mybatis-3-mapper.dtd">
<!-- namespace 指用户自定义的命名空间。 -->
<mapper namespace="org.fkit.mapper.UserMapper">
  <insert id="saveUser" parameterType="user" useGeneratedKeys="true">
     INSERT INTO TB_USER(name,sex,age)
     VALUES(#{name},#{sex},#{age})
  </insert>
  <select id="selectUser" parameterType="int" resultType="user">
     SELECT * FROM TB_USER WHERE id = #{id}
  </select>
  <update id="modifyUser" parameterType="user">
     UPDATE TB_USER
     SET name = #{name},sex = #{sex},age = #{age}
     WHERE id = #{id}
  </update>
  <delete id="removeUser" parameterType="int">
     DELETE FROM TB_USER WHERE id = #{id}
  </delete>
</mapper>
```

在 UserMapper.xml 中定义了 insert、update、delete 和 select 4 个元素，分别对应插入、更新、删除和查询 4 个数据库操作。

因为每次测试都需要读取 mybatis-config.xml 根配置文件，根据根配置文件信息创建 SqlSessionFactory 对象，再获取 SqlSession 对象，使得该操作比较频繁，所以开发一个 FKSqlSessionFactory 工厂类封装以上操作的重复代码。

程序清单：codes/09/DMLTest/src/org/fkit/factory/FKSqlSessionFactory.java

```
import java.io.InputStream;
import org.apache.ibatis.io.Resources;
import org.apache.ibatis.session.SqlSession;
import org.apache.ibatis.session.SqlSessionFactory;
import org.apache.ibatis.session.SqlSessionFactoryBuilder;
public class FKSqlSessionFactory {
   private static SqlSessionFactory sqlSessionFactory = null;
   // 初始化创建 SqlSessionFactory 对象
   static{

try (// 读取 mybatis-config.xml 文件
   InputStream is = Resources.getResourceAsStream("mybatis-config.xml");
   ){
         sqlSessionFactory = new SqlSessionFactoryBuilder()
               .build(is);
      } catch (Exception e) {
         e.printStackTrace();
      }
   }
   // 获取 SqlSession 对象的静态方法
   public static SqlSession getSqlSession(){
      return sqlSessionFactory.openSession();
   }
   // 获取 SqlSessionFactory 对象的静态方法
   public static SqlSessionFactory getSqlSessionFactory() {
```

```
            return sqlSessionFactory;
        }
    }
```

首先测试<insert.../>元素。

程序清单：codes/09/DMLTest/src/org/fkit/test/InsertTest.java

```
import org.apache.ibatis.session.SqlSession;
import org.fkit.domain.User;
import org.fkit.factory.FKSqlSessionFactory;

public class InsertTest {

    public static void main(String[] args)  {
        // 定义 SqlSession 变量
        SqlSession sqlSession = null;
        try {
            // 创建 SqlSession 实例
            sqlSession = FKSqlSessionFactory.getSqlSession();
            // 创建 User 对象
            User user = new User("jack", "男", 22);
            // 插入数据
            sqlSession.insert("org.fkit.mapper.UserMapper.saveUser", user);
            // 提交事务
            sqlSession.commit();
        } catch (Exception e) {
                                    // 回滚事务
                                    sqlSession.rollback();
            e.printStackTrace();
        }finally {
            // 关闭 SqlSession
            if(sqlSession != null)
                sqlSession.close();
        }

    }

}
```

运行 InsertTest 类的 main 方法，创建 User 对象，并将 User 对象作为参数调用 SqlSession 的 insert 方法，insert 方法的第一个参数是 org.fkit.mapper.UserMapper.saveUser，MyBatis 会找到 org.fkit.mapper.UserMapper 命名空间下 id="saveUser"的元素，执行该元素中的 SQL 语句。

```
    <insert id="saveUser" parameterType="user" useGeneratedKeys="true" keyProperty=
"id"、>
        INSERT INTO TB_USER(name,sex,age)
        VALUES(#{name},#{sex},#{age})
    </insert>
```

<insert.../>元素中的 parameterType="user"表示该插入语句需要一个 User 对象作为参数；useGeneratedKeys="true"表示使用数据库的自动增长的主键，该操作需要底层数据库的支持；keyProperty="id"表示将插入数据生成的主键设置到 user 对象的 id 当中。元素中的 SQL 语句是一条标准的 INSERT INTO 语句，需要注意的是，#{name}使用了 MyBatis 的表达式，其会查找参数 user 当中的 name 属性作为值并将其设置到 SQL 语句中；如果传入的参数是一个 Map，则会以 name 作为 key 查找 Map 当中的值并将其设置到 SQL 语句中。#{sex}、#{age}和#{name}操作相同。

运行 InsertTest 类的 main 方法，将会插入一条数据到数据库当中。控制台结果如下所示：

```
    DEBUG [main] ==>  Preparing: INSERT INTO TB_USER(name,sex,age) VALUES(?,?,?)
```

```
    DEBUG [main] ==> Parameters: jack(String), 男(String), 22(Integer)
    DEBUG [main] <==    Updates: 1
```

插入数据后数据库表数据如图 9.2 所示。

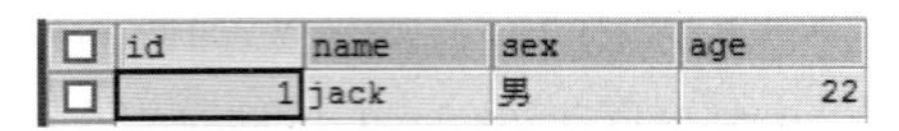

id	name	sex	age
1	jack	男	22

图 9.2　测试<insert.../>元素

接下来测试<select.../>元素。

程序清单：codes/09/DMLTest/src/org/fkit/test/SelectTest.java

```
import org.apache.ibatis.session.SqlSession;
import org.fkit.domain.User;
import org.fkit.factory.FKSqlSessionFactory;

public class SelectTest {

    public static void main(String[] args) {
        // 定义 SqlSession 变量
        SqlSession sqlSession = null;
        try {
            // 创建 SqlSession 实例
            sqlSession = FKSqlSessionFactory.getSqlSession();
            // 根据 id 查询 User 对象
            User user = sqlSession.selectOne("org.fkit.mapper.UserMapper.selectUser",1);
            System.out.println(user);
            // 提交事务
            sqlSession.commit();
        } catch (Exception e) {
                                    // 回滚事务
                                    sqlSession.rollback();
            e.printStackTrace();
        }finally {
            // 关闭 SqlSession
            if(sqlSession != null)
                sqlSession.close();
        }
    }

}
```

运行 SelectTest 类的 main 方法，调用 SqlSession 的 select 方法，selectOne 方法的第一个参数是 org.fkit.mapper.UserMapper.selectUser，MyBatis 会找到 org.fkit.mapper.UserMapper 命名空间下 id="selectUser "的元素，执行该元素中的 SQL 语句。

```
    <select id="selectUser" parameterType="int" resultType="user">
        SELECT * FROM TB_USER WHERE id = #{id}
    </select>
```

<select.../>元素中的 parameterType="int"表示该插入新语句需要一个 int 类型的值作为参数；resultType="user"表示该条查询语句需要返回一个 User 对象。元素中的 SQL 语句是一条标准的 SELECT 语句，该语句需要的参数 id 值正是调用时传入的 int 值。

运行 SelectTest 类的 main 方法，程序将会到数据库当中查询 id 为 1 的一条数据并封装成 User 类型的对象返回。控制台结果如下所示：

```
    DEBUG [main] ==>  Preparing: SELECT * FROM TB_USER WHERE id = ?
    DEBUG [main] ==> Parameters: 1(Integer)
    DEBUG [main] <==      Total: 1
    User [id=1, name=jack, sex=男, age=22]
```

接下来测试<update.../>元素。

程序清单：codes/09/DMLTest/src/org/fkit/test/UpdateTest.java

```
import org.apache.ibatis.session.SqlSession;
import org.fkit.domain.User;
import org.fkit.factory.FKSqlSessionFactory;

public class UpadeTest {

    public static void main(String[] args) {
        // 定义 SqlSession 变量
        SqlSession sqlSession = null;
        try {
            // 创建 SqlSession 实例
            sqlSession = FKSqlSessionFactory.getSqlSession();
            // 根据 id 查询 User 对象
            User user = sqlSession.selectOne("org.fkit.mapper.UserMapper.selectUser",1);
            // 修改 User 对象的属性值
            user.setName("tom");
            user.setAge(25);
            // 修改 User 对象
            sqlSession.update("org.fkit.mapper.UserMapper.modifyUser", user);
            // 提交事务
            sqlSession.commit();
        } catch (Exception e) {
                                    // 回滚事务
                                    sqlSession.rollback();
            e.printStackTrace();
        }finally {
            // 关闭 SqlSession
            if(sqlSession != null)
                sqlSession.close();
        }
    }
}
```

运行 UpdateTest 类的 main 方法，首先调用 SqlSession 的 selectOne 方法，查询出 id 为 1 的数据返回给 User 对象。接下来修改该 User 对象的属性值，最后调用 SqlSession 的 update 方法修改该 User 对象。update 方法的第一个参数是 org.fkit.mapper.UserMapper.updateUser ，MyBatis 会找到 org.fkit.mapper.UserMapper 命名空间下 id="modifyUser "的元素，执行该元素中的 SQL 语句。

```
<update id="modifyUser" parameterType="user">
    UPDATE TB_USER
    SET name = #{name},sex = #{sex},age = #{age}
    WHERE id = #{id}
</update>
```

<update../>元素中的 parameterType="user"表示该更新语句需要一个 user 对象作为参数。元素中的 SQL 语句是一条标准的 UPDATE 语句，该语句根据传入的 User 对象的属性更新表数据。

运行 UpdateTest 类的 main 方法，我们将会看到数据库中的更新数据。控制台结果如下所示：

```
    DEBUG [main] ==>  Preparing: SELECT * FROM TB_USER WHERE id = ?
    DEBUG [main] ==> Parameters: 1(Integer)
    DEBUG [main] <==      Total: 1
    DEBUG [main] ==>  Preparing: UPDATE TB_USER SET name = ?,sex = ?,age = ? WHERE id
= ?
```

```
DEBUG [main] ==> Parameters: tom(String), 男(String), 25(Integer), 1(Integer)
DEBUG [main] <==    Updates: 1
```

更新数据后数据库表数据如图 9.3 所示。

id	name	sex	age
1	tom	男	25

图 9.3 测试<update.../>元素

接下来测试<delete.../>元素。

程序清单：codes/09/DMLTest/src/org/fkit/test/DeleteTest.java

```
import org.apache.ibatis.session.SqlSession;
import org.fkit.factory.FKSqlSessionFactory;

public class DeleteTest {

    public static void main(String[] args)  {
        // 定义 SqlSession 变量
        SqlSession sqlSession = null;
        try {
            // 创建 SqlSession 实例
            sqlSession = FKSqlSessionFactory.getSqlSession();
            // 删除 id 为 1 的 User 对象
            sqlSession.delete("org.fkit.mapper.UserMapper.removeUser", 1);
            // 提交事务
            sqlSession.commit();
        } catch (Exception e) {
                                    // 回滚事务
                                    sqlSession.rollback();
            e.printStackTrace();
        }finally {
            // 关闭 SqlSession
            if(sqlSession != null)
                sqlSession.close();
        }
    }

}
```

运行 DeleteTest 类的 main 方法，调用 SqlSession 的 delete 方法。delete 方法的第一个参数是 org.fkit.mapper.UserMapper.removeUser，MyBatis 会找到 org.fkit.mapper.UserMapper 命名空间下的 id="removeUser "的元素，执行该元素中的 SQL 语句。

```
<delete id="removeUser" parameterType="int">
    DELETE FROM TB_USER WHERE id = #{id}
</delete>
```

<delete.../>元素中的 parameterType="int"表示该删除语句需要一个 int 类型的值作为参数。元素中的 SQL 语句是一条标准的 DELETE 语句，该语句需要的参数 id 值正是调用时传入的 int 值。

运行 DeleteTest 类的 main 方法，会将数据库当中 id 为 1 的数据删除。控制台结果如下所示：

```
DEBUG [main] ==>  Preparing: DELETE FROM TB_USER WHERE id = ?
DEBUG [main] ==> Parameters: 1(Integer)
DEBUG [main] <==    Updates: 1
```

9.4.5 ResultMaps

resultMap 元素是 MyBatis 中最重要最强大的元素。它的作用是告诉 MyBatis 将从结果集中取出的数据转换成开发者所需要的对象。

下面是最简单的映射语句示例：

```
<select id="selectUser" resultType="map">
    SELECT * FROM TB_USER
</select>
```

selectUser 的<select.../>元素执行一条查询语句，查询 TB_USER 表的所有数据。resultType="map"表示返回的数据是一个 Map 集合（使用列名作为 key，列值作为 value）。

示例：测试 ResultMaps

程序清单：codes/09/ResultMapTest /src/org/fkit/test/SelectMapTest.java

```
import java.util.List;
import java.util.Map;
import org.apache.ibatis.session.SqlSession;
import org.fkit.factory.FKSqlSessionFactory;

public class SelectMapTest {

    public static void main(String[] args) {
        // 定义 SqlSession 变量
        SqlSession sqlSession = null;
        try {
            // 创建 SqlSession 实例
            sqlSession = FKSqlSessionFactory.getSqlSession();
            // 查询 TB_USER 表所有数据返回 List 集合,集合中的每个元素都是一个 Map
            List<Map<String,Object>> list
                = sqlSession.selectList("org.fkit.mapper.UserMapper.selectUser");
            // 遍历 List 集合，打印每一个 Map 对象
                                        list.forEach(row -> System.out.println(row));
            // 提交事务
            sqlSession.commit();
        } catch (Exception e) {
                                        // 回滚事务
                                        sqlSession.rollback();
            e.printStackTrace();
        }finally {
            // 关闭 SqlSession
            if(sqlSession != null)
                sqlSession.close();
        }
    }
}
```

执行之前先往 TB_USER 表中插入几条测试数据，SQL 脚本如下：

```
INSERT INTO TB_USER(NAME,sex,age) VALUES('jack','男',22);
INSERT INTO TB_USER(NAME,sex,age) VALUES('rose','女',18);
INSERT INTO TB_USER(NAME,sex,age) VALUES('tom','男',25);
INSERT INTO TB_USER(NAME,sex,age) VALUES('mary','女',20);
```

运行 SelectMapTest 类的 main 方法，控制台显示如下：

```
DEBUG [main] ==>  Preparing: SELECT * FROM TB_USER
DEBUG [main] ==> Parameters:
DEBUG [main] <==      Total: 4
{sex=男, name=jack, id=1, age=22}
{sex=女, name=rose, id=2, age=18}
{sex=男, name=tom, id=3, age=25}
{sex=女, name=mary, id=4, age=20}
```

可以看到，查询语句返回的每一条数据都被封装成一个 Map 集合，列名作为 Map 集合的 key，而列的值作为 Map 的 value。

虽然数据被封装成 Map 集合返回，但是 Map 集合并不能很好地描述一个领域模型。在实际项目开发中更加建议使用 JavaBean 或 POJO（Plain Old Java Object，普通 Java 对象）作为领域模型描述数据。例如：

```
<select id="selectUser" resultType="org.fkit.domain.User ">
    SELECT * FROM TB_USER
  </select>
```

默认情况下，MyBatis 会将查询到的数据的列和需要返回的对象（User）的属性逐一进行匹配赋值，但是如果查询到的数据的列和需要返回的对象（User）的属性不一致，则 MyBatis 就不会自动赋值了，这时，可以使用 resultMap 进行处理。

进入 mybatis 数据库，创建一个表 TB_USER2，并插入几条测试数据。

```
# 如果存在 tb_user2 表则删除
DROP TABLE IF EXISTS tb_user2;
CREATE TABLE TB_USER2(
user_id INT PRIMARY KEY AUTO_INCREMENT,
user_name VARCHAR(18),
user_sex VARCHAR(18),
user_age INT
);
INSERT INTO TB_USER2(user_name,user_sex,user_age) VALUES('jack','男',22);
INSERT INTO TB_USER2(user_name,user_sex,user_age) VALUES('rose','女',18);
INSERT INTO TB_USER2(user_name,user_sex,user_age) VALUES('tom','男',25);
INSERT INTO TB_USER2(user_name,user_sex,user_age) VALUES('mary','女',20);
```

接下来创建一个 User 对象映射 TB_USER2 表。

程序清单：codes/09/ResultMapTest/src/org/fkit/domain/User.java

```
import java.io.Serializable;
public class User implements Serializable{
   private Integer id;
   private String name;
   private String sex;
   private Integer age;
   // 省略 set 和 get 方法......
}
```

程序清单：codes/09/ResultMapTest/src/org/fkit/mapper/UserMapper.xml

```
<resultMap id="userResultMap" type="org.fkit.domain.User" >
     <id property="id" column="user_id" />
     <result property="name" column="user_name"/>
     <result property="sex" column="user_sex"/>
     <result property="age" column="user_age"/>
   </resultMap>
  <!-- resultMap="userResultMap"表示引用上面的 resultMap 进行数据库表和返回类型对象的映射 -->
  <select id="selectUser2" resultMap="userResultMap">
     SELECT * FROM TB_USER2
  </select>
```

上面使用了一个新的元素<resultMap.../>，该元素常用属性如下：

➢ id。resultMap 的唯一标识符。

➢ type。resultMap 实际返回的类型。

上面使用了<resultMap.../>的两个子元素 id 和 result。

- id。表示数据库表的主键，其中，column 属性表示数据库表的列名，property 表示数据库列映射到返回类型的属性。
- result。表示数据库表的普通列，其中，column 属性表示数据库表的列名，property 表示数据库列映射到返回类型的属性。

程序清单：codes/09/ResultMapTest /src/org/fkit/test/ResultMapTest.java

```
import java.util.List;
import org.apache.ibatis.session.SqlSession;
import org.fkit.domain.User;
import org.fkit.factory.FKSqlSessionFactory;

public class ResultMapTest {

    public static void main(String[] args) {
        // 定义 SqlSession 变量
        SqlSession sqlSession = null;
        try {
            // 创建 SqlSession 实例
            sqlSession = FKSqlSessionFactory.getSqlSession();
            // 查询 TB_USER 表所有数据返回 List 集合,集合中的每个元素都是一个 Map
            List<User> user_list
                = sqlSession.selectList("org.fkit.mapper.UserMapper.selectUser2");
            // 遍历 List 集合，打印每一个 Map 对象
                                  user_list.forEach(user -> System.out.println(user));
            // 提交事务
            sqlSession.commit();
        } catch (Exception e) {
                                  // 回滚事务
                                  sqlSession.rollback();
            e.printStackTrace();
        }finally {
            // 关闭 SqlSession
            if(sqlSession != null)
                sqlSession.close();
        }
    }
}
```

运行 ResultMapTest 类的 main 方法，控制台显示如下：

```
DEBUG [main] ==>  Preparing: SELECT * FROM TB_USER2
DEBUG [main] ==> Parameters:
DEBUG [main] <==      Total: 4
User [id=1, name=jack, sex=男, age=22]
User [id=2, name=rose, sex=女, age=18]
User [id=3, name=tom, sex=男, age=25]
User [id=4, name=mary, sex=女, age=20]
```

可以看到，TB_USER2 的列名虽然和 User 对象的属性名不一致，数据依然被正确封装到 User 对象当中。

在实际项目开发中，还有更加复杂的情况，例如执行的是一个多表查询语句，而返回的对象关联到另一个对象，此时简单地映射已经无法解决问题，必须使用<resultMap.../>元素来完成关联映射。

进入 mybatis 数据库，创建两个表 TB_CLAZZ 和 TB_STUDENT，并分别插入几条测试数据。

```
# 如果存在 TB_CLAZZ 表则删除
DROP TABLE IF EXISTS TB_CLAZZ;
```

```
CREATE TABLE TB_CLAZZ(
id INT PRIMARY KEY AUTO_INCREMENT,
CODE VARCHAR(18)
);
INSERT INTO TB_CLAZZ(CODE) VALUES('j1601');
INSERT INTO TB_CLAZZ(CODE) VALUES('j1602');
# 如果存在 TB_STUDENT 表则删除
CREATE TABLE TB_STUDENT(
id INT PRIMARY KEY AUTO_INCREMENT,
NAME VARCHAR(18),
sex CHAR(3),
age INT,
clazz_id INT,
FOREIGN KEY (clazz_id) REFERENCES TB_CLAZZ(id)
);
INSERT INTO TB_STUDENT(NAME,sex,age,clazz_id) VALUES('jack','男',22,1);
INSERT INTO TB_STUDENT(NAME,sex,age,clazz_id) VALUES('rose','女',18,1);
INSERT INTO TB_STUDENT(NAME,sex,age,clazz_id) VALUES('tom','男',25,2);
INSERT INTO TB_STUDENT(NAME,sex,age,clazz_id) VALUES('mary','女',20,2);
```

以上 SQL 语句插入了两个班级记录和 4 个学生记录，两个学生分配在 1 班，两个学生分配在 2 班。需要指出的是，TB_STUDENT 表中的 clazz_id 列作为外键引用 TB_CLAZZ 表的 id 列，表示学生对应的班级。

接下来我们要做的是查询出所有的学生信息，同时关联查询出学生对应的班级信息。

创建一个 Clazz 对象和 Student 对象并分别映射 TB_CLAZZ 表和 TB_STUDENT 表。

程序清单：codes/09/ResultMapTest/src/org/fkit/domain/Clazz.java

```
import java.io.Serializable;
import java.util.List;
public class Clazz implements Serializable{
   private Integer id;
   private String code;
   // 省略 set 和 get 方法.....
}
```

程序清单：codes/09/ResultMapTest/src/org/fkit/domain/Student.java

```
import java.io.Serializable;
public class Student implements Serializable{
   private Integer id;
   private String name;
   private String sex;
   private Integer age;
   // 关联的 Clazz 对象
   private Clazz clazz;
   // 省略 set 和 get 方法.....
}
```

需要注意的是，Student 中的属性 Clazz 是一个对象，该对象包括 Clazz 的 id 和 code。这是现代开发中最常用的对象关联方式。

程序清单：codes/09/ResultMapTest /src/org/fkit/mapper/UserMapper.xml

```
<!-- 映射学生对象的 resultMap -->
  <resultMap id="studentResultMap" type="org.fkit.domain.Student">
     <id property="id" column="id" />
     <result property="name" column="name"/>
     <result property="sex" column="sex"/>
     <result property="age" column="age"/>
     <!-- 关联映射 -->
     <association property="clazz" column="clazz_id"
```

```
        javaType="org.fkit.domain.Clazz"
        select="selectClazzWithId"/>
    </resultMap>
  <!-- 根据班级 id 查询班级 -->
  <select id="selectClazzWithId" resultType="org.fkit.domain.Clazz">
    SELECT * FROM TB_CLAZZ where id = #{id}
  </select>
  <!-- 查询所有学生信息 -->
  <select id="selectStudent" resultMap="studentResultMap">
    SELECT * FROM TB_STUDENT
  </select>
```

上面的映射相对之前复杂了一些，具体解释如下：

① 首先执行 id 为 selectStudent 的<select.../>元素，查询所有的学生数据，此时返回的不是简单的 Student 对象，因为 Student 对象中还包含了 Clazz 对象，所以使用 resultMap 去映射返回类型。

② id 为 studentResultMap 的<resultMap.../>元素返回类型为 org.fkit.domain.Student，其中，id、name、sex 和 age 都是简单的属性映射，而查询的班级 id 列 clazz_id 则使用了关联映射<association.../>。

<association.../>元素的解释如下：

- column。表示数据库表的列名。
- property。表示返回类型 Student 的属性名 clazz。
- javaType。表示该属性对应的类型名称，本示例是一个 Clazz 类型。
- select。表示执行一条查询语句，将查询到的数据封装到 property 所代表的类型对象当中。上面的 selectClazzWithId 执行一条 SQL 语句，将学生的 clazz_id 作为参数查询对应的班级信息。

程序清单：codes/09/ResultMapTest /src/org/fkit/test/SelectStudentTest.java

```
import java.util.List;
import org.apache.ibatis.session.SqlSession;
import org.fkit.domain.Student;
import org.fkit.factory.FKSqlSessionFactory;

public class SelectStudentTest {

  public static void main(String[] args) {
    // 定义 SqlSession 变量
    SqlSession sqlSession = null;
    try {
      // 创建 SqlSession 实例
      sqlSession = FKSqlSessionFactory.getSqlSession();
      // 查询 TB_USER 表所有数据返回 List 集合,集合中的每个元素都是一个 Student 对象
      List<Student> student_list
        = sqlSession.selectList("org.fkit.mapper.UserMapper.selectStudent");
      // 遍历 List 集合，打印每一个 Student 对象，该对象包含关联的 Clazz 对象
                        student_list.forEach(stu -> System.out.println(stu));
      // 提交事务
      sqlSession.commit();
    } catch (Exception e) {
                        // 回滚事务
                        sqlSession.rollback();
      e.printStackTrace();
    }finally {
      // 关闭 SqlSession
      if(sqlSession != null)
```

```
            sqlSession.close();
        }
    }
}
```

运行 SelectStudentTest 类的 main 方法，控制台显示如下：

```
DEBUG [main] ==>  Preparing: SELECT * FROM TB_STUDENT
DEBUG [main] ==> Parameters:
DEBUG [main] ====>  Preparing: SELECT * FROM TB_CLAZZ where id = ?
DEBUG [main] ====> Parameters: 1(Integer)
DEBUG [main] <====      Total: 1
DEBUG [main] ====>  Preparing: SELECT * FROM TB_CLAZZ where id = ?
DEBUG [main] ====> Parameters: 2(Integer)
DEBUG [main] <====      Total: 1
DEBUG [main] <==     Total: 4
Student [id=1, name=jack, sex=男, age=22, clazz=Clazz [id=1, code=j1601]]
Student [id=2, name=rose, sex=女, age=18, clazz=Clazz [id=1, code=j1601]]
Student [id=3, name=tom, sex=男, age=25, clazz=Clazz [id=2, code=j1602]]
Student [id=4, name=mary, sex=女, age=20, clazz=Clazz [id=2, code=j1602]]
```

可以看到，因为使用了关联映射，查询学生信息时学生对应的班级对象也被查询出来了。

现在查询所有学生时可以关联查询出班级信息了，那如果反过来，查询所有班级时需要查询出班级中的所有学生对象，应该如何映射呢？

学生通常只对应一个班级，但是班级中会有多个学生存在，所以首先在 Clazz.java 类中增加一个字段 students，该字段是一个 List 集合，表示班级的多个学生。

程序清单：codes/09/ResultMapTest/src/org/fkit/domain/Clazz.java

```
private List<Student> students;
public List<Student> getStudents() {
      return students;
}
public void setStudents(List<Student> students) {
      this.students = students;
}
```

程序清单：codes/09/ResultMapTest /src/org/fkit/mapper/UserMapper.xml

```
<!-- 映射班级对象的 resultMap -->
  <resultMap id="clazzResultMap" type="org.fkit.domain.Clazz">
     <id property="id" column="id" />
     <result property="code" column="code"/>
     <!-- 班级的学生属性，因为一个班级有多个学生，所以该属性是一个集合 -->
     <collection property="students" javaType="ArrayList"
     column="id" ofType="org.fkit.domain.Student"
     select="selectStudentWithId"/>
    </resultMap>
    <!-- 根据班级 id 查询学生 -->
    <select id="selectStudentWithId" resultType="org.fkit.domain.Student">
        SELECT * FROM TB_STUDENT where clazz_id = #{id}
     </select>
    <!-- 查询所有班级信息 -->
  <select id="selectClazz" resultMap="clazzResultMap">
     SELECT * FROM TB_CLAZZ
  </select>
```

上面的映射和查询学生关联班级类似，具体解释如下：

① 首先执行 id 为 selectClazz 的<select.../>元素，查询所有的班级数据，此时返回的不是简单的 Clazz 对象，因为 Clazz 对象中还包含了学生的集合对象，所以使用 resultMap 去映射返回类型。

② id 为 clazzResultMap 的<resultMap.../>元素返回类型为 org.fkit.domain.Clazz，其中，id 和 code 都是简单的属性映射，而查询班级所有学生时则使用了集合映射<collection.../>。

<collection.../>元素的解释如下：

- property。表示返回类型 Clazz 的属性名 students。
- javaType。表示该属性对应的类型名称，本示例中是一个 ArrayList 集合。
- ofType。表示集合当中的类型，本示例中是 Student 类型。
- column。表示使用 id 作为参数进行之后的 select 语句查询。
- select。表示执行一条查询语句，将查询到的数据封装到 property 所代表的类型对象当中。上面的 selectStudentWithId 执行一条 SQL 语句，将班级的 id 作为参数查询班级对应的所有学生信息。

程序清单：codes/09/ResultMapTest /src/org/fkit/test/SelectClazzTest.java

```
import java.util.List;
import org.apache.ibatis.session.SqlSession;
import org.fkit.domain.Clazz;
import org.fkit.domain.Student;
import org.fkit.factory.FKSqlSessionFactory;

public class SelectClazzTest {

   public static void main(String[] args) {
      // 定义 SqlSession 变量
      SqlSession sqlSession = null;
      try {
         // 创建 SqlSession 实例
         sqlSession = FKSqlSessionFactory.getSqlSession();
         // 查询 TB_CLAZZ 表所有数据返回 List 集合,集合中的每个元素都是一个 Clazz 对象
         List<Clazz> clazz_list
            = sqlSession.selectList("org.fkit.mapper.UserMapper.selectClazz");
         // 遍历 List 集合，打印每一个 Clazz 对象和该 Clazz 关联的所有 Student 对象
         for(Clazz clazz : clazz_list){
            System.out.println(clazz);
            List<Student> student_list = clazz.getStudents();
            for(Student stu : student_list){
               System.out.println(stu.getId() + " " + stu.getName() + " " + stu.getSex()
                + " " + stu.getAge());
            }
         }
         // 提交事务
         sqlSession.commit();
      } catch (Exception e) {
                           // 回滚事务
                           sqlSession.rollback();
         e.printStackTrace();
      }finally {
         // 关闭 SqlSession
         if(sqlSession != null)
            sqlSession.close();
      }

   }
}
```

运行 SelectClazzTest 类的 main 方法，控制台显示如下：

```
    DEBUG [main] ==> Preparing: SELECT * FROM TB_CLAZZ
    DEBUG [main] ==> Parameters:
    DEBUG [main] ====> Preparing: SELECT * FROM TB_STUDENT where clazz_id = ?
```

```
DEBUG [main] ====> Parameters: 1(Integer)
DEBUG [main] <====      Total: 2
DEBUG [main] ====>  Preparing: SELECT * FROM TB_STUDENT where clazz_id = ?
DEBUG [main] ====> Parameters: 2(Integer)
DEBUG [main] <====      Total: 2
DEBUG [main] <==      Total: 2
Clazz [id=1, code=j1601]
1 jack 男 22
2 rose 女 18
Clazz [id=2, code=j1602]
3 tom 男 25
4 mary 女 20
```

可以看到，因为使用了集合映射，所以查询班级信息时班级对应的所有学生对象也被查询出来了。

9.5　本章小结

本章首先介绍了 MyBatis 最重要的两个类 SqlSessionFactory 和 SqlSession；接下来详细介绍了 MyBatis 的根配置文件 mybatis-config.xml 中的元素及使用方法，包括 MyBatis 的日志信息配置；最后重点介绍了 Mapper XML 映射文件的元素使用方法，包括 insert、delete、update、select 及强大的 ResultMaps。

第 10 章将深入 MyBatis 内部，重点介绍 MyBatis 的一对一、一对多、多对多、动态 SQL、调用存储过程和缓存机制。

CHAPTER

10

第10章 深入使用MyBatis

本章要点

- 一对一映射
- 一对多映射
- 多对多映射
- 动态SQL映射
- 调用存储过程
- 事务管理
- 缓存机制

通过 MyBatis 的支持，应用程序可以从底层的 JDBC 中抽离出来，以面向对象的方式进行数据库访问。但面向对象远不止这些内容，比如对象和对象之间的关联关系，这对于客观世界的建模是非常重要的。本章将深入介绍 MyBatis 的关联映射，也会详细介绍 MyBatis 的动态 SQL 查询。

10.1 MyBatis 关联映射

客观世界中的对象很少有孤立存在的，例如班级，往往与班级的学生存在关联关系，如果得到某个班级的实例，那么应该可以直接获取班级对应的全部学生。反过来，如果已经得到一个学生的实例，那么也应该可以访问该学生对应的班级。这种实例之间的互相访问就是关联关系。

关联关系是面向对象分析、面向对象设计最重要的知识，MyBatis 完全可以理解这种关联关系，如果映射得当，MyBatis 的关联映射将可以大大简化持久层数据的访问。关联关系大致有如下分类。

- 一对一
- 一对多
- 多对多

10.1.1 一对一

在实际项目开发中，经常存在一对一关系，比如一个人只能有一个身份证，一个身份证只能给一个人使用，这就是一对一的关系。一对一关系推荐使用唯一主外键关联，即两张表使用外键关联关系，由于是一对一关联，因此还需要给外键列增加 unique 唯一约束。下面我们就用一个示例来看看 MyBatis 怎么处理一对一关系。

示例：OneToOneTest

首先，给之前创建的 mybatis 数据库创建两个表 tb_card 和 tb_person，并插入测试数据。SQL 脚本如下：

```
# 如果存在 tb_user 表则删除
DROP TABLE IF EXISTS tb_card;
CREATE TABLE tb_card (
id INT PRIMARY KEY AUTO_INCREMENT,
CODE VARCHAR(18)
);
INSERT INTO tb_card (CODE) VALUES('432801198009191038');
# 如果存在 tb_user 表则删除
DROP TABLE IF EXISTS tb_person;
CREATE TABLE tb_person (
id INT PRIMARY KEY AUTO_INCREMENT,
NAME VARCHAR(18),
sex VARCHAR(18),
age INT,
card_id INT UNIQUE,
FOREIGN KEY (card_id) REFERENCES tb_card (id)
);
INSERT INTO TB_PERSON (NAME,sex,age,card_id) VALUES('jack','男',23,1);
```

提示：

tb_person 表的 card_id 作为外键参照 tb_card 表的主键 id，因为是一对一关系，即一个 card 只能让一个 person 使用，所以把 card_id 做成了唯一键约束。如此一来，当一个 person 使用了一个 card 之后，其他的 person 就不能使用该 card 了。

在 mybatis 数据库中执行 SQL 脚本，完成创建数据库和表的操作。

接下来，创建一个 Card 对象和一个 Person 对象分别映射 tb_card 和 tb_peson 表。

程序清单：codes/10/OneToOneTest/src/org/fkit/domain/Card.java

```
import java.io.Serializable;
public class Card implements Serializable {
    private Integer id;  // 主键 id
    private String code; // 身份证编号
    // 省略构造器和 set/get 方法……
}
```

程序清单：codes/10/OneToOneTest/src/org/fkit/domain/Person.java

```
import java.io.Serializable;
public class Person implements Serializable {
    private Integer id;  // 主键 id
    private String name; // 姓名
    private String sex;  // 性别
    private Integer age; // 年龄
    // 人和身份证号码之间是一对一的关系，即一个人只有一个身份证号码
    private Card card;
    // 省略构造器和 set/get 方法……
}
```

人和身份证号码之间是一对一的关系，即一个人只有一个身份证。在 Person 类中定义了一个 card 属性，该属性是一个 Card 类型，用来映射一对一的关联关系，表示这个人的身份证。

再接下来是 XML 映射文件。

程序清单：codes/10/OneToOneTest/src/org/fkit/mapper/CardMapper.xml

```
<mapper namespace="org.fkit.mapper.CardMapper">
    <!-- 根据 id 查询 Card，返回 Card 对象 -->
  <select id="selectCardById" parameterType="int" resultType="org.fkit.domain.Card">
      SELECT * from tb_card where id = #{id}
  </select>
</mapper>
```

程序清单：codes/10/OneToOneTest/src/org/fkit/mapper/PersonMapper.xml

```
<mapper namespace="org.fkit.mapper.PersonMapper">
    <!-- 根据 id 查询 Person，返回 resultMap -->
  <select id="selectPersonById" parameterType="int"
      resultMap="personMapper">
      SELECT * from tb_person where id = #{id}
  </select>
  <!-- 映射 Peson 对象的 resultMap -->
    <resultMap type="org.fkit.domain.Person" id="personMapper">
        <id property="id" column="id"/>
        <result property="name" column="name"/>
        <result property="sex" column="sex"/>
        <result property="age" column="age"/>
        <!-- 一对一关联映射:association   -->
        <association property="card" column="card_id"
        select="org.fkit.mapper.CardMapper.selectCardById"
        javaType="org.fkit.domain.Card"/>
    </resultMap>
</mapper>
```

在 PersonMapper.xml 中定义了一个<select.../>，其根据 id 查询 Peson 信息，由于 Peson 类除了简单的属性 id、name、sex 和 age 之外，还有一个关联对象 card，所以返回的是一个名为 personMapper 的 resultMap。personMapper 中使用了<association.../>元素映射一对一的关联关系，

select 属性表示会使用 column 属性的 card_id 值作为参数执行 CardMapper 中定义的 selectCardById 查询对应的 Card 数据，查询出的数据将被封装到 property 表示的 Card 对象当中。

之前的测试都是使用 SqlSession 对象调用 insert、update、delete 和 select 方法进行测试。实际上，Mybatis 官方手册建议通过 mapper 接口的代理对象访问 Mybatis，该对象关联了 SqlSession 对象，开发者可以通过该对象直接调用方法操作数据库。下面定义一个 mapper 接口对象，需要注意的是，mapper 接口对象的类名必须和之前的 XML 文件中的 mapper 的 namespace 一致，而方法名和参数也必须和 XML 文件中的<select.../>元素的 id 属性和 parameterType 属性一致。

```
<mapper namespace="org.fkit.mapper.PersonMapper">
    <select id="selectPersonById" parameterType="int".../>
</mapper>
```

程序清单：codes/10/OneToOneTest/src/org/fkit/mapper/PersonMapper.java

```
package org.fkit.mapper;
import org.fkit.domain.Person;
public interface PersonMapper {
    /**
     * 根据 id 查询 Person,
     * 方法名和参数必须和 XML 文件中的<select.../>元素的 id 属性和 parameterType 属性一致
     * @param id
     * @return Person 对象
     * */
    Person selectPersonById(Integer id);
}
```

最后，完成测试类。

程序清单：codes/10/OneToOneTest/src/org/fkit/test/OneToOneTest.java

```
import org.apache.ibatis.session.SqlSession;
import org.fkit.domain.Person;
import org.fkit.factory.FKSqlSessionFactory;
import org.fkit.mapper.PersonMapper;

public class OneToOneTest {

    public static void main(String[] args) {
        // 定义 SqlSession 变量
        SqlSession sqlSession = null;
        try {
            // 创建 SqlSession 实例
            sqlSession = FKSqlSessionFactory.getSqlSession();
            // 获得 mapper 接口的代理对象
            PersonMapper pm = sqlSession.getMapper(PersonMapper.class);
            // 直接调用接口的方法，查询 id 为 1 的 Peson 数据
            Person p = pm.selectPersonById(1);
            // 打印 Peson 对象
            System.out.println(p);
            // 打印 Person 对象关联的 Card 对象
            System.out.println(p.getCard());
            // 提交事务
            sqlSession.commit();
        } catch (Exception e) {
            // 回滚事务
            sqlSession.rollback();
            e.printStackTrace();
        }finally {
            // 关闭 SqlSession
```

```
        if(sqlSession != null)
            sqlSession.close();
        }
    }

}
```

项目的 mybatis-config.xml 和 log4j.xml 文件内容请参考配套资源文件，此处不再赘述。

运行 OneToOneTest 类的 main 方法，通过 SqlSession 的 getMapper(Class<T> type)方法获得 mapper 接口的代理对象 PersonMapper，调用 selectPersonById 方法时会执行 PersonMapper.xml 中<select.../>元素中定义的 SQL 语句。控制台显示如下：

```
DEBUG [main] ==>  Preparing: SELECT * from tb_person where id = ?
DEBUG [main] ==> Parameters: 1(Integer)
DEBUG [main] ====>  Preparing: SELECT * from tb_card where id = ?
DEBUG [main] ====> Parameters: 1(Integer)
DEBUG [main] <====      Total: 1
DEBUG [main] <==      Total: 1
Person [id=1, name=jack, sex=男, age=23]
Card [id=1, code=432801198009191038]
```

可以看到，查询 Person 信息时 Person 对应的 Card 对象也被查询出来了。

10.1.2 一对多

在实际项目开发中，一对多是非常常见的关系，比如，一个班级可以有多个学生，一个学生只能属于一个班级，班级和学生之间是一对多的关系，而学生和班级之间是多对一的关系。在数据库中一对多关系通常使用主外键关联，外键列应该在多方，即多方维护关系。下面我们就用一个示例来看看 MyBatis 怎么处理一对多关系。

示例：OneToManyTest

首先，给之前创建的 mybatis 数据库创建两个表 tb_clazz 和 tb_student，并插入测试数据。SQL 脚本如下：

```
# 如果存在 tb_clazz 表则删除
DROP TABLE IF EXISTS tb_clazz;
CREATE TABLE tb_clazz(
id INT PRIMARY KEY AUTO_INCREMENT,
CODE VARCHAR(18),
NAME VARCHAR(18)
);
INSERT INTO tb_clazz(CODE,NAME) VALUES('j1601','Java 就业班');
# 如果存在 tb_student 表则删除
DROP TABLE IF EXISTS tb_student;
CREATE TABLE tb_student(
id INT PRIMARY KEY AUTO_INCREMENT,
NAME VARCHAR(18),
sex VARCHAR(18),
age INT,
clazz_id INT,
FOREIGN KEY (clazz_id) REFERENCES tb_clazz(id)
);
INSERT INTO tb_student(NAME,sex,age,clazz_id) VALUES('jack','男',23,1);
INSERT INTO tb_student(NAME,sex,age,clazz_id) VALUES('rose','女',18,1);
INSERT INTO tb_student(NAME,sex,age,clazz_id) VALUES('tom','男',21,1);
INSERT INTO tb_student(NAME,sex,age,clazz_id) VALUES('alice','女',20,1);
```

提示： tb_student 表的 clazz_id 作为外键参照 tb_clazz 表的主键 id。

在 mybatis 数据库中执行 SQL 脚本，完成创建数据库和表的操作。

接下来，创建一个 Clazz 对象和一个 Student 对象分别映射 tb_clazz 和 tb_student 表。

程序清单：codes/10/OneToManyTest/src/org/fkit/domain/Clazz.java

```
import java.io.Serializable;
import java.util.List;
public class Clazz implements Serializable {
    private Integer id; // 班级 id，主键
    private String code; // 班级编号
    private String name; // 班级名称
    // 班级和学生之间是一对多的关系，即一个班级可以有多个学生
    private List<Student> students;
    // 省略构造器和 set/get 方法......
}
```

班级和学生之间是一对多的关系，即一个班级可以有多个学生。在 Clazz 类当中定义了一个 students 属性，该属性是一个 List 集合，用来映射一对多的关联关系，表示一个班级有多个学生。

程序清单：codes/10/OneToManyTest/src/org/fkit/domain/Student.java

```
import java.io.Serializable;
public class Student implements Serializable {
    private Integer id; // 学生 id，主键
    private String name; // 姓名
    private String sex;  // 性别
    private Integer age; // 年龄
    // 学生和班级之间是多对一的关系，即一个学生只属于一个班级
    private Clazz clazz;
    // 省略构造器和 set/get 方法......
}
```

学生和班级之间是多对一的关系，即一个学生只属于一个班级。在 Student 类当中定义了一个 clazz 属性，该属性是一个 Clazz 类型，用来映射多对一的关联关系，表示该学生所属的班级。

再接下来是 XML 映射文件。

程序清单：codes/10/OneToManyTest/src/org/fkit/mapper/ClazzMapper.xml

```
<mapper namespace="org.fkit.mapper.ClazzMapper">
    <!-- 根据 id 查询班级信息，返回 resultMap -->
      <select id="selectClazzById" parameterType="int" resultMap="clazzResultMap">
         SELECT * FROM tb_clazz  WHERE id = #{id}
      </select>
       <!-- 映射 Clazz 对象的 resultMap -->
    <resultMap type="org.fkit.domain.Clazz" id="clazzResultMap">
        <id property="id" column="id"/>
        <result property="code" column="code"/>
        <result property="name" column="name"/>
        <!-- 一对多关联映射:collection fetchType="lazy"表示懒加载  -->
        <collection property="students" javaType="ArrayList"
      column="id" ofType="org.fkit.domain.Student"
      select="org.fkit.mapper.StudentMapper.selectStudentByClazzId"
      fetchType="lazy">
          <id property="id" column="id"/>
          <result property="name" column="name"/>
          <result property="sex" column="sex"/>
```

```
            <result property="age" column="age"/>
        </collection>
    </resultMap>
</mapper>
```

ClazzMapper.xml 中定义了一个<select.../>，其根据 id 查询班级信息。由于 Clazz 类除了简单的属性 id、code、name 以外，还有一个关联对象 students，所以返回的是一个名为 clazzResultMap 的 resultMap。由于 students 是一个 List 集合，所以 clazzResultMap 中使用了<collection.../>元素映射一对多的关联关系，select 属性表示会使用 column 属性的 id 值作为参数执行 StudentMapper 中定义的 selectStudentByClazzId 查询该班级对应的所有学生数据，查询出的数据将被封装到 property 表示的 students 对象当中。

还使用了一个新的属性 fetchType，该属性的取值有 eager 和 lazy，eager 表示立即加载，即查询 Clazz 对象的时候，会立即执行关联的 selectStudentByClazzId 中定义的 SQL 语句去查询班级的所有学生；lazy 表示懒加载，其不会立即发送 SQL 语句去查询班级的所有学生，而是等到需要使用到班级的 students 属性时，才会发送 SQL 语句去查询班级的所有学生。fetch 机制更多的是为了性能考虑，如果查询班级时确定会访问班级的所有学生，则该属性应该被设置为 eager；如果查询班级时只是查询班级信息，有可能不会访问班级的所有学生，则该属性应该被设置为 lazy。正常情况下，一对多所关联的集合对象，都应该被设置成 lazy。

使用懒加载还需要在 mybatis-config.xml 中增加如下配置：

```
<settings>
    <!-- 要使延迟加载生效必须配置下面两个属性 -->
    <setting name="lazyLoadingEnabled" value="true"/>
    <setting name="aggressiveLazyLoading" value="false"/>
</settings>
```

- lazyLoadingEnabled 属性表示延迟加载的全局开关。当开启时，所有关联对象都会延迟加载。默认为 false。
- aggressiveLazyLoading 属性启用时，会使带有延迟加载属性的对象立即加载；反之，每种属性将会按需加载。默认为 true，所以这里需要设置成 false。

程序清单：codes/10/OneToManyTest/src/org/fkit/mapper/StudentMapper.xml

```
<mapper namespace="org.fkit.mapper.StudentMapper">
    <!-- 根据 id 查询学生信息，多表连接，返回 resultMap
    注意：多表联合查询时，tb_clazz 表和 tb_student 表中有同名字段，所以给同名字段 id 和 name
    起别名
-->
  <select id="selectStudentById" parameterType="int" resultMap="studentResultMap">
      SELECT c.id as cid,c.code,c.name as cname,s.* FROM
      tb_clazz c,tb_student s
      WHERE c.id = s.clazz_id
       AND s.id = #{id}
  </select>
  <!-- 根据班级 id 查询学生信息，返回 resultMap -->
  <select id="selectStudentByClazzId" parameterType="int"
  resultMap="studentResultMap">
      SELECT * FROM tb_student WHERE clazz_id = #{id}
  </select>
   <!-- 映射 Student 对象的 resultMap -->
    <resultMap type="org.fkit.domain.Student" id="studentResultMap">
        <id property="id" column="id"/>
          <result property="name" column="name"/>
          <result property="sex" column="sex"/>
          <result property="age" column="age"/>
        <!-- 多对一关联映射:association
```

```
        注意：此处映射时使用的是多表连接查询的别名 cid 和 cname 进行映射
        -->
        <association property="clazz" javaType="org.fkit.domain.Clazz">
            <id property="id" column="cid"/>
            <result property="code" column="code"/>
            <result property="name" column="cname"/>
        </association>
    </resultMap>
</mapper>
```

StudentMapper.xml 中定义了一个<select id="selectStudentById".../>，其会根据学生 id 查询学生信息，由于 Student 类除了简单的属性 id、name、sex 和 age 之外，还有一个关联对象 clazz，所以它返回的是一个名为 clazzResultMap 的 resultMap。clazzResultMap 中使用了< association.../>元素映射多对一的关联关系，因为<select id="selectStudentById".../>的 SQL 语句是一条多表连接，关联 tb_clazz 表的同时查询班级数据，所以< association.../>只是简单地装载数据。需要注意的是，多表联合查询时，tb_clazz 表和 tb_student 表中有同名字段，需要给同名字段 id 和 name 起别名，同时在 association 映射时也要使用多表连接查询的别名 cid 和 cname 进行映射，否则会出现映射混乱问题。

提示

在实际开发中，一对多关系通常映射为集合对象，而由于多方的数据量可能很大，所以通常使用懒加载；而多对一只是关联到一个对象，所以通常使用多表连接直接把数据提取出来。

StudentMapper.xml 中还定义了一个<select id=" selectStudentByClazzId ".../>，其会根据班级 id 查询所有学生信息，该查询用于 ClazzMapper.xml 中的关联查询。

再接下来是 mapper 接口对象。

程序清单：codes/10/OneToManyTest/src/org/fkit/mapper/ClazzMapper.java

```java
import org.fkit.domain.Clazz;
public interface ClazzMapper {
    // 根据 id 查询班级信息
    Clazz selectClazzById(Integer id);
}
```

程序清单：codes/10/OneToManyTest/src/org/fkit/mapper/StudentMapper.java

```java
import org.fkit.domain.Student;
public interface StudentMapper {
    // 根据 id 查询学生信息
    Student selectStudentById(Integer id);
}
```

最后，完成测试类。

程序清单：codes/10/OneToManyTest/src/org/fkit/test/OneToManyTest.java

```java
import java.util.List;
import org.apache.ibatis.session.SqlSession;
import org.fkit.domain.Clazz;
import org.fkit.domain.Student;
import org.fkit.factory.FKSqlSessionFactory;
import org.fkit.mapper.ClazzMapper;
import org.fkit.mapper.StudentMapper;

public class OneToManyTest {

   public static void main(String[] args) {
```

```
        // 定义 SqlSession 变量
        SqlSession sqlSession = null;
        try {
            // 创建 SqlSession 实例
            sqlSession = FKSqlSessionFactory.getSqlSession();
            OneToManyTest t = new OneToManyTest();
            t.testSelectClazzById(sqlSession);
            // 提交事务
            sqlSession.commit();
        } catch (Exception e) {
            // 回滚事务
            sqlSession.rollback();
            e.printStackTrace();
        }finally {
            // 关闭 SqlSession
            if(sqlSession != null)
                sqlSession.close();
        }
    }

    // 测试一对多关系，查询班级 Clazz（一）的时候关联查询学生 Student（多）
    public void testSelectClazzById(SqlSession sqlSession){
        // 获得 ClazzMapper 接口的代理对象
        ClazzMapper cm = sqlSession.getMapper(ClazzMapper.class);
        // 调用 selectClazzById 方法
        Clazz clazz = cm.selectClazzById(1);
        // 查看查询到的 clazz 对象信息
        System.out.println(clazz.getId() + " "+ clazz.getCode() + " "+clazz.getName());
    }
}
```

项目的 mybatis-config.xml 和 log4j.xml 文件内容请参考配套资源文件，此处不再赘述。

在 OneToManyTest 类中定义了一个 testSelectClazzById()方法，该方法用于测试一对多关系，查询班级 Clazz（一）的时候关联查询学生 Student（多）的信息。

运行 OneToManyTest 类的 main 方法，其通过 SqlSession 的 getMapper(Class<T> type)方法获得 mapper 接口的代理对象 ClazzMapper，调用 selectClazzById 方法时会执行 ClazzMapper.xml 中<select id="selectClazzById".../>的元素中定义的 SQL 语句。控制台显示如下：

```
DEBUG [main] ==>  Preparing: SELECT * FROM tb_clazz WHERE id = ?
DEBUG [main] ==> Parameters: 1(Integer)
DEBUG [main] <==      Total: 1
1 j1601 Java 就业班
```

可以看到，MyBatis 只是执行了查询班级的 SQL 语句，由于<select id="selectClazzById".../>中的<collection fetchType="lazy"/>使用的是懒加载，因此，当没有使用到关联的学生对象时，并没有发送查询学生的 SQL 语句。

修改 testSelectClazzById()方法，增加访问关联学生的代码：

```
// 测试一对多关系，查询班级 Clazz（一）的时候关联查询学生 Student（多）
    public void testSelectClazzById(SqlSession sqlSession){
        // 获得 ClazzMapper 接口的代理对象
        ClazzMapper cm = sqlSession.getMapper(ClazzMapper.class);
        // 调用 selectClazzById 方法
        Clazz clazz = cm.selectClazzById(1);
        // 查看查询到的 clazz 对象信息
        System.out.println(clazz.getId() + " "+ clazz.getCode() + " "+clazz.getName());
        // 查看 clazz 对象关联的学生信息
        List<Student> students = clazz.getStudents();
```

```
        students.forEach(stu -> System.out.println(stu));
    }
```

再次运行 OneToManyTest 类的 main 方法。控制台显示如下：

```
    DEBUG [main] ==>  Preparing: SELECT * FROM tb_clazz WHERE id = ?
    DEBUG [main] ==> Parameters: 1(Integer)
    DEBUG [main] <==      Total: 1
    1 j1601 Java 就业班
    DEBUG [main] ==>  Preparing: SELECT * FROM tb_student WHERE clazz_id = ?
    DEBUG [main] ==> Parameters: 1(Integer)
    DEBUG [main] <==      Total: 4
    Student [id=1, name=jack, sex=男, age=23]
    Student [id=2, name=rose, sex=女, age=18]
    Student [id=3, name=tom, sex=男, age=21]
    Student [id=4, name=alice, sex=女, age=20]
```

可以看到，MyBatis 执行了查询班级的 SQL 语句之后，又执行了根据 clazz_id 查询学生信息的 SQL 语句。这就是所谓的懒加载。

增加一个 testSelectStudentById()方法，测试多对一关系。

```
// 测试多对一关系，查询学生 Student（多）的时候关联查询班级 Clazz（一）
    public void testSelectStudentById(SqlSession sqlSession){
        // 获得 StudentMapper 接口的代理对象
        StudentMapper sm = sqlSession.getMapper(StudentMapper.class);
        // 调用 selectStudentById 方法
        Student stu = sm.selectStudentById(1);
        // 查看查询到的 Student 对象信息
        System.out.println(stu);
        // 查看 Student 对象关联的班级信息
        System.out.println(stu.getClazz());
    }
```

在 main 方法中运行 testSelectStudentById()方法，控制台显示如下：

```
    DEBUG [main] ==>  Preparing: SELECT * FROM tb_clazz c,tb_student s WHERE c.id =
s.clazz_id AND s.id = ?
    DEBUG [main] ==> Parameters: 1(Integer)
    DEBUG [main] <==      Total: 1
    Student [id=1, name=Java 就业班, sex=男, age=23]
    Clazz [id=1, code=j1601, name=Java 就业班]
```

可以看到，MyBatis 执行了一个多表查询语句，并且将查询到的班级信息封装到了学生对象的关联属性中。

10.1.3 多对多

在实际项目开发中，多对多关系也是非常常见的关系，比如，一个购物系统中，一个用户可以有多个订单，这是一对多的关系；一个订单中可以购买多种商品，一种商品也可以属于多个不同的订单，订单和商品之间就是多对多的关系。对于数据库中多对多的关系建议使用一个中间表来维护关系，中间表中的订单 id 作为外键参照订单表的 id，商品 id 作为外键参照商品表的 id。下面我们就用一个示例来看看 MyBatis 怎么处理多对多关系。

示例：ManyToManyTest

首先，给之前创建的 mybatis 数据库创建三个表 tb_user、tb_article 和 tb_order，再创建一个中间表维护 tb_article 和 tb_order 的关系，并插入测试数据。SQL 脚本如下：

```
    # 如果存在 tb_user 表则删除
    DROP TABLE IF EXISTS tb_user;
```

```
#创建用户表
CREATE TABLE tb_user(
id INT PRIMARY KEY AUTO_INCREMENT,
username VARCHAR(18),
loginname VARCHAR(18),
PASSWORD VARCHAR(18),
phone VARCHAR(18),
address VARCHAR(18)
);
#插入用户表测试数据
INSERT INTO tb_user(username,loginname,PASSWORD,phone,address)
VALUES('杰克','jack','123456','13920001616','广州');
# 如果存在 tb_article 表则删除
DROP TABLE IF EXISTS tb_article;
#创建商品表
CREATE TABLE tb_article(
id INT PRIMARY KEY AUTO_INCREMENT,
NAME VARCHAR(18),
price DOUBLE,
remark VARCHAR(18)
);
#插入商品表测试数据
INSERT INTO tb_article(NAME,price,remark)
VALUES('疯狂 Java 讲义',108.9,'李刚老师经典著作');
INSERT INTO tb_article(NAME,price,remark)
VALUES('疯狂 Android 讲义',99.9,'李刚老师经典著作');
INSERT INTO tb_article(NAME,price,remark)
VALUES('疯狂 iOS 讲义',89.9,'李刚老师经典著作');
INSERT INTO tb_article(NAME,price,remark)
VALUES('SpringMVC+MyBatis 企业开发',69.9,'肖文吉老师经典著作');
# 如果存在 tb_order 表则删除
DROP TABLE IF EXISTS tb_order;
#创建订单表
CREATE TABLE tb_order(
id INT PRIMARY KEY AUTO_INCREMENT,
CODE VARCHAR(32),
total DOUBLE,
user_id INT,
FOREIGN KEY (user_id) REFERENCES tb_user(id)
);
#插入订单表测试数据
INSERT INTO tb_order(CODE,total,user_id)
VALUES('6aa3fa359ff14619b77fab5990940a2d',388.6,1);
INSERT INTO tb_order(CODE,total,user_id)
VALUES('6aa3fa359ff14619b77fab5990940b3c',217.8,1);
# 如果存在 tb_item 表则删除
DROP TABLE IF EXISTS tb_item;
#创建中间表
CREATE TABLE tb_item(
order_id INT,
article_id INT,
amount INT,
PRIMARY KEY(order_id,article_id),
FOREIGN KEY (order_id) REFERENCES tb_order(id),
FOREIGN KEY (article_id) REFERENCES tb_article(id)
);
#创建插入中间表数据
INSERT INTO tb_item(order_id,article_id,amount)
VALUES(1,1,1);
INSERT INTO tb_item(order_id,article_id,amount)
VALUES(1,2,1);
INSERT INTO tb_item(order_id,article_id,amount)
VALUES(1,3,2);
```

```
INSERT INTO tb_item(order_id,article_id,amount)
VALUES(2,4,2);
INSERT INTO tb_item(order_id,article_id,amount)
VALUES(2,1,1);
```

提示： tb_order表的user_id作为外键参照tb_user表的主键id。tb_item表作为中间表，用来维护 tb_article 和 tb_order 的多对多关系，tb_imte 表的 order_id 作为外键参照 tb_order 表的主键 id，article_id 作为外键参照 tb_article 表的主键 id

在 mybatis 数据库中执行 SQL 脚本，完成创建数据库和表的操作。

接下来，创建一个 User 对象、一个 Article 对象和一个 Order 对象分别映射 tb_user、tb_article 和 tb_order 表。

程序清单：codes/10/ManyToManyTest/src/org/fkit/domain/User.java

```
import java.io.Serializable;
import java.util.List;
public class User implements Serializable{
   private Integer id;  // 用户 id，主键
   private String username;  // 用户名
   private String loginname; // 登录名
   private String password;  // 密码
   private String phone;    // 联系电话
   private String address;  // 收货地址
   // 用户和订单之间是一对多的关系，即一个用户可以有多个订单
   private List<Order> orders;
   // 省略构造器和 set/get 方法……
}
```

用户和订单之间是一对多的关系，即一个用户可以有多个订单。在 User 类中定义了一个 orders 属性，该属性是一个 List 集合，用来映射一对多的关联关系，表示一个用户有多个订单。

程序清单：codes/10/ManyToManyTest/src/org/fkit/domain/Order.java

```
import java.io.Serializable;
import java.util.List;
public class Order implements Serializable {
   private Integer id;  // 订单 id，主键
   private String code;  // 订单编号
   private Double total; // 订单总金额
   // 订单和用户之间是多对一的关系，即一个订单只属于一个用户
   private User user;
   // 订单和商品之间是多对多的关系，即一个订单可以包含多种商品
   private List<Article> articles;
   // 省略构造器和 set/get 方法……
}
```

订单和用户之间是多对一的关系，一个订单只属于一个用户，在 Order 类中定义了一个 user 属性，用来映射多对一的关联关系，表示该订单的用户；订单和商品之间是多对多的关系，即一个订单中可以包含多种商品，在 Order 类中定义了一个 articles 属性，该属性是一个 List 集合，用来映射多对多的关联关系，表示一个订单中包含多种商品。

程序清单：codes/10/ManyToManyTest/src/org/fkit/domain/Articlejava

```
import java.io.Serializable;
import java.util.List;
public class Article implements Serializable {
   private Integer id;        // 商品 id，主键
```

```
    private String name;   // 商品名称
    private Double price;    // 商品价格
    private String remark;    // 商品描述
    // 商品和订单之间是多对多的关系，即一种商品可以出现在多个订单中
    private List<Order> orders;
    // 省略构造器和 set/get 方法……
}
```

商品和订单之间是多对多的关系，即一种商品可以出现在多个订单中。在 Article 类中定义了一个 orders 属性，该属性是一个 List 集合，用来映射多对多的关联关系，表示该商品关联的多个订单。

再接下来是 XML 映射文件。

程序清单：codes/10/ManyToManyTest/src/org/fkit/mapper/UserMapper.xml

```
<mapper namespace="org.fkit.mapper.UserMapper">
   <resultMap type="org.fkit.domain.User" id="userResultMap">
      <id property="id" column="id"/>
      <result property="username" column="username"/>
      <result property="loginname" column="loginname"/>
      <result property="password" column="password"/>
      <result property="phone" column="phone"/>
      <result property="address" column="address"/>
      <!-- 一对多关联映射:collection   -->
      <collection property="orders" javaType="ArrayList"
     column="id" ofType="org.fkit.domain.User"
     select="org.fkit.mapper.OrderMapper.selectOrderByUserId"
     fetchType="lazy">
        <id property="id" column="id"/>
        <result property="code" column="code"/>
        <result property="total" column="total"/>
     </collection>
   </resultMap>
  <select id="selectUserById" parameterType="int" resultMap="userResultMap">
     SELECT * FROM tb_user  WHERE id = #{id}
  </select>
</mapper>
```

UserMapper.xml 中定义了一个<select.../>，其根据 id 查询用户信息。由于 User 类除了简单的属性 id、username、loginname、password、phone 和 address 之外，还有一个关联对象 orders，所以返回的是一个名为 userResultMap 的 resultMap。由于 orders 是一个 List 集合，因此 userResultMap 中使用了<collection.../>元素映射一对多的关联关系，select 属性表示会使用 column 属性的 id 值作为参数执行 OrderMapper 中定义的 selectOrderByUserId 查询该用户下的所有订单，查询出的数据将被封装到 property 表示的 orders 对象当中。注意，一对多使用的都是 lazy（懒加载）。

程序清单：codes/10/ManyToManyTest/src/org/fkit/mapper/OrderMapper.xml

```
<mapper namespace="org.fkit.mapper.OrderMapper">
   <resultMap type="org.fkit.domain.Order" id="orderResultMap">
      <id property="id" column="oid"/>
        <result property="code" column="code"/>
        <result property="total" column="total"/>
      <!-- 多对一关联映射:association   -->
      <association property="user" javaType="org.fkit.domain.User">
         <id property="id" column="id"/>
         <result property="username" column="username"/>
         <result property="loginname" column="loginname"/>
         <result property="password" column="password"/>
         <result property="phone" column="phone"/>
```

```
            <result property="address" column="address"/>
        </association>
        <!-- 多对多映射的关键:collection  -->
        <collection property="articles" javaType="ArrayList"
      column="oid" ofType="org.fkit.domain.Article"
      select="org.fkit.mapper.ArticleMapper.selectArticleByOrderId"
      fetchType="lazy">
          <id property="id" column="id"/>
          <result property="name" column="name"/>
          <result property="price" column="price"/>
          <result property="remark" column="remark"/>
      </collection>
    </resultMap>
    <!-- 注意，如果查询出来的列同名，例如 tb_user 表的 id 和 tb_order 表的 id 都是 id，同名，则需要使用别名区分 -->
    <select id="selectOrderById" parameterType="int" resultMap="orderResultMap">
      SELECT u.*,o.id AS oid,CODE,total,user_id
       FROM tb_user u,tb_order o
      WHERE u.id = o.user_id
       AND o.id = #{id}
    </select>
    <!-- 根据 userid 查询订单 -->
    <select id="selectOrderByUserId" parameterType="int" resultType="org.fkit.domain.Order">
      SELECT * FROM tb_order WHERE user_id = #{id}
    </select>
  </mapper>
```

OrderMapper.xml 中定义了一个<select id=" selectOrderByUserId".../>，其根据用户 id 查询订单信息，返回的是简单的 Order 对象。

同时定义了一个<select id=" selectOrderById ".../>，其根据订单 id 查询订单信息，由于 Order 类和用户是多对一关系，和商品是多对多关系，而多对一通常都是立即加载，因此 SQL 语句是一条关联了 tb_user 和 tb_order 的多表查询语句。查询结果返回一个名为 orderResultMap 的 resultMap。orderResultMap 中使用了< association.../>元素映射多对一的关联关系，其将查询到的用户信息装载到 Order 对象的 user 属性当中；orderResultMap 中还使用了<collection.../>元素映射多对多的关联关系，select 属性表示会使用 column 属性的 oid 值作为参数执行 ArticleMapper 中定义的 selectArticleByOrderId 查询该订单中的所有商品，查询出的数据将被封装到 property 表示的 articles 对象当中。注意，一对多使用的都是 lazy（懒加载）。

提示

因为多表查询返回的结果集中 tb_user 有个 id 列，tb_order 也有个 id 列，当列同名时，MyBatis 使用的元素中的 column 属性如果是 id，则 MyBatis 会默认使用查询出的第一个 id 列。为了区分同名的列，最好的方法是给列取一个别名。SQL 语句中的 o.id AS oid，resultMap 中的 column="oid"就是指使用的是 tb_order 表的 id 值。

程序清单：codes/10/ManyToManyTest/src/org/fkit/mapper/ArticleMapper.xml

```
<mapper namespace="org.fkit.mapper.ArticleMapper">
  <select id="selectArticleByOrderId" parameterType="int"
  resultType="org.fkit.domain.Article">
     SELECT * FROM tb_article WHERE id IN (
       SELECT article_id FROM tb_item WHERE order_id = #{id}
    )
  </select>
</mapper>
```

ArticleMapper.xml 中定义了一个<select id="selectArticleByOrderId ".../>，其根据订单 id 查询订单关联的所有商品，由于订单和商品是多对多的关系，数据库使用了一个中间表 tb_item

维护多对多的关系，故此处使用了一个子查询，首先根据订单 id 到中间表中查询出所有的商品，之后根据所有商品的 id 查询出所有的商品信息，并将这些信息封装到 Article 对象当中。

再接下来是 mapper 接口对象。

程序清单：codes/10/ManyToManyTest/src/org/fkit/mapper/UserMapper.java

```
import org.fkit.domain.User;
public interface UserMapper {
    User selectUserById(int id);
}
```

程序清单：codes/10/ManyToManyTest/src/org/fkit/mapper/OrderMapper.java

```
import org.fkit.domain.Order;
public interface OrderMapper {
    Order selectOrderById(int id);
}
```

最后，完成测试类。

程序清单：codes/10/OneToManyTest/src/org/fkit/test/ManyToManyTest.java

```
import java.util.List;
import org.apache.ibatis.session.SqlSession;
import org.fkit.domain.Article;
import org.fkit.domain.Order;
import org.fkit.domain.User;
import org.fkit.factory.FKSqlSessionFactory;
import org.fkit.mapper.OrderMapper;
import org.fkit.mapper.UserMapper;


public class ManyToManyTest {

    public static void main(String[] args) {
        // 定义 SqlSession 变量
        SqlSession sqlSession = null;
        try {
            // 创建 SqlSession 实例
            sqlSession = FKSqlSessionFactory.getSqlSession();
            ManyToManyTest t = new ManyToManyTest();
            // 根据用户 id 查询用户，测试一对多关系
            t.testSelectUserById(sqlSession);
            // 根据订单 id 查询订单，测试多对多关系
            // t.testSelectOrderById(sqlSession);
            // 提交事务
            sqlSession.commit();
        } catch (Exception e) {
            // 回滚事务
            sqlSession.rollback();
            e.printStackTrace();
        }finally {
            // 关闭 SqlSession
            if(sqlSession != null)
                sqlSession.close();
        }
    }

    // 测试一对多关系，查询班级 User（一）的时候关联查询订单 Order（多）
    public void testSelectUserById(SqlSession sqlSession){
        // 获得 UserMapper 接口的代理对象
        UserMapper um = sqlSession.getMapper(UserMapper.class);
        // 调用 selectUserById 方法
```

```
        User user = um.selectUserById(1);
        // 查看查询到的 user 对象信息
        System.out.println(user.getId() + " " + user.getUsername());
        // 查看 user 对象关联的订单信息
        List<Order> orders = user.getOrders();
        orders.forEach(order -> System.out.println(order));
    }

    // 测试多对多关系，查询订单 Order（多）的时候关联查询订单的商品 Article（多）
    public void testSelectOrderById(SqlSession sqlSession){
        // 获得 OrderMapper 接口的代理对象
        OrderMapper om = sqlSession.getMapper(OrderMapper.class);
        // 调用 selectOrderById 方法
        Order order = om.selectOrderById(2);
        // 查看查询到的 order 对象信息
        System.out.println(order.getId() + " " + order.getCode() + " " +
order.getTotal());
        // 查看 order 对象关联的用户信息
        User user = order.getUser();
        System.out.println(user);
        // 查看 order 对象关联的商品信息
        // List<Article> articles = order.getArticles();
        // articles.forEach(article -> System.out.println(article));
    }

}
```

项目的 mybatis-config.xml 和 log4j.xml 文件内容请参考配套资源文件，此处不再赘述。

运行 ManyToManyTest 类的 main 方法，首先测试 testSelectUserById()方法，根据用户 id 查询用户。控制台显示如下：

```
    DEBUG [main] ==>  Preparing: SELECT * FROM tb_user WHERE id = ?
    DEBUG [main] ==> Parameters: 1(Integer)
    DEBUG [main] <==      Total: 1
    1 杰克
    DEBUG [main] ==>  Preparing: SELECT * FROM tb_order WHERE user_id = ?
    DEBUG [main] ==> Parameters: 1(Integer)
    DEBUG [main] <==      Total: 2
    Order [id=1, code=6aa3fa359ff14619b77fab5990940a2d, total=388.6]
    Order [id=2, code=6aa3fa359ff14619b77fab5990940b3c, total=217.8]
```

可以看到，MyBatis 执行了根据用户 id 查询用户的 SQL 语句，查询出了用户信息；由于在测试方法中立即又获取了用户的订单集合，所以 MyBatis 又执行了根据用户 id 查询订单的 SQL 语句，查询出了该用户的两个订单。

接下来测试 testSelectOrderById()方法，根据订单 id 查询订单信息。控制台显示如下：

```
    DEBUG [main] ==>  Preparing: SELECT u.*,o.id AS oid,CODE,total,user_id FROM tb_user
u,tb_order o WHERE u.id = o.user_id AND o.id = ?
    DEBUG [main] ==> Parameters: 2(Integer)
    DEBUG [main] <==      Total: 1
    2 6aa3fa359ff14619b77fab5990940b3c 217.8
    User [id=1, username=杰克, loginname=jack, password=123456, phone=13920001616,
address=广州]
```

可以看到，MyBatis 执行了一个多表连接查询，同时查询出了订单信息和用户信息，由于测试方法中注释了查询订单中的商品代码，故 MyBatis 采用了懒加载机制，没有立即查询商品信息。

取消 testSelectOrderById()方法中查询订单中的商品的代码注释，再次执行。控制台显示如下：

```
    DEBUG [main] ==>  Preparing: SELECT u.*,o.id AS oid,CODE,total,user_id FROM tb_user
u,tb_order o WHERE u.id = o.user_id AND o.id = ?
    DEBUG [main] ==> Parameters: 2(Integer)
    DEBUG [main] <==      Total: 1
    2 6aa3fa359ff14619b77fab5990940b3c 217.8
    User [id=1, username=杰克, loginname=jack, password=123456, phone=13920001616,
address=广州]
    DEBUG [main] ==>  Preparing: SELECT * FROM tb_article WHERE id IN ( SELECT article_id
FROM tb_item WHERE order_id = ? )
    DEBUG [main] ==> Parameters: 2(Integer)
    DEBUG [main] <==      Total: 2
    Article [id=1, name=疯狂Java讲义, price=108.9, remark=李刚老师经典著作]
    Article [id=4, name=SpringMVC+MyBatis企业开发, price=69.9, remark=肖文吉老师经典著作]
```

可以看到，MyBatis 执行了 ArticleMapper.xml 中定义的子查询，查询出了订单所关联的所有商品信息。

提示

多对多查询因为关联到中间表查询，所以读者需要对数据库的 SQL 知识有一定的了解。

10.2 MyBatis 动态 SQL

MyBatis 还有一个强大特性就是它的动态 SQL。

在实际项目开发中，经常需要根据不同条件拼接 SQL 语句，拼接时一定不能忘了必要的空格，有时候还要注意省略掉列名列表最后的逗号，等等。在使用 JDBC 或其他类似持久层框架操作数据库时，处理这种情况是非常麻烦的，甚至可以用痛苦来形容，而在 MyBatis 中利用动态 SQL 这一特性可以很简单地解决这个问题。

动态 SQL 元素和使用 JSTL 或其他类似基于 XML 的文本处理器相似，MyBatis 采用功能强大的基于 OGNL 的表达式来完成动态 SQL。OGNL 的表达式可以用在任意的 SQL 映射语句中。

常用的动态 SQL 元素包括：

- if
- choose (when、otherwise)
- where
- set
- foreach
- bind

下面我们就用一个简单示例来看看在 MyBatis 中怎么使用动态 SQL。

示例：DynamicSQLTest

首先，给之前创建的 mybatis 数据库创建一个表 tb_employee，并插入测试数据。SQL 脚本如下：

```
# 如果存在 tb_employee 表则删除
DROP TABLE IF EXISTS tb_employee;
CREATE TABLE tb_employee (
  ID INT(11) PRIMARY KEY AUTO_INCREMENT,
```

```
  loginname VARCHAR(18),
  PASSWORD VARCHAR(18),
  NAME VARCHAR(18) DEFAULT NULL,
  SEX CHAR(2) DEFAULT NULL,
  AGE INT(11) DEFAULT NULL,
  phone VARCHAR(21),
  sal DOUBLE,
  state VARCHAR(18)
 );
INSERT INTO tb_employee(loginname,PASSWORD,NAME,sex,age,phone,sal,state)
 VALUES('jack','123456','杰克','男',26,'13902019999',9800,'ACTIVE');
  INSERT INTO tb_employee(loginname,PASSWORD,NAME,sex,age,phone,sal,state)
 VALUES('rose','123456','露丝','女',21,'13902018888',6800,'ACTIVE');
 INSERT INTO tb_employee(loginname,PASSWORD,NAME,sex,age,phone,sal,state)
 VALUES('tom','123456','汤姆','男',25,'13902017777',8800,'ACTIVE');
  INSERT INTO tb_employee(loginname,PASSWORD,NAME,sex,age,phone,sal,state)
 VALUES('alice','123456','爱丽丝','女',20,'13902016666',5800,'ACTIVE');
```

在 mybatis 数据库中执行 SQL 脚本，完成创建数据库和表的操作。

接下来，创建一个 Employee 对象映射 tb_ employee 表。

程序清单：codes/10/DynamicSQLTest/src/org/fkit/domain/Employee.java

```
import java.io.Serializable;
public class Employee implements Serializable {
   private Integer id;            // 主键 id
   private String loginname;     // 登录名
   private String password;     // 密码
   private String name;        // 真实姓名
   private String sex;           // 性别
   private Integer age;        // 年龄
   private String phone;        // 电话
   private Double sal;           // 薪水
   private String state;        // 状态
   // 省略构造器和 set/get 方法……
}
```

10.2.1 if

动态 SQL 通常会做的事情是有条件地包含 where 子句的一部分。比如:

程序清单：codes/10/DynamicSQLTest/src/org/fkit/mapper/EmployeeMapper.xml

```
<mapper namespace="org.fkit.mapper.EmployeeMapper">
   <select id="selectEmployeeByIdLike"
        resultType="org.fkit.domain.Employee">
        SELECT * FROM tb_employee WHERE state = 'ACTIVE'
      <!-- 可选条件，如果传进来的参数有 id 属性，则加上 id 查询条件 -->
        <if test="id != null ">
           and id = #{id}
        </if>
    </select>
</mapper>
```

以上语句提供了一个可选的根据 id 查找 Employee 的功能。如果没有传入 id，那么所有处于“ACTIVE”状态的 Employee 都会被返回；反之若传入了 id，那么就会把查找 id 内容的 Employee 结果返回。

程序清单：codes/10/DynamicSQLTest/src/org/fkit/mapper/EmployeeMapper.java

```
import java.util.HashMap;
```

```
import java.util.List;
import org.fkit.domain.Employee;
public interface EmployeeMapper {
    List<Employee> selectEmployeeByIdLike(HashMap<String, Object> params);
}
```

以上代码提供了一个和 EmployeeMapper.xml 中的 select 元素的 id 同名的方法，需要注意的是，selectEmployeeByIdLike 接受一个 HashMap 作为参数。

在 MyBatis 中，#{id}表达式获取参数有两种方式：一是从 HashMap 中获取集合中的 property 对象；二是从 JavaBean 中获取 property 对象。

程序清单：codes/10/DynamicSQLTest/src/org/fkit/mapper/DynamicSQLTest.java

```
import java.util.ArrayList;
import java.util.HashMap;
import java.util.List;
import org.apache.ibatis.session.SqlSession;
import org.fkit.domain.Employee;
import org.fkit.factory.FKSqlSessionFactory;
import org.fkit.mapper.EmployeeMapper;

public class DynamicSQLTest {

    public static void main(String[] args) {

        // 定义 SqlSession 变量
        SqlSession sqlSession = null;
        try {
            // 创建 SqlSession 实例
            sqlSession = FKSqlSessionFactory.getSqlSession();

            DynamicSQLTest t = new DynamicSQLTest();

            t.testSelectEmployeeByIdLike(sqlSession);

            // 提交事务
            sqlSession.commit();
        } catch (Exception e) {
            // 回滚事务
            sqlSession.rollback();
            e.printStackTrace();
        }finally {
            // 关闭 SqlSession
            if(sqlSession != null)
                sqlSession.close();
        }

    }

// 测试<select id="selectEmployeeByIdLike" ...>
public void testSelectEmployeeByIdLike(SqlSession sqlSession){
    // 获得 EmployeeMapper 接口的代理对象
    EmployeeMapper em = sqlSession.getMapper(EmployeeMapper.class);
    // 创建一个 HashMap 存储参数
    HashMap<String, Object> params = new HashMap<String, Object>();
    // 设置 id 属性
    params.put("id", 1);
    // 调用 EmployeeMapper 接口的 selectEmployeeByIdLike 方法
    List<Employee> list = em.selectEmployeeByIdLike(params);
    // 查看查询结果
    list.forEach(employee -> System.out.println(employee));
}
```

项目的 mybatis-config.xml 和 log4j.xml 文件内容请参考配套资源文件，此处不再赘述。

运行 DynamicSQLTest 类的 main 方法，其通过 SqlSession 的 getMapper(Class<T> type)方法获得 mapper 接口的代理对象 EmployeeMapper。调用 selectEmployeeByIdLike 方法时会执行 EmployeeMapper.xml 中<select id="selectEmployeeByIdLike".../>元素中定义的 SQL 语句。控制台显示如下：

```
DEBUG [main] ==>  Preparing: SELECT * FROM tb_employee WHERE state = 'ACTIVE'
DEBUG [main] ==> Parameters:
DEBUG [main] <==      Total: 2
Employee [id=1, loginname=jack, password=123456, name=杰克, sex=男, age=26, phone=13902019999, sal=9800.0, state=ACTIVE]
```

可以看到，执行的 SQL 语句中因为传入了 id 属性，所以 SQL 语句中包含了“and id = ?”，查询返回的 Emplyee 对象就是 id 为 1 的对象。

接下来注释以下代码：

```
// params.put("id", 1);
```

再次执行 main 方法，控制台显示如下：

```
DEBUG [main] ==>  Preparing: SELECT * FROM tb_employee WHERE state = 'ACTIVE'
DEBUG [main] ==> Parameters:
DEBUG [main] <==      Total: 4
Employee [id=1, loginname=jack, password=123456, name=杰克, sex=男, age=26, phone=13902019999, sal=9800.0, state=ACTIVE]
Employee [id=2, loginname=rose, password=123456, name=露丝, sex=女, age=21, phone=13902018888, sal=6800.0, state=ACTIVE]
Employee [id=3, loginname=tom, password=123456, name=汤姆, sex=男, age=25, phone=13902017777, sal=8800.0, state=ACTIVE]
Employee [id=4, loginname=alice, password=123456, name=爱丽丝, sex=女, age=20, phone=13902016666, sal=5800.0, state=ACTIVE]
```

可以看到，由于传递的 HashMap 中没有 id 属性，故执行的 SQL 语句中不再包含“and id = ?”，查询返回了所有 state = ACTIVE 的数据。

如果想通过两个或多个条件搜索该怎么办呢？很简单，只要多加入一个或多个条件即可。

程序清单：codes/10/DynamicSQLTest/src/org/fkit/mapper/EmployeeMapper.xml

```
<select id="selectEmployeeByLoginLike"
     resultType="org.fkit.domain.Employee">
     SELECT * FROM tb_employee WHERE state = 'ACTIVE'
   <!-- 两个可选条件，例如登录功能的登录名和密码查询 -->
     <if test="loginname != null and password != null">
        and loginname = #{loginname} and password = #{password}
     </if>
  </select>
```

程序清单：codes/10/DynamicSQLTest/src/org/fkit/mapper/EmployeeMapper.java

```
List<Employee> selectEmployeeByLoginLike(HashMap<String, Object> params);
```

程序清单：codes/10/DynamicSQLTest/src/org/fkit/mapper/DynamicSQLTest.java

```
public void testSelectEmployeeByLoginLike(SqlSession sqlSession){
        EmployeeMapper em = sqlSession.getMapper(EmployeeMapper.class);
        HashMap<String, Object> params = new HashMap<String, Object>();
        // 设置 loginname 和 password 属性
        params.put("loginname", "jack");
        params.put("password", "123456");
        List<Employee> list = em.selectEmployeeByLoginLike(params);
        list.forEach(employee -> System.out.println(employee));
}
```

运行 DynamicSQLTest 类的 main 方法，测试 selectEmployeeByLoginLike()方法，控制台显示如下：

```
    DEBUG [main] ==>  Preparing: SELECT * FROM tb_employee WHERE state = 'ACTIVE' and
loginname = ? and password = ?
    DEBUG [main] ==> Parameters: jack(String), 123456(String)
    DEBUG [main] <==      Total: 1
    Employee  [id=1,  loginname=jack,  password=123456,  name=杰克,  sex=男,  age=26,
phone=13902019999, sal=9800.0, state=ACTIVE]
```

可以看到，执行的 SQL 语句中因为传入了 loginname 和 password 属性，故 SQL 语句中包含了“and loginname = ? and password = ?”，查询返回的 Emplyee 对象就是 loginname 是 jack，并且 password 是 123456 的对象。

10.2.2　choose（when、otherwise）

有些时候，我们不想用所有的条件语句，而只想从中择其一二。针对这种情况，MyBatis 提供了 choose 元素，它有点像 Java 中的 switch 语句。

还是上面的例子，但是这次变为提供了 id 就按 id 查找，提供了 loginname 和 password 就按 loginname 和 password 查找，若两者都没有提供，就返回所有 sex 等于男的 Employee。

程序清单：codes/10/DynamicSQLTest/src/org/fkit/mapper/EmployeeMapper.xml

```
    <select id="selectEmployeeChoose"
         parameterType="hashmap"
         resultType="org.fkit.domain.Employee">
         SELECT * FROM tb_employee WHERE state = 'ACTIVE'
         <!-- 如果传入了id，就根据id查询，没有传入id就根据loginname和password查询，否则查
询sex等于男的数据 -->
         <choose>
            <when test="id != null">
               and id = #{id}
            </when>
            <when test="loginname != null and password != null">
               and loginname = #{loginname} and password = #{password}
            </when>
            <otherwise>
               and sex = '男'
            </otherwise>
         </choose>
      </select>
```

程序清单：codes/10/DynamicSQLTest/src/org/fkit/mapper/EmployeeMapper.java

```
    List<Employee> selectEmployeeChoose(HashMap<String, Object> params);
```

程序清单：codes/10/DynamicSQLTest/src/org/fkit/mapper/DynamicSQLTest.java

```
    public void testSelectEmployeeChoose(SqlSession sqlSession){
          EmployeeMapper em = sqlSession.getMapper(EmployeeMapper.class);
          HashMap<String, Object> params = new HashMap<String, Object>();
          // 设置id属性
          params.put("id", 1);
          params.put("loginname", "jack");
          params.put("password", "123456");
          List<Employee> list = em.selectEmployeeChoose(params);
          list.forEach(employee -> System.out.println(employee));
      }
```

运行 DynamicSQLTest 类的 main 方法，测试 selectEmployeeChoose()方法，控制台显示如下：

```
DEBUG [main] ==>  Preparing: SELECT * FROM tb_employee WHERE state = 'ACTIVE' and id = ?
```

```
DEBUG [main] ==> Parameters: 1(Integer)
DEBUG [main] <==      Total: 1
Employee [id=1, loginname=jack, password=123456, name=杰克, sex=男, age=26,
phone=13902019999, sal=9800.0, state=ACTIVE]
```

可以看到，HashMap 里面传递的参数包括了 id、loginname 和 password，执行的 SQL 语句中因为使用了 choose 标签，所以 SQL 语句执行的是按照 id 查询，查询返回的 Emplyee 对象就是 id 为 1 的对象。

接下来注释以下代码：

```
    // params.put("id", 1);
```

再次执行 main 方法，控制台显示如下：

```
    DEBUG [main] ==>  Preparing: SELECT * FROM tb_employee WHERE state = 'ACTIVE' and
loginname = ? and password = ?
    DEBUG [main] ==> Parameters: jack(String), 123456(String)
    DEBUG [main] <==      Total: 1
    Employee [id=1, loginname=jack, password=123456, name=杰克, sex=男, age=26,
phone=13902019999, sal=9800.0, state=ACTIVE]
```

可以看到，HashMap 里面传递的参数只包括 loginname 和 password，所以 SQL 语句是按照 loginname 和 password 查找，查询返回的 Emplyee 对象就是 loginname 是 jack，并且 password 是 123456 的对象。

接下来注释以下代码：

```
    // params.put("id", 1);
    // params.put("loginname", "jack");
    // params.put("password", "123456");
```

再次执行 main 方法，控制台显示如下：

```
    DEBUG [main] ==>  Preparing: SELECT * FROM tb_employee WHERE state = 'ACTIVE' and
sex = '男'
    DEBUG [main] ==> Parameters:
    DEBUG [main] <==      Total: 2
    Employee [id=1, loginname=jack, password=123456, name=杰克, sex=男, age=26,
phone=13902019999, sal=9800.0, state=ACTIVE]
    Employee [id=3, loginname=tom, password=123456, name=汤姆, sex=男, age=25,
phone=13902017777, sal=8800.0, state=ACTIVE]
```

可以看到，由于没有传递任何参数，故 SQL 语句执行的是< otherwise >元素里面的查询条件，即“and sex = ‘男’”，查询返回的 Emplyee 对象就是“sex= ‘男’”的对象。

10.2.3 where

前面几个例子已经很好地解决了动态 SQL 问题。现在回到之前的 if 示例，这次我们将 state = 'ACTIVE'也设置成动态的条件，看看会发生什么。

程序清单：codes/10/DynamicSQLTest/src/org/fkit/mapper/EmployeeMapper.xml

```
    <select id="selectEmployeeByIdLike"
        resultType="org.fkit.domain.Employee">
        SELECT * FROM tb_employee WHERE
        <if test="state != null ">
           state = #{state}
        </if>
        <if test="id != null ">
           and id = #{id}
        </if>
      </select>
```

如果传入 state 参数，则执行正常。

如果没有传入参数，则会执行 SQL 语句：

```
SELECT * FROM tb_employee WHERE
```

如果只是传入 id，则会执行 SQL 语句：

```
SELECT * FROM tb_employee WHERE and id = ?
```

也就是说，如果没有传入 state 参数，会导致执行失败。这个问题不能简单地用条件语句来解决。MyBatis 有一个简单的处理方法，只要简单地修改就能得到想要的效果：

程序清单：codes/10/DynamicSQLTest/src/org/fkit/mapper/EmployeeMapper.xml

```
<select id="selectEmployeeLike"
      resultType="org.fkit.domain.Employee">
      SELECT * FROM tb_employee
      <where>
         <if test="state != null ">
            state = #{state}
         </if>
         <if test="id != null ">
            and id = #{id}
         </if>
         <if test="loginname != null and password != null">
            and loginname = #{loginname} and password = #{password}
         </if>
      </where>
   </select>
```

where 元素知道只有在一个以上的 if 条件有值的情况下才会插入 WHERE 子句。而且，若最后的内容是“AND”或“OR”开头，则 where 元素也知道如何将它们去除。

程序清单：codes/10/DynamicSQLTest/src/org/fkit/mapper/EmployeeMapper.java

```
List<Employee> selectEmployeeLike (HashMap<String, Object> params);
```

程序清单：codes/10/DynamicSQLTest/src/org/fkit/mapper/DynamicSQLTest.java

```
public void testSelectEmployeeLike(SqlSession sqlSession){
   EmployeeMapper em = sqlSession.getMapper(EmployeeMapper.class);
   HashMap<String, Object> params = new HashMap<String, Object>();
   // 设置 id、loginname 和 password 属性
   params.put("id", 1);
   params.put("loginname", "jack");
   params.put("password", "123456");
   List<Employee> list = em.selectEmployeeLike(params);
   list.forEach(employee -> System.out.println(employee));
}
```

运行 DynamicSQLTest 类的 main 方法，测试 selectEmployeeLike ()方法，控制台显示如下：

```
DEBUG [main] ==>  Preparing: SELECT * FROM tb_employee WHERE id = ? and loginname
= ? and password = ?
DEBUG [main] ==> Parameters: 1(Integer), jack(String), 123456(String)
DEBUG [main] <==      Total: 1
Employee [id=1, loginname=jack, password=123456, name=杰克, sex=男, age=26,
phone=13902019999, sal=9800.0, state=ACTIVE]
```

可以看到，HashMap 里面传递的参数只包括 id、loginname 和 password，所以 SQL 语句是按照 id、loginname 和 password 查找，同时自动在 SQL 语句中添加了 where 关键字，查询返回的 Emplyee 对象就是 id 是 1，loginname 是 jack，并且 password 是 123456 的对象。

接下来注释以下代码：

```
// params.put("id", 1);
```

```
    // params.put("loginname", "jack");
    // params.put("password", "123456");
```

再次执行 main 方法，控制台显示如下：

```
    DEBUG [main] ==>  Preparing: SELECT * FROM tb_employee
    DEBUG [main] ==> Parameters:
    DEBUG [main] <==      Total: 4
    Employee [id=1, loginname=jack, password=123456, name=杰克, sex=男, age=26,
phone=13902019999, sal=9800.0, state=ACTIVE]
    Employee [id=2, loginname=rose, password=123456, name=露丝, sex=女, age=21,
phone=13902018888, sal=6800.0, state=ACTIVE]
    Employee [id=3, loginname=tom, password=123456, name=汤姆, sex=男, age=25,
phone=13902017777, sal=8800.0, state=ACTIVE]
    Employee [id=4, loginname=alice, password=123456, name=爱丽丝, sex=女, age=20,
phone=13902016666, sal=5800.0, state=ACTIVE]
```

当没有传入参数时，MyBatis 查询返回了所有数据。

10.2.4 set

关于动态更新语句还可以使用 set 元素。set 元素可以用于动态包含需要更新的列，而舍去其他的。

程序清单：codes/10/DynamicSQLTest/src/org/fkit/mapper/EmployeeMapper.xml

```
    <!-- 根据 id 查询员工信息 -->
    <select  id="selectEmployeeWithId"  parameterType="int"  resultType="org.fkit.
domain.Employee">
        SELECT * FROM tb_employee where id = #{id}
    </select>
    <!-- 动态更新员工信息 -->
    <update id="updateEmployeeIfNecessary" parameterType="org.fkit.domain.Employee">
      update tb_employee
        <set>
          <if test="loginname != null">loginname=#{loginname},</if>
          <if test="password != null">password=#{password},</if>
          <if test="name != null">name=#{name},</if>
          <if test="sex != null">sex=#{sex},</if>
          <if test="age != null">age=#{age},</if>
          <if test="phone != null">phone=#{phone},</if>
          <if test="sal != null">sal=#{sal},</if>
          <if test="state != null">state=#{state}</if>
        </set>
      where id=#{id}
    </update>
```

set 元素会动态前置 SET 关键字，同时也会消除无关的逗号，因为使用了条件语句之后很可能就会在生成的赋值语句的后面留下这些逗号。

程序清单：codes/10/DynamicSQLTest/src/org/fkit/mapper/EmployeeMapper.java

```
    // 根据 id 查询员工
    Employee selectEmployeeWithId(Integer id);
    // 动态更新员工
    void updateEmployeeIfNecessary(Employee employee);
```

需要注意的是，updateEmployeeIfNecessary 方法传递的参数不是之前使用的 HashMap，而是一个 Employee 对象，因为通常在进行更新操作时都是先查询出一个实体对象再进行更新操作。

程序清单：codes/10/DynamicSQLTest/src/org/fkit/mapper/DynamicSQLTest.java

```
    public void testUpdateEmployeeIfNecessary(SqlSession sqlSession){
```

```
        EmployeeMapper em = sqlSession.getMapper(EmployeeMapper.class);
        // 查询 id 为 4 的员工信息
        Employee employee = em.selectEmployeeWithId(4);
        // 设置需要修改的属性
        employee.setLoginname("mary");
        employee.setPassword("123");
        employee.setName("玛丽");
        em.updateEmployeeIfNecessary(employee);
    }
```

运行 DynamicSQLTest 类的 main 方法，测试 updateEmployeeIfNecessary ()方法，控制台显示如下：

```
    DEBUG [main] ==>  Preparing: SELECT * FROM tb_employee where id = ?
    DEBUG [main] ==> Parameters: 4(Integer)
    DEBUG [main] <==      Total: 1
    DEBUG [main] ==>  Preparing: update tb_employee SET loginname=?, password=?, name=?,
sex=?, age=?, phone=?, sal=?, state=? where id=?
    DEBUG [main] ==> Parameters: mary(String), 123(String), 玛丽(String), 女(String),
20(Integer), 13902016666(String), 5800.0(Double), ACTIVE(String), 4(Integer)
    DEBUG [main] <==    Updates: 1
```

可以看到，首先执行了一条查询语句，查询 id 为 4 的员工，之后执行了一条 update 语句，根据传入的 Employee 对象更新员工信息。

切换到数据库，可以看到 id 为 4 的员工信息已经更新。

10.2.5 foreach

关于动态 SQL 另外一个常用的操作就是需要对一个集合进行遍历，通常发生在构建 IN 条件语句时。

程序清单：codes/10/DynamicSQLTest/src/org/fkit/mapper/EmployeeMapper.xml

```
<select id="selectEmployeeIn" resultType="org.fkit.domain.Employee">
  SELECT *
  FROM tb_employee
  WHERE ID in
  <foreach item="item" index="index" collection="list"
      open="(" separator="," close=")">
    #{item}
  </foreach>
</select>
```

foreach 元素的功能非常强大，它允许指定一个集合，声明可以用在元素体内的集合项和索引变量。它也允许指定开闭匹配的字符串以及在迭代中间放置分隔符。这个元素是很智能的，因此它不会随机地附加多余的分隔符。

程序清单：codes/10/ DynamicSQLTest /src/org/fkit/mapper/ EmployeeMapper.java

```
// 根据传入的 id 集合查询员工
List<Employee> selectEmployeeIn(List<Integer> ids);
```

selectEmployeeIn 方法传入的参数是一个 List 集合，该集合中的每一个 Integer 元素表示需要查询的员工的 id。

程序清单：codes/10/DynamicSQLTest/src/org/fkit/mapper/DynamicSQLTest.java

```
public void testSelectEmployeeIn(SqlSession sqlSession){
    EmployeeMapper em = sqlSession.getMapper(EmployeeMapper.class);
    // 创建 List 集合
    List<Integer> ids = new ArrayList<Integer>();
```

```
    // 往 List 集合中添加两个测试数据
    ids.add(1);
    ids.add(2);
    List<Employee> list = em.selectEmployeeIn(ids);
    list.forEach(employee -> System.out.println(employee));
}
```

运行 DynamicSQLTest 类的 main 方法，测试 selectEmployeeIn ()方法，控制台显示如下：

```
DEBUG [main] ==>  Preparing: SELECT * FROM tb_employee WHERE ID in ( ? , ? )
DEBUG [main] ==> Parameters: 1(Integer), 2(Integer)
DEBUG [main] <==      Total: 2
Employee [id=1, loginname=jack, password=123456, name=杰克, sex=男, age=26, phone=13902019999, sal=9800.0, state=ACTIVE]
Employee [id=2, loginname=rose, password=123456, name=露丝, sex=女, age=21, phone=13902018888, sal=6800.0, state=ACTIVE]
```

可以看到，执行的 SQL 语句是一个 in 条件语句，返回的是 List 集合中的 id 为 1 和 2 的员工数据。

10.2.6 bind

bind 元素可以从 OGNL 表达式创建一个变量并将其绑定到上下文。

程序清单：codes/10/DynamicSQLTest/src/org/fkit/mapper/EmployeeMapper.xml

```
<select id="selectEmployeeLikeName"  resultType="org.fkit.domain.Employee">
  <bind name="pattern" value="'%' + _parameter.getName() + '%'" />
   SELECT * FROM tb_employee
   WHERE loginname LIKE #{pattern}
</select>
```

程序清单：codes/10/DynamicSQLTest/src/org/fkit/mapper/EmployeeMapper.java

```
// 根据传入的参数进行模糊查询
List<Employee> selectEmployeeLikeName(Employee employee);
```

程序清单：codes/10/DynamicSQLTest/src/org/fkit/mapper/DynamicSQLTest.java

```
public void testSelectEmployeeLikeName(SqlSession session){
    EmployeeMapper em = session.getMapper(EmployeeMapper.class);
    Employee employee = new Employee();
    // 设置模糊查询的参数
    employee.setName("o");
    List<Employee> list = em.selectEmployeeLikeName(employee);
    System.out.println(list);
}
```

运行 DynamicSQLTest 类的 main 方法，测试 selectEmployeeLikeName ()方法，控制台显示如下：

```
DEBUG [main] ==>  Preparing: SELECT * FROM tb_employee WHERE loginname LIKE ?
DEBUG [main] ==> Parameters: %o%(String)
DEBUG [main] <==      Total: 2
Employee [id=2, loginname=rose, password=123456, name=露丝, sex=女, age=21, phone=13902018888, sal=6800.0, state=ACTIVE]
Employee [id=3, loginname=tom, password=123456, name=汤姆, sex=男, age=25, phone=13902017777, sal=8800.0, state=ACTIVE]
```

可以看到，执行的 SQL 语句是一个 like 模糊查询语句，返回的是“loginname”包含“o”的所有数据。

10.3　MyBatis 调用存储过程

SQL 语句在执行的时候需要先编译，然后执行。数据库的存储过程（Procedure）是一组为了完成特定功能的 SQL 语句，编译后存储在数据库中，用户通过指定存储过程的名字并给定参数（如果该存储过程带有参数）来调用存储过程。使用存储过程可提高数据库执行速度，并且存储过程可以重复使用，从而减少数据库开发人员的工作量。

现如今大多数数据库都支持存储过程，包括 Oracle、MySQL 等，很多企业在项目开发中都使用存储过程，特别是金融行业和银行系统。下面重点介绍如何使用 MyBatis 调用存储过程。

示例：MyBatis 调用存储过程

本示例直接使用第 8 章创建的 tb_user 表、数据库脚本、User.java、mybatis-config.xml 和 log4j.xml，具体请参考第 8 章内容，此处不再赘述。

10.3.1　插入数据

在之前创建的 mybatis 数据库中创建一个插入 tb_user 表数据的存储过程。SQL 脚本如下：

```
# 插入数据的存储过程
DROP PROCEDURE IF EXISTS insert_user;
DELIMITER //
CREATE PROCEDURE insert_user
(OUT v_id INTEGER,IN v_name VARCHAR(18),IN v_sex VARCHAR(19),IN v_age INTEGER)
BEGIN
# 执行 insert 操作，values 中的值是传入的参数
INSERT INTO tb_user(NAME,sex,age) VALUES(v_name,v_sex,v_age);
# 调用 LAST_INSERT_ID 函数将插入成功自动生成的主键 id 值设置到变量 v_id 中
SET v_id = LAST_INSERT_ID();
END
//
```

以上存储过程代码解释如下：

（1）DELIMITER 是 MySQL 解释器，该段命令是否已经结束了，MySQL 是否可以执行了，因为在 MySQL 当中，默认情况下一遇到分号，它就要自动执行。

（2）在存储过程中使用参数时，除了参数名和参数类型之外，还建议指定参数的 mode（模式），可选值为 IN、OUT 和 INOUT 三种。入参使用 IN，出参（返回）使用 OUT，输入输出参数使用 INOUT。

（3）LAST_INSERT_ID()是 MySQL 的函数，和 AUTO_INCREMENT 属性一起使用，当往带有 AUTO_INCREMENT 属性字段的表中新增数据时，LAST_INSERT_ID()函数返回该字段的值。

提示

关于 MySQL 存储过程的知识，已超出了本书的范围，更多内容请读者自行参考 MySQL 的官方文档。

程序清单：codes/10/procedureTest/src/org/fkit/mapper/UserMapper.xml

```
<?xml version="1.0" encoding="UTF-8"?>
<!DOCTYPE mapper PUBLIC "-//mybatis.org//DTD Mapper 3.0//EN"
"http://mybatis.org/dtd/mybatis-3-mapper.dtd">
<mapper namespace="org.fkit.mapper.UserMapper">
  <insert id="saveUser" parameterType="org.fkit.domain.User" statementType=
"CALLABLE">
```

```
      {call insert_user(
      #{id,mode=OUT,jdbcType=INTEGER},
    #{name,mode=IN},#{sex,mode=IN},
    #{age,mode=IN}
      )}
  </insert>
</mapper>
```

<insert…/>元素调用名为“insert_user”的存储过程完成一个插入数据的操作。调用存储过程时，需要把 statementType 属性的值设置为“CALLABLE”。“call”是用来调用存储过程的关键字，需要注意的是，OUT 模式的参数必须指定 jdbcType，这是因为在 IN 模式下，MyBatis 提供了默认的 jdbcType，而在 OUT 模式下没有提供。

程序清单：codes/10/procedureTest/src/org/fkit/mapper/UserMapper.java

```
import java.util.List;
import org.fkit.domain.User;

public interface UserMapper {
    /**
     * 插入数据
     * @param user 包含要查询的 User 信息
     */
    void saveUser(User user);
}
```

程序清单：codes/10/procedureTest/src/org/fkit/test/ProcedureTest.java

```
import java.util.List;
import org.apache.ibatis.session.SqlSession;
import org.fkit.domain.User;
import org.fkit.factory.FKSqlSessionFactory;
import org.fkit.mapper.UserMapper;

public class ProcedureTest {

    public static void main(String[] args) {
        // 定义 SqlSession 变量
        SqlSession sqlSession = null;
        try {
            // 创建 SqlSession 实例
            sqlSession = FKSqlSessionFactory.getSqlSession();
            ProcedureTest t = new ProcedureTest();
            // 测试插入数据
            t.testSaveUser(sqlSession);
            // 提交事务
            sqlSession.commit();
        } catch (Exception e) {
            // 回滚事务
            sqlSession.rollback();
            e.printStackTrace();
        }finally {
            // 关闭 SqlSession
            if(sqlSession != null)
                sqlSession.close();
        }
    }

    // 测试插入数据
    public void testSaveUser(SqlSession sqlSession){
        // 获得 UserMapper 接口的代理对象
        UserMapper um = sqlSession.getMapper(UserMapper.class);
```

```
        User user = new User("test","男",23);
                // 调用保存 User 的方法
        um.saveUser(user);
        System.out.println("插入成功，返回的 id = " + user.getId());
    }

}
```

运行 ProcedureTest 类的 main 方法，测试 testSaveUser ()方法，插入一个用户数据并输出自动生成的 id 值。控制台显示如下：

```
DEBUG [main] ==>  Preparing: {call insert_user( ?,?,?,? )}
DEBUG [main] ==> Parameters: test(String), 男(String), 23(Integer)
DEBUG [main] <==    Updates: 1
插入成功，返回的 id = 1
```

可以看到，MyBatis 调用了名为“insert_user”的存储过程，并且传入了 4 个参数，插入成功后返回的 id 值为 1，打开数据库，可以看到 tb_user 表中插入了一条记录，如图 10.1 所示。

再使用“insert_user”存储过程向 tb_user 表中插入几条数据，用于接下来的测试。数据插入结果如图 10.2 所示。

ID	NAME	SEX	AGE
1	test	男	23
(Auto)	(NULL)	(NULL)	(NULL)

图 10.1 存储过程插入数据结果

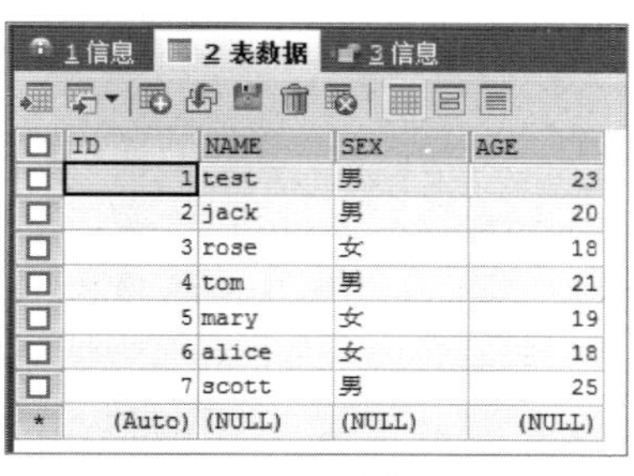

ID	NAME	SEX	AGE
1	test	男	23
2	jack	男	20
3	rose	女	18
4	tom	男	21
5	mary	女	19
6	alice	女	18
7	scott	男	25
(Auto)	(NULL)	(NULL)	(NULL)

图 10.2 测试数据

10.3.2 查询数据返回集合

在之前创建的 mybatis 数据库中创建一个查询 tb_user 表所有数据的存储过程。SQL 脚本如下：

```
#查询所有数据的存储过程
DROP PROCEDURE IF EXISTS select_user;
DELIMITER //
CREATE PROCEDURE select_user()
BEGIN
SELECT id,NAME,sex,age FROM tb_user;
END
//
```

在 UserMapper.xml 中加入调用名为“select_user”存储过程的<select.../>元素。

程序清单：codes/10/procedureTest/src/org/fkit/mapper/UserMapper.xml

```
<select id="selectUser" resultType="org.fkit.domain.User" statementType="CALLABLE">
  {call select_user()}
</select>
```

<select.../>元素调用名为“select_user”的存储过程查询所有 User 数据并返回 List，查询到的每一条数据会被封装到 User 对象，这和之前执行 SQL 语句返回数据的方式完全一致。

在 UserMapper.java 接口中加入抽象方法。

程序清单：codes/10/procedureTest/src/org/fkit/mapper/UserMapper.java

```
/**
 * 查询所有数据
 * @return List<User> 包含所有 User 数据的 List 集合
 */
List<User> selectUser();
```

在 procedureTest 类中加入测试方法。

程序清单：codes/10/procedureTest/src/org/fkit/test/ProcedureTest.java

```
// 测试查询所有数据
public void testSelectUser(SqlSession sqlSession){
    // 获得 UserMapper 接口的代理对象
    UserMapper um = sqlSession.getMapper(UserMapper.class);
    // 调用查询所有 User 数据的方法
    List<User> users = um.selectUser();
    // 遍历打印每一个 User 对象
    users.forEach(user -> System.out.println(user));
}
```

修改 main 方法测试查询所有数据。

```
// 测试插入数据
// t.testSaveUser(sqlSession);
// 测试查询所有数据
t.testSelectUser(sqlSession);
```

运行 ProcedureTest 类的 main 方法，测试 testSelectUser ()方法，调用存储过程查询所有用户信息。控制台显示如下：

```
DEBUG [main] ==>  Preparing: {call select_user()}
DEBUG [main] ==> Parameters:
DEBUG [main] <==      Total: 7
DEBUG [main] <==    Updates: 0
User [id=1, name=test, sex=男, age=23]
User [id=2, name=jack, sex=男, age=20]
User [id=3, name=rose, sex=女, age=18]
User [id=4, name=tom, sex=男, age=21]
User [id=5, name=mary, sex=女, age=19]
User [id=6, name=alice, sex=女, age=18]
User [id=7, name=scott, sex=男, age=25]
```

10.3.3　根据 id 查询数据返回对象

在之前创建的 mybatis 数据库中创建一个根据 id 查询 tb_user 表对应数据的存储过程。SQL 脚本如下：

```
# 根据 id 查询数据的存储过程
DROP PROCEDURE IF EXISTS select_user_by_id;
DELIMITER //
CREATE PROCEDURE select_user_by_id(IN v_id INTEGER)
BEGIN
SELECT id,NAME,sex,age FROM tb_user WHERE id = v_id;
END
//
```

在 UserMapper.xml 中加入调用名为“select_user_by_id”的存储过程的<select…/>元素。

```
<select id="selectUserById" parameterType="int" resultType=
"org.fkit.domain.User" statementType="CALLABLE">
```

```
        {call select_user_by_id(#{id,mode=IN})}
    </select>
```

<select.../>元素调用名为“select_user_by_id”的存储过程根据 id 查询对应的数据并返回 User 对象，这和之前执行 SQL 语句返回数据的方式完全一致。

在 UserMapper.java 接口中加入抽象方法。

程序清单：codes/10/procedureTest/src/org/fkit/mapper/UserMapper.java

```
/**
 * 根据 id 查询对应的数据
 * @param id 用户的 id
 * @return User 包含用户信息的 User 对象
 */
User selectUserById(Integer id);
```

在 procedureTest 类中加入测试方法。

程序清单：codes/10/procedureTest/src/org/fkit/test/ProcedureTest.java

```
// 测试根据 id 查询对应的数据
public void testSelectUserById(SqlSession sqlSession){
    // 获得 UserMapper 接口的代理对象
    UserMapper um = sqlSession.getMapper(UserMapper.class);
    // 调用根据 id 查询 User 数据的方法
    User user = um.selectUserById(1);
    System.out.println(user);
}
```

修改 main 方法测试查询所有数据。

```
// 测试插入数据
// t.testSaveUser(sqlSession);
// 测试查询所有数据
// t.testSelectUser(sqlSession);
// 测试根据 id 查询对应的数据
t.testSelectUserById(sqlSession);
```

运行 ProcedureTest 类的 main 方法，测试 testSelectUserById ()方法，调用存储过程根据 id 查询对应的用户信息。控制台显示如下：

```
DEBUG [main] ==>  Preparing: {call select_user_by_id(?)}
DEBUG [main] ==> Parameters: 1(Integer)
DEBUG [main] <==      Total: 1
DEBUG [main] <==    Updates: 0
User [id=1, name=test, sex=男, age=23]
```

10.3.4 修改数据

在之前创建的 mybatis 数据库中创建一个修改 tb_user 表数据的存储过程。SQL 脚本如下：

```
# 修改数据的存储过程
DROP PROCEDURE IF EXISTS update_user;
DELIMITER //
CREATE PROCEDURE update_user
(IN v_id INTEGER,IN v_name VARCHAR(18),IN v_sex VARCHAR(19),IN v_age INTEGER)
BEGIN
UPDATE tb_user SET NAME = v_name,sex = v_sex,age = v_age
WHERE id = v_id;
END
//
```

在 UserMapper.xml 中加入调用名为“update_user”的存储过程的<update.../>元素。

程序清单：codes/10/procedureTest/src/org/fkit/mapper/UserMapper.xml

```
    <update id="modifyUser" parameterType="org.fkit.domain.User" statementType=
"CALLABLE">
         {call update_user(
         #{id,mode=IN},#{name,mode=IN},#{sex,mode=IN},#{age,mode=IN}
         )}
      </update>
```

<update…/>元素调用名为“update_user”的存储过程修改 User 数据。

在 UserMapper.java 接口中加入抽象方法。

程序清单：codes/10/procedureTest/src/org/fkit/mapper/UserMapper.java

```
/**
 * 修改数据
 * @param user 包含要修改的 User 信息
 */
void modifyUser(User user);
```

在 procedureTest 类中加入测试方法。

程序清单：codes/10/procedureTest/src/org/fkit/test/ProcedureTest.java

```
// 测试修改数据
public void testModifyUser(SqlSession sqlSession){
    // 获得 UserMapper 接口的代理对象
    UserMapper um = sqlSession.getMapper(UserMapper.class);
    // 查询出 id 为 1 的 User 对象
    User user = um.selectUserById(1);
    // 修改对象的属性
    user.setName("fkit");
    user.setSex("男");
    user.setAge(19);
    // 调用修改数据的方法
    um.modifyUser(user);
}
```

修改 main 方法测试查询所有数据。

```
// 测试插入数据
// t.testSaveUser(sqlSession);
// 测试查询所有数据
// t.testSelectUser(sqlSession);
// 测试根据 id 查询对应的数据
// t.testSelectUserById(sqlSession);
// 测试修改数据
t.testModifyUser(sqlSession);
```

运行 ProcedureTest 类的 main 方法，测试 testModifyUser ()方法，调用存储过程修改数据。控制台显示如下：

```
DEBUG [main] ==>  Preparing: {call select_user_by_id(?)}
DEBUG [main] ==> Parameters: 1(Integer)
DEBUG [main] <==      Total: 1
DEBUG [main] <==    Updates: 0
DEBUG [main] ==>  Preparing: {call update_user( ?,?,?,? )}
DEBUG [main] ==> Parameters: 1(Integer), fkit(String), 男(String), 19(Integer)
DEBUG [main] <==    Updates: 1
```

执行 testModifyUser()方法时，先调用“select_user_by_id”存储过程查询 id 为 1 的 User 数据，之后调用“update_user”存储过程修改该 User 数据。打开数据库，可以看到执行修改存储过程之后 tb_user 表的记录，如图 10.3 所示。

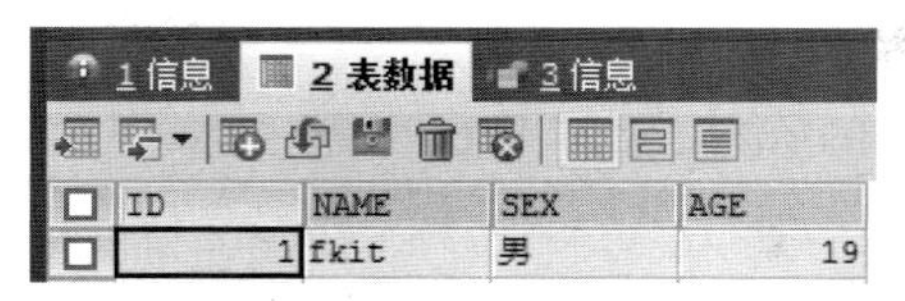

ID	NAME	SEX	AGE
1	fkit	男	19

图 10.3　修改后的记录

10.3.5　删除数据

在之前创建的 mybatis 数据库中创建一个删除 tb_user 表数据的存储过程。SQL 脚本如下：

```
# 删除数据的存储过程
DROP PROCEDURE IF EXISTS delete_user_by_id;
DELIMITER //
CREATE PROCEDURE delete_user_by_id(IN v_id INTEGER)
BEGIN
DELETE FROM tb_user WHERE id = v_id;
END
//
```

在 UserMapper.xml 中加入调用名为"delete_user_by_id"的存储过程的<delete…/>元素。

程序清单：codes/10/procedureTest/src/org/fkit/mapper/UserMapper.xml

```
<delete id="removeUser" parameterType="org.fkit.domain.User" statementType=
"CALLABLE">
    {call delete_user_by_id(#{id,mode=IN})}
</delete>
```

<delete…/>元素调用名为"delete_user_by_id"的存储过程根据 id 删除 User 数据。

在 UserMapper.java 接口中加入抽象方法。

程序清单：codes/10/procedureTest/src/org/fkit/mapper/UserMapper.java

```
/**
 * 删除数据
 * @param user 包含要修改的 User 信息
 */
void removeUser(User user);
```

在 procedureTest 类中加入测试方法。

程序清单：codes/10/procedureTest/src/org/fkit/test/ProcedureTest.java

```
// 测试删除数据
public void testRemoveUser(SqlSession sqlSession){
    // 获得 UserMapper 接口的代理对象
    UserMapper um = sqlSession.getMapper(UserMapper.class);
    // 查询出 id 为 1 的 User 对象
    User user = um.selectUserById(1);
    // 调用删除数据的方法
    um.removeUser(user);
}
```

修改 main 方法测试查询所有数据。

```
// 测试插入数据
// t.testSaveUser(sqlSession);
// 测试查询所有数据
// t.testSelectUser(sqlSession);
// 测试根据 id 查询对应的数据
// t.testSelectUserById(sqlSession);
// 测试修改数据
// t.testModifyUser(sqlSession);
// 测试删除数据
```

```
    t.testRemoveUser(sqlSession);
```

运行 ProcedureTest 类的 main 方法，测试 testRemoveUser ()方法，调用存储过程删除数据。控制台显示如下：

```
    DEBUG [main] ==>  Preparing: {call select_user_by_id(?)}
    DEBUG [main] ==> Parameters: 1(Integer)
    DEBUG [main] <==      Total: 1
    DEBUG [main] <==    Updates: 0
    DEBUG [main] ==>  Preparing: {call delete_user_by_id(?)}
    DEBUG [main] ==> Parameters: 1(Integer)
    DEBUG [main] <==    Updates: 1
```

执行 testRemoveUser ()方法时，先调用“select_user_by_id”存储过程查询 id 为 1 的 User 数据，之后调用“delete_user_by_id”存储过程将该 User 数据删除。打开数据库，可以看到 tb_user 表中 id 为 1 的数据已经被删除了。

常见的存储过程大概就是以上几种，包括插入、查询、修改、删除，读者掌握了这几种 MyBatis 调用存储过程的方法后，在实际项目开发中几乎可以应对所有的开发场景了。

10.4 MyBatis 事务管理

使用 MyBatis 可以很方便地以面向对象的方式进行数据库访问。在所有的 Java 语言数据库框架中，数据库的事务管理都是非常重要的一个方面。同时我们也经常需要合理地利用缓存来加快数据库的查询，进而有效地提升数据库的性能。下面将重点介绍 MyBatis 的事务管理。

10.4.1 事务的概念

每个业务逻辑都是由一系列数据库访问完成的，这一系列数据库访问可能会修改多条数据记录，这一系列修改应该是一个整体，绝不能仅修改其中的几条数据记录。也就是说，多个数据库原子访问应该被绑定成一个整体，这就是事务。事务是一个最小的逻辑执行单元，整个事务不能分开执行，要么同时执行，要么同时放弃执行。

事务是一步或几步操作组成的逻辑执行单元，这些基本操作作为一个整体执行单元，它们要么全部执行，要么全部取消执行，绝不能仅仅执行一部分。一般而言，一个用户请求对应一个业务逻辑方法，一个业务逻辑方法往往具有逻辑上的原子性，此时应该使用事务。例如一个转账操作，对应修改两个账户的余额，这两个账户的修改要么同时生效，要么同时取消，同时生效是转账成功，同时取消是转账失败；但不可只修改其中一个账户，那将破坏数据库的完整性。

通常来讲，事务具备 4 个特性：原子性（Atomicity）、一致性（Consistency）、隔离性（Isolation）和持续性（Durability）。这 4 个特性也简称为 ACID 特性，具体描述如下：

- **原子性（Atomicity）**。事务是应用中最小的执行单位，就如原子是自然界最小颗粒，具有不可再分的特征一样。事务是应用中不可再分的最小逻辑执行体。
- **一致性（Consistency）**。事务执行的结果，必须使数据库从一种一致性状态，变到另一种一致性状态。当数据库只包含事务成功提交的结果时，数据库处于一致性状态。如果系统运行发生中断，某个事务尚未完成而被迫中断，而该未完成的事务对数据库所做的修改已被写入数据库，此时，数据库就处于一种不正确的状态。比如银行在两个账户之间转账：从 A 账户向 B 账户转入 1000 元。系统先减少 A 账户的 1000 元，然后再为 B 账户增加 1000 元。如果全部执行成功，数据库就处于一致性状态。如果仅执行完 A 账户金额的修改，而没有增加 B 账户的金额，则数据库就处于不一致性状态。

因此，一致性是通过原子性来保证的。
- **隔离性（Isolation）**。各个事务的执行互不干扰，任意一个事务的内部操作对其他并发的事务，都是隔离的，即并发执行的事务之间不能互相影响。
- **持续性（Durability）**。持续性也被称为持久性（Persistence），指事务一旦提交，对数据所做的任何改变都要记录到永久存储器中，通常就是保存到物理数据库。

10.4.2 Transaction 接口

MyBatis 事务设计的重点是 org.apache.ibatis.transaction.Transaction 接口，Transaction 接口有两个实现类，分别是 org.apache.ibatis.transaction.jdbc.JdbcTransaction 和 org.apache.ibatis.transaction.managed.ManagedTransaction。同时 MyBatis 还设计了 org.apache.ibatis.transaction.TransactionFactory 接口和两个实现类 org.apache.ibatis.transaction.jdbc.JdbcTransactionFactory 和 org.apache.ibatis.transaction.managed. ManagedTransactionFactory 用来获取事务的实例对象。

对数据库的事务而言，应该具有以下几个动作：创建（create）、提交（commit）、回滚（rollback）、关闭（close）。对应地，MyBatis 将事务抽象成了 Transaction 接口。该接口源代码如下：

```
public interface Transaction {

  // 获取数据库连接
  Connection getConnection() throws SQLException;

  // 提交
  void commit() throws SQLException;

  // 回滚
  void rollback() throws SQLException;

  // 关闭数据库连接
  void close() throws SQLException;

}
```

MyBatis 的事务管理分为两种形式：
- **使用 JDBC 的事务管理机制**。即利用 java.sql.Connection 对象完成对事务的提交（commit()）、回滚（rollback()）和关闭（close()）等操作。
- **使用 MANAGED 的事务管理机制**。对于这种机制，MyBatis 自身不会去实现事务管理，而是让容器如 WebLogic、JBoss 等来实现对事务的管理。

10.4.3 事务的配置创建和使用

1. 事务的配置

我们在使用 MyBatis 时，一般会在 MyBatis 的根配置文件 mybatis-config.xml 中定义类似如下的信息：

```
<environment id="mysql">
    <!-- 指定事务管理类型，type="JDBC"指直接简单使用了 JDBC 的提交和回滚设置
         type="MANAGED"指让容器实现对事务的管理
    -->
     <transactionManager type="JDBC"/>
      <dataSource type="POOLED">
        <property name="driver" value="${driver}"/>
        <property name="url" value="${url}"/>
```

```
        <property name="username" value="${username}"/>
        <property name="password" value="${password}"/>
      </dataSource>
    </environment>
```

<environment >元素定义了连接某个数据库的信息，其子元素< transactionManager > 的 type 决定我们用什么类型的事务管理机制。

2. 事务工厂的创建

MyBatis 事务的创建是交给 org.apache.ibatis.transaction.TransactionFactory 事务工厂来完成的。如果我们将<transactionManager>的 type 配置为 JDBC，那么，在 MyBatis 初始化解析<environment>节点时，会根据 type="JDBC"创建一个 JdbcTransactionFactory 工厂，其源码如下：

```
// 解析<transactionManager>节点，创建对应的 TransactionFactory
private TransactionFactory transactionManagerElement(XNode context) throws Exception {
/*
   在 Configuration 初始化的时候，会通过以下语句，注册 JDBC 和 MANAGED 对应的工厂类
   typeAliasRegistry.registerAlias("JDBC", JdbcTransactionFactory.class);
   typeAliasRegistry.registerAlias("MANAGED", ManagedTransactionFactory.class);
  */
     if (context != null) {
         String type = context.getStringAttribute("type");
         Properties props = context.getChildrenAsProperties();
         // 创建对应的工厂实例，返回的是 JdbcTransactionFactory 或者
            MangedTransactionFactory 其中之一
          TransactionFactory factory = (TransactionFactory)
            resolveClass(type).newInstance();
         factory.setProperties(props);
         return factory;
    }
    throw new BuilderException("Environment declaration requires a TransactionFactory.");
  }
```

如上代码所示，如果 type = "JDBC"，则 MyBatis 会创建一个 JdbcTransactionFactory 的实例；如果 type="MANAGED"，则 MyBatis 会创建一个 MangedTransactionFactory 的实例。

3. 事务工厂 TransactionFactory

通过事务工厂 TransactionFactory 很容易获取到 Transaction 对象实例。我们以 JdbcTransaction 为例，看一下 JdbcTransactionFactory 是怎样生成 JdbcTransaction 的。JdbcTransaction 源代码如下：

```
public class JdbcTransactionFactory implements TransactionFactory {
  @Override
  public void setProperties(Properties props) {
  }
  // 根据给定的数据库连接 Connection 创建 Transaction
  @Override
  public Transaction newTransaction(Connection conn) {
    return new JdbcTransaction(conn);
  }
  // 根据 DataSource、隔离级别和是否自动提交创建 Transacion
  @Override
  public Transaction newTransaction(DataSource ds, TransactionIsolationLevel level,
boolean autoCommit) {
    return new JdbcTransaction(ds, level, autoCommit);
  }
}
```

如上代码所示，JdbcTransactionFactory 会创建 JDBC 类型的 Transaction，即 JdbcTransaction。类似地，ManagedTransactionFactory 也会创建 ManagedTransaction。下面我们分别深入解析

JdbcTransaction 和 ManagedTransaction，看它们到底是怎样实现事务管理的。

4. JdbcTransaction

JdbcTransaction 可直接使用 JDBC 的提交和回滚事务管理机制。它依赖于从 dataSource 中取得的连接 connection 来管理 transaction 的作用域，connection 对象的获取被延迟到调用 getConnection()方法时。如果将 autocommit 设置为 on，开启状态的话，则它会忽略 commit 和 rollback。

也就是说，JdbcTransaction 是使用 java.sql.Connection 上的 commit 和 rollback 功能来完成事务操作的，JdbcTransaction 只是相当于对 java.sql.Connection 事务处理进行了再次封装，Transaction 的事务管理都是通过 java.sql.Connection 实现的。JdbcTransaction 的代码实现如下：

```
    public class JdbcTransaction implements Transaction {
      private static final Log log = LogFactory.getLog(JdbcTransaction.class);
      // 数据库连接
      protected Connection connection;
      // 数据源
      protected DataSource dataSource;
      // 隔离级别
      protected TransactionIsolationLevel level;
      // 是否为自动提交
      protected boolean autoCommmit;
      public JdbcTransaction(DataSource ds, TransactionIsolationLevel desiredLevel,
boolean desiredAutoCommit) {
        dataSource = ds;
        level = desiredLevel;
        autoCommmit = desiredAutoCommit;
      }
      public JdbcTransaction(Connection connection) {
        this.connection = connection;
      }
      @Override
      public Connection getConnection() throws SQLException {
        if (connection == null) {
          openConnection();
        }
        return connection;
      }
      // 使用 connection 的 commit()
      @Override
      public void commit() throws SQLException {
        if (connection != null && !connection.getAutoCommit()) {
          if (log.isDebugEnabled()) {
            log.debug("Committing JDBC Connection [" + connection + "]");
          }
          connection.commit();
        }
      }
      // 使用 connection 的 rollback()
      @Override
      public void rollback() throws SQLException {
        if (connection != null && !connection.getAutoCommit()) {
          if (log.isDebugEnabled()) {
            log.debug("Rolling back JDBC Connection [" + connection + "]");
          }
          connection.rollback();
        }
      }
```

```
    // 使用 connection 的 close()
    @Override
    public void close() throws SQLException {
      if (connection != null) {
        resetAutoCommit();
        if (log.isDebugEnabled()) {
          log.debug("Closing JDBC Connection [" + connection + "]");
        }
        connection.close();
      }
    }
    // 省略部分代码……
  }
```

从 JdbcTransaction 类的源代码可以看出，JdbcTransaction 就是使用 java.sql.Connection 上的 commit、rollback 功能来完成事务操作的。

5. ManagedTransaction

ManagedTransaction 让容器来管理事务 Transaction 的整个生命周期，意思就是说，使用 ManagedTransaction 的 commit 和 rollback 功能不会对事务有任何的影响，它什么都不会做，它将事务管理的权力移交给了容器。ManagedTransaction 的代码实现如下：

```
  /**
   *
   * 让容器管理事务 transaction 的整个生命周期
   * connection 的获取延迟到 getConnection()方法调用时
   * 忽略所有的 commit 和 rollback 操作
   * 默认情况下，可以关闭一个连接 connection，也可以配置它不可以关闭一个连接
   * 让容器来管理 transaction 的整个生命周期
   * @see ManagedTransactionFactory
   */
  public class ManagedTransaction implements Transaction {
    private static final Log log = LogFactory.getLog(ManagedTransaction.class);
    private DataSource dataSource;
    private TransactionIsolationLevel level;
    private Connection connection;
    private boolean closeConnection;
    public ManagedTransaction(Connection connection, boolean closeConnection) {
      this.connection = connection;
      this.closeConnection = closeConnection;
    }
    public ManagedTransaction(DataSource ds, TransactionIsolationLevel level, boolean
closeConnection) {
      this.dataSource = ds;
      this.level = level;
      this.closeConnection = closeConnection;
    }
    @Override
    public Connection getConnection() throws SQLException {
      if (this.connection == null) {
        openConnection();
      }
      return this.connection;
    }
    @Override
    public void commit() throws SQLException {
      // Does nothing
    }
    @Override
    public void rollback() throws SQLException {
      // Does nothing
```

```
    }
    @Override
    public void close() throws SQLException {
      if (this.closeConnection && this.connection != null) {
        if (log.isDebugEnabled()) {
          log.debug("Closing JDBC Connection [" + this.connection + "]");
        }
        this.connection.close();
      }
    }
    protected void openConnection() throws SQLException {
      if (log.isDebugEnabled()) {
        log.debug("Opening JDBC Connection");
      }
      this.connection = this.dataSource.getConnection();
      if (this.level != null) {
        this.connection.setTransactionIsolation(this.level.getLevel());
      }
    }
  }
```

从 ManagedTransaction 类的源代码可以看出，提交和回滚时它什么都没有做，所以，当使用 ManagedTransaction 时 MyBatis 的事务是交给容器来操作管理的。

10.5 MyBatis 缓存机制

在实际项目开发中，通常对数据库查询的性能要求很高，而 MyBatis 提供了查询缓存来缓存数据，从而达到提高查询性能的要求。MyBatis 的查询缓存分为一级缓存和二级缓存。一级缓存是 SqlSession 级别的缓存，二级缓存是 mapper 级别的缓存，二级缓存是多个 SqlSession 共享的。MyBatis 通过缓存机制减轻数据压力，提高数据库性能。

10.5.1 一级缓存（SqlSession 级别）

MyBatis 的一级缓存是 SqlSession 级别的缓存。在操作数据库时需要构造 SqlSession 对象，在 SqlSession 对象中有一个 HashMap 用于存储缓存数据。不同的 SqlSession 之间的缓存数据区域（HashMap）是互相不影响的。

一级缓存的作用域是 SqlSession 范围的，当在同一个 SqlSession 中执行两次相同的 SQL 语句时，第一次执行完毕会将从数据库中查询的数据写到缓存（内存），第二次查询时会从缓存中获取数据，不再去底层数据库查询，从而提高查询效率。需要注意的是，如果 SqlSession 执行了 DML 操作（insert、update 和 delete），并提交到数据库，MyBatis 则会清空 SqlSession 中的一级缓存，这样做的目的是为了保证缓存中存储的是最新的信息，避免出现脏读现象。

当一个 SqlSession 结束后该 SqlSession 中的一级缓存也就不存在了。Mybatis 默认开启一级缓存，不需要进行任何配置。

注意：

MyBatis 的缓存机制是基于 id 进行缓存的，也就是说，MyBatis 使用 HashMap 缓存数据时，是使用对象的 id 作为 key，对象作为 value 保存的。

示例：OneLevelCacheTest

接下来我们测试 MyBatis 的一级缓存。

本示例使用了第 8 章创建的 tb_user 表、数据库脚本，User.java、mybatis-config.xml 和 log4j.xml 文件请参考第 8 章内容，此处不再赘述。

执行以下 SQL 脚本，向 tb_user 表插入几条测试数据：

```
INSERT INTO tb_user(NAME,sex,age) VALUES('jack','男',22);
INSERT INTO tb_user(NAME,sex,age) VALUES('rose','女',19);
INSERT INTO tb_user(NAME,sex,age) VALUES('tom','男',25);
INSERT INTO tb_user(NAME,sex,age) VALUES('mary','女',20);
```

程序清单：codes/10/OneLevelCacheTest/src/org/fkit/mapper/UserMapper.xml

```
<mapper namespace="org.fkit.mapper.UserMapper">
  <!-- 根据 id 查询 User -->
  <select id="selectUserById" parameterType="int"
  resultType="org.fkit.domain.User">
     SELECT * FROM TB_USER WHERE id = #{id}
  </select>
  <!-- 查询所有 User -->
  <select id="selectAllUser" resultType="org.fkit.domain.User">
     SELECT * FROM TB_USER
  </select>
<!-- 根据 id 删除 User -->
  <delete id="deleteUserById" parameterType="int">
     DELETE FROM TB_USER WHERE id = #{id}
  </delete>
</mapper>
```

程序清单：codes/10/OneLevelCacheTest/src/org/fkit/mapper/UserMapper.java

```
import java.util.List;
import org.fkit.domain.User;
public interface UserMapper {
    // 根据 id 查询 User
    User selectUserById(Integer id);
    // 查询所有 User
    List<User> selectAllUser();
    // 根据 id 删除 User
    void deleteUserById(Integer id);
}
```

程序清单：codes/10/OneLevelCacheTest/src/org/fkit/test/OneLevelCacheTest.java

```
import org.apache.ibatis.session.SqlSession;
import org.fkit.domain.User;
import org.fkit.factory.FKSqlSessionFactory;
import org.fkit.mapper.UserMapper;

public class OneLevelCacheTest {

    public static void main(String[] args) {

        // 定义 SqlSession 变量
        SqlSession sqlSession = null;
        try {
            // 创建 SqlSession 实例
            sqlSession = FKSqlSessionFactory.getSqlSession();
            TestOneLevelCache t = new TestOneLevelCache();

            //t.testCache1(sqlSession);
            //t.testCache2(sqlSession);
            t.testCache3(sqlSession);

            // 提交事务
            sqlSession.commit();
```

```
        } catch (Exception e) {
            // 回滚事务
            sqlSession.rollback();
            e.printStackTrace();
        }finally {
            // 关闭 SqlSession
            if(sqlSession != null)
                sqlSession.close();
        }
    }

    /*
     * 一级缓存: 也即 Session 级的缓存(默认开启)
     */
    public void testCache1 (SqlSession sqlSession){
        // 获得 UserMapping 对象
        UserMapper um = sqlSession.getMapper(UserMapper.class);
        // 查询 id 为 1 的 User 对象，会执行 select 语句
        User user = um.selectUserById(1);
        System.out.println(user);
        // 再次查询 id 为 1 的 User 对象,因为是同一个 SqlSession,所以会从之前的一级缓存中查找数据
        User user2 = um.selectUserById(1);
        System.out.println(user2);
    }

}
```

运行 OneLevelCacheTest 类的 main 方法，测试 testCache1 ()方法，控制台显示如下：

```
DEBUG [main] ==>  Preparing: SELECT * FROM TB_USER WHERE id = ?
DEBUG [main] ==> Parameters: 1(Integer)
DEBUG [main] <==      Total: 1
User [id=1, name=jack, sex=男, age=22]
User [id=1, name=jack, sex=男, age=22]
```

仔细观察 MyBatis 的执行结果，在第一次查询 id 为 1 的 User 对象时执行了一条 select 语句，但是第二次获取 id 为 1 的 User 对象时并没有执行 select 语句。因为此时一级缓存也就是 SqlSession 缓存中已经缓存了 id 为 1 的 User 对象，MyBatis 直接从缓存中将对象取出来，并没有再次去数据库查询，所以第二次没有再执行 select 语句。

程序清单：codes/10/OneLevelCacheTest /src/org/fkit/test/OneLevelCacheTest.java

```
// 测试一级缓存执行 DML 语句并提交
public void testCache2 (SqlSession sqlSession){
    // 获得 UserMapping 对象
    UserMapper um = sqlSession.getMapper(UserMapper.class);
    // 查询 id 为 1 的 User 对象，会执行 select 语句
    User user = um.selectUserById(1);
    System.out.println(user);
    // 执行 delete 操作
    um.deleteUserById(5);
    // commit 提交
    sqlSession.commit();
    // 再次查询 id 为 1 的 User 对象,因为 DML 操作会清空 SqlSession 缓存,所以会再次执行 select 语句
    User user2 = um.selectUserById(1);
    System.out.println(user2);
}
```

运行 OneLevelCacheTest 类的 main 方法，测试 testCache2 ()方法，控制台显示如下：

```
DEBUG [main] ==>  Preparing: SELECT * FROM TB_USER WHERE id = ?
DEBUG [main] ==> Parameters: 1(Integer)
DEBUG [main] <==      Total: 1
```

```
User [id=1, name=jack, sex=男, age=22]
DEBUG [main] ==>  Preparing: DELETE FROM TB_USER WHERE id = ?
DEBUG [main] ==> Parameters: 5(Integer)
DEBUG [main] ==>  Preparing: SELECT * FROM TB_USER WHERE id = ?
DEBUG [main] ==> Parameters: 1(Integer)
DEBUG [main] <==      Total: 1
User [id=1, name=jack, sex=男, age=22]
```

仔细观察 MyBatis 的执行结果，在第一次查询 id 为 1 的 User 对象时执行了一条 select 语句，接下来执行了一个 delete 操作，MyBatis 为了保证缓存中存储的是最新的信息，会清空 SqlSession 缓存。当第二次获取 id 为 1 的 User 对象时一级缓存也就是 SqlSession 缓存中并没有缓存任何对象，所以 MyBatis 再次执行 select 语句去查询 id 为 1 的 User 对象。

如果注释下面代码：

```
// 执行 delete 操作
// um.deleteUserById(5);
// commit 提交
// sqlSession.commit();
```

由于并没有执行 DML 操作并将操作提交到数据库，故此时 MyBatis 不会清空 SqlSession 缓存，当再次查询 id 为 1 的 User 对象时不会执行 select 语句。

程序清单：codes/10/OneLevelCacheTest/src/org/fkit/test/OneLevelCacheTest.java

```
// 测试一级缓存 clearCache
public void testCache3 (SqlSession sqlSession){
    // 获得 UserMapping 对象
    UserMapper um = sqlSession.getMapper(UserMapper.class);
    // 查询 id 为 1 的 User 对象，会执行 select 语句
    User user = um.selectUserById(1);
    System.out.println(user);
    // 清空一级缓存
    sqlSession.clearCache();
    // 再次获得 UserMapping 对象
    um = sqlSession.getMapper(UserMapper.class);
    // 再次访问，因为 SqlSession 一级缓存已经被清空，所以会再次执行 select 语句
    User user2 = um.selectUserById(1);
    System.out.println(user2);
}
```

运行 OneLevelCacheTest 类的 main 方法，测试 testCache3 ()方法，控制台显示如下：

```
DEBUG [main] ==>  Preparing: SELECT * FROM TB_USER WHERE id = ?
DEBUG [main] ==> Parameters: 1(Integer)
DEBUG [main] <==      Total: 1
User [id=1, name=jack, sex=男, age=22]
DEBUG [main] ==>  Preparing: SELECT * FROM TB_USER WHERE id = ?
DEBUG [main] ==> Parameters: 1(Integer)
DEBUG [main] <==      Total: 1
User [id=1, name=jack, sex=男, age=22]
```

仔细观察 MyBatis 的执行结果，在第一次查询 id 为 1 的 User 对象时执行了一条 select 语句，接下来调用 SqlSession 的 clearCache ()方法，该方法会清空 SqlSession 缓存。当第二次获取 id 为 1 的 User 对象时一级缓存中并没有缓存任何对象，所以 MyBatis 再次执行 select 语句去查询 id 为 1 的 User 对象。

程序清单：codes/10/OneLevelCacheTest/src/org/fkit/test/OneLevelCacheTest.java

```
// 测试一级缓存 close
public void testCache4 (){
```

```
        // 使用工厂类获得 SqlSession 对象
        SqlSession sqlSession = FKSqlSessionFactory.getSqlSession();
        // 获得 UserMapping 对象
        UserMapper um = sqlSession.getMapper(UserMapper.class);
        // 查询 id 为 1 的 User 对象，会执行 select 语句
        User user = um.selectUserById(1);
        System.out.println(user);
        // 关闭一级缓存
        sqlSession.close();
        // 再次访问,重新获取一级缓存，然后才能查找数据，否则会抛出异常
        SqlSession sqlSession2 = FKSqlSessionFactory.getSqlSession();
        // 再次获得 UserMapping 对象
        um = sqlSession2.getMapper(UserMapper.class);
        // 再次访问，因为现在使用的是一个新的 SqlSession 对象,所以会再次执行 select 语句
        User user2 = um.selectUserById(1);
        System.out.println(user2);
        // 关闭 SqlSession 对象
        sqlSession2.close();
    }
```

运行 OneLevelCacheTest 类的 main 方法，测试 testCache34()方法，控制台显示如下：

```
DEBUG [main] ==>  Preparing: SELECT * FROM TB_USER WHERE id = ?
DEBUG [main] ==> Parameters: 1(Integer)
DEBUG [main] <==      Total: 1
User [id=1, name=jack, sex=男, age=22]
DEBUG [main] ==>  Preparing: SELECT * FROM TB_USER WHERE id = ?
DEBUG [main] ==> Parameters: 1(Integer)
DEBUG [main] <==      Total: 1
User [id=1, name=jack, sex=男, age=22]
```

仔细观察 MyBatis 的执行结果，在第一次查询 id 为 1 的 User 对象时执行了一条 select 语句，接下来调用 SqlSession 的 close ()方法，该方法会关闭 SqlSession 缓存。当第二次获取 id 为 1 的 User 对象时一级缓存也就是 SqlSession 缓存是一个全新的对象，一级缓存中并没有缓存任何对象，所以 MyBatis 再次执行 select 语句去查询 id 为 1 的 User 对象。

10.5.2 二级缓存（mapper 级别）

二级缓存是 mapper 级别的缓存。使用二级缓存时，多个 SqlSession 使用同一个 mapper 的 SQL 语句去操作数据库，得到的数据会存在二级缓存区域，它同样是使用 HashMap 进行数据存储的。相比一级缓存 SqlSession，二级缓存的范围更大，多个 SqlSession 可以共享二级缓存中的数据，二级缓存是跨 SqlSession 的。

二级缓存是多个 SqlSession 共享的，其作用域是 mapper 的同一个 namespace。不同的 SqlSession 两次执行相同的 namespace 下的 SQL 语句，且向 SQL 中传递的参数也相同，即最终执行相同的 SQL 语句，当第一个 SqlSession 调用 close()方法关闭一级缓存时，第一次从数据库中查询到的数据会被保存到二级缓存，第二次查询时会从二级缓存中获取数据，不再去底层数据库查询，从而提高查询效率。

MyBatis 默认没有开启二级缓存，需要在 setting 全局参数中配置开启二级缓存。

示例：TwoLevelCacheTest

接下来测试 MyBatis 的二级缓存，所有代码和测试一级缓存的代码完全一样，只是需要在配置文件中开启二级缓存。

程序清单：codes/10/TwoLevelCacheTest /src/mybatis-config.xml

```
<settings>
    <!-- 开启二级缓存 -->
    <setting name="cacheEnabled" value="true"/>
</settings>
```

cacheEnabled 的 value 为 true 时表示在此配置文件下开启二级缓存，该属性默认为 false。

MyBatis 的二级缓存是和命名空间绑定的，即二级缓存需要配置在 Mapper.xml 映射文件或者 Mapper 接口中。在映射文件中，命名空间就是 XML 根节点 mapper 的 namespace 属性。在 Mapper 接口中，命名空间就是接口的全限定名称。

开启默认的二级缓存代码如下：

```
<cach />
```

默认的二级缓存会有如下作用：

- 映射语句文件中的所有 SELECT 语句将会被缓存。
- 映射语句文件中的所有 INSERT、UPDATE、DELETE 语句会刷新缓存。
- 缓存会使用 Least Recently Used（LRU 最近最少使用）策略来收回。
- 根据时间表（如 no Flush Interval，没有刷新间隔），缓存不会以任何时间顺序来刷新。
- 缓存会存储集合或对象（无论查询方法返回什么类型的值）的 1024 个引用。
- 缓存会被视为 read/write（可读/可写）的，这意味着对象检索不是共享的，而且可以安全地被调用者修改，而不干扰其他调用者或线程所做的潜在修改。

<cache/>元素中所有这些行为都可以通过 cache 元素的属性来进行修改。

程序清单：codes/10/TwoLevelCacheTest/src/org/fkit/mapper/UserMapper.xml

```
<!-- 开启当前 mapper 的 namespace 下的二级缓存 -->
<cache eviction="LRU"  flushInterval="60000"
   size="512" readOnly="true"/>
```

以上配置创建了一个 LRU 缓存，并每隔 60s 刷新，最大存储 512 个对象，而且返回的对象被认为是只读的。

cache 元素用来开启当前 mapper 的 namespace 下的二级缓存。该元素的属性设置如下：

- flushInterval。刷新间隔。可以被设置为任意的正整数，而且它们代表一个合理的毫秒形式的时间段。默认情况下是不设置，也就是没有刷新间隔，缓存仅仅调用语句时刷新。
- size。缓存数目。可以被设置为任意正整数，要记住你缓存的对象数目和你运行环境的可用内存资源数目。默认值是 1024。
- readOnly。只读。该属性可以被设置为 true 或 false。只读的缓存会给所有调用者返回缓存对象的相同实例，因此这些对象不能被修改。这提供了很重要的性能优势。可读写的缓存会返回缓存对象的拷贝（通过序列化）。这种方式会慢一些，但是安全，因此默认是 false。
- eviction。收回策略，默认为 LRU。有如下几种：
    - LRU。最近最少使用的策略，移除最长时间不被使用的对象。
    - FIFO。先进先出策略，按对象进入缓存的顺序来移除它们。
    - SOFT。软引用策略，移除基于垃圾回收器状态和软引用规则的对象。
    - WEAK。弱引用策略，更积极地移除基于垃圾收集器状态和弱引用规则的对象。

提示

使用二级缓存时，与查询结果映射的 Java 对象必须实现 java.io.Serializable 接口的序列化和反序列化操作，如果存在父类，其成员都需要实现序列化接口。实现序列化接口是为了对缓存数据进行序列化和反序列化操作，因为二级缓存数据存储介质多种多样，不一定在内存，有可能是硬盘或者远程服务器。

程序清单：codes/10/TwoLevelCacheTest/src/org/fkit/mapper/UserMapper.java

```
import org.fkit.domain.User;
public interface UserMapper {
    // 根据 id 查询 User
    User selectUserById(Integer id);
    // 根据 id 删除 User
    void deleteUserById(Integer id);
}
```

程序清单：codes/10/TwoLevelCacheTest /src/org/fkit/test/TwoLevelCacheTest.java

```
import org.apache.ibatis.session.SqlSession;
import org.fkit.domain.User;
import org.fkit.factory.FKSqlSessionFactory;
import org.fkit.mapper.UserMapper;

public class TwoLevelCacheTest {

    public static void main(String[] args)  {
        // 使用工厂类获得 SqlSession 对象
        SqlSession sqlSession1 = FKSqlSessionFactory.getSqlSession();
        // 获得 UserMapping 对象
        UserMapper um = sqlSession1.getMapper(UserMapper.class);
        // 查询 id 为 1 的 User 对象，会执行 select 语句
        User user = um.selectUserById(1);
        System.out.println(user);
        // 关闭一级缓存
        sqlSession1.close();
        // 重新获取一级缓存
        SqlSession sqlSession2 = FKSqlSessionFactory.getSqlSession();
        // 再次获得 UserMapping 对象
        um = sqlSession2.getMapper(UserMapper.class);
        // 再次查询 id 为 1 的 User 对象，虽然现在使用的是一个新的 SqlSession 对象，
        // 但是二级缓存中缓存了数据，所以不会再次执行 select 语句
        User user2 = um.selectUserById(1);
        System.out.println(user2);
        // 关闭 SqlSession 对象
        sqlSession2.close();
    }
}
```

运行 TwoLevelCacheTest 类的 main 方法，控制台显示如下：

```
DEBUG [main] Cache Hit Ratio [org.fkit.mapper.UserMapper]: 0.0
DEBUG [main] ==>  Preparing: SELECT * FROM TB_USER WHERE id = ?
DEBUG [main] ==> Parameters: 1(Integer)
DEBUG [main] <==      Total: 1
User [id=1, name=jack, sex=男, age=22]
DEBUG [main] - Cache Hit Ratio [org.fkit.mapper.UserMapper]: 0.5
User [id=1, name=jack, sex=男, age=22]
```

仔细观察 MyBatis 的执行结果，日志中有几条以 Cache Hit Ration 开头的语句，这行日志后面输出的值为当前执行方法的缓存命中率。在第一次查询 id 为 1 的 User 对象时执行了一条 select 语句，接下来调用 SqlSession 的 close ()方法，该方法会关闭 SqlSession 一级缓存，同时会将查询数据保存到二级缓存中。当第二次获取 id 为 1 的 User 对象时重新获得的一级缓存 SqlSession 中并没有缓存任何对象，但是因为启用了二级缓存，当 MyBatis 在一级缓存中没有找到 id 为 1 的 User 对象时，会去二级缓存中查找，所以不会再次执行 select 语句。

10.6 本章小结

本章首先介绍了 MyBatis 的关联映射，包括最常见的一对一、一对多、多对多关联映射，动态 SQL 机制和存储过程调用。这些都是在实际项目开发时最常用的操作，我们需要重点掌握。

接下来介绍了 MyBatis 的事务管理，包括 Transaction 和该接口的两个实现类 JdbcTransaction 和 ManagedTransaction，以及它们如何处理事务。

最后介绍了 MyBatis 的缓存机制，包括一级缓存 SqlSession 和二级缓存 mapper。使用缓存可以最大程度地减轻数据查询压力，提高数据库性能

第 11 章将重点介绍 MyBatis 的注解配置。

CHAPTER

11

第 11 章 MyBatis 的注解配置

本章要点

- MyBatis 注解插入、修改、删除和查询操作
- MyBatis 注解一对一、一对多和多对多操作
- MyBatis 注解动态 SQL
- MyBatis 注解调用存储过程
- MyBatis 注解使用二级缓存

前面的章节介绍了 MyBatis 的基本用法、关联映射、动态 SQL 和缓存机制等知识，其所有的配置都使用 XML 完成，但是大量的 XML 配置文件的编写是非常烦琐的，因此 MyBatis 也提供了更加简便的基于注解（annotation）的配置方式。本章将重点介绍 MyBatis 的注解配置。

11.1 常用注解

MyBatis 的注解位于 org.apache.ibatis.annotations 包下。常用的注解如下：

- Select。映射查询的 SQL 语句。
- SelectProvider。Select 语句的动态 SQL 映射。允许指定一个类名和一个方法在执行时返回运行的查询语句。有两个属性：type 和 method，type 属性是类的完全限定名，method 是该类中的方法名。
- Insert。映射插入的 SQL 语句。
- InsertProvider。Insert 语句的动态 SQL 映射。允许指定一个类名和一个方法在执行时返回运行的插入语句。有两个属性：type 和 method，type 属性是类的完全限定名，method 是该类中的方法名。
- Update。映射更新的 SQL 语句。
- UpdateProvider。Update 语句的动态 SQL 映射。允许指定一个类名和一个方法在执行时返回运行的更新语句。有两个属性：type 和 method，type 属性是类的完全限定名，method 是该类中的方法名。
- Delete。映射删除的 SQL 语句。
- DeleteProvider。Delete 语句的动态 SQL 映射。允许指定一个类名和一个方法在执行时返回运行的删除语句。有两个属性：type 和 method，type 属性是类的完全限定名，method 是该类中的方法名。
- Result。在列和属性之间的单独结果映射。属性包括：id、column、property、javaType、jdbcType、type Handler、one、many。id 属性是一个布尔值，表示是否被用于主键映射。one 属性是单独的映射，和 XML 配置中的 <association> 相似，而 many 属性是对集合而言的，和 XML 配置的<collection>相似。
- Results。多个结果映射（Result）列表。
- Options。提供配置选项的附加值，它们通常在映射语句上作为附加功能配置出现。
- One。复杂类型的单独属性值映射。必须指定 select 属性，表示已映射的 SQL 语句的完全限定名。
- Many。复杂类型的集合属性映射。必须指定 select 属性，表示已映射的 SQL 语句的完全限定名。
- Param。当映射器方法需要多个参数时，这个注解可以被应用于映射器方法参数来给每个参数取一个名字。否则，多参数将会以它们的顺序位置和 SQL 语句中的表达式进行映射，这是默认的。使用@Param("id")时，SQL 中参数应该被命名为#{id}。

11.2 注解的使用

@select、@insert、@update 和@delete 可以完成常见的 CRUD（增删改查）SQL 语句映射。

示例：测试 select、insert、update 和 delete 操作

本示例使用了第 8 章创建的 tb_user 表、数据库脚本，User.java、mybatis-config.xml 和 log4j.xml 文件请参考第 8 章内容，此处不再赘述。

执行以下 SQL 脚本，向 tb_user 表插入几条测试数据：

```
INSERT INTO tb_user(NAME,sex,age) VALUES('jack','男',22);
INSERT INTO tb_user(NAME,sex,age) VALUES('rose','女',19);
INSERT INTO tb_user(NAME,sex,age) VALUES('tom','男',25);
INSERT INTO tb_user(NAME,sex,age) VALUES('mary','女',20);
```

程序清单：codes/11/ADMLTest/src/org/fkit/mapper/UserMapper.java

```
import org.apache.ibatis.annotations.Delete;
import org.apache.ibatis.annotations.Insert;
import org.apache.ibatis.annotations.Options;
import org.apache.ibatis.annotations.Param;
import org.apache.ibatis.annotations.Result;
import org.apache.ibatis.annotations.Results;
import org.apache.ibatis.annotations.Select;
import org.apache.ibatis.annotations.Update;
import org.fkit.domain.User;
public interface UserMapper {
    @Insert("INSERT INTO TB_USER(name,sex,age) VALUES(#{name},#{sex},#{age})")
    @Options(useGeneratedKeys=true,keyProperty="id")
    int saveUser(User user);

    @Delete("DELETE FROM TB_USER WHERE id = #{id}")
    int removeUser(@Param("id") Integer id);

    @Update("UPDATE TB_USER SET name = #{name},sex = #{sex},age = #{age} WHERE id
     = #{id}")
    void modifyUser(User user);

    @Select("SELECT * FROM TB_USER WHERE id = #{id}")
    @Results({
        @Result(id=true,column="id",property="id"),
        @Result(column="name",property="name"),
        @Result(column="sex",property="sex"),
        @Result(column="age",property="age")
    })
    User selectUserById(Integer id);

    @Select("SELECT * FROM TB_USER ")
    List<User> selectAllUser();
}
```

在 UserMapper.java 中定义了 saveUser、removeUser、modifyUser、selectUserById 和 selectAllUser 5 个方法，分别对应插入、更新、删除、根据 id 查询用户和查询所有用户 5 个数据库操作，并使用注解代替了之前的 XML 配置。注解配置中的 SQL 语句和 XML 配置中的 SQL 语句用法一致，此处不再赘述。

程序清单：codes/11/ADMLTest/src/org/fkit/test/InsertTest.java

```
import org.apache.ibatis.session.SqlSession;
import org.fkit.domain.User;
import org.fkit.factory.FKSqlSessionFactory;
import org.fkit.mapper.UserMapper;
public class InsertTest {

   public static void main(String[] args) {
      // 定义 SqlSession 变量
```

```
        SqlSession sqlSession = null;
        try {
            // 创建 SqlSession 实例
            sqlSession = FKSqlSessionFactory.getSqlSession();

            // 创建 UserMapper 实例
            UserMapper um = sqlSession.getMapper(UserMapper.class);
            // 创建 User 对象并设置属性
            User user = new User();
            user.setName("test");
            user.setSex("男");
            user.setAge(18);
            // 插入数据
            um.saveUser(user);
            // 查看插入数据生成的主键
            System.out.println("插入数据生成的主键 id 为: " + user.getId());

            // 提交事务
            sqlSession.commit();
        } catch (Exception e) {
            // 回滚事务
            sqlSession.rollback();
            e.printStackTrace();
        }finally {
            // 关闭 SqlSession
            if(sqlSession != null)
                sqlSession.close();
        }
    }

}
```

运行 InsertTest 类的 main 方法，测试 saveUser ()方法，控制台显示如下：

```
DEBUG [main] ==>  Preparing: INSERT INTO TB_USER(name,sex,age) VALUES(?,?,?)
DEBUG [main] ==> Parameters: test(String), 男(String), 18(Integer)
DEBUG [main] <==    Updates: 1
插入数据生成的主键 id 为: 5
```

saveUser 方法的声明如下：

```
@Insert("INSERT INTO TB_USER(name,sex,age) VALUES(#{name},#{sex},#{age})")
@Options(useGeneratedKeys=true,keyProperty="id")
int saveUser(User user);
```

调用 saveUser 方法，会执行@insert 注解中的 SQL 语句。需要注意的是，saveUser 方法还使用了@Options 注解，属性 useGeneratedKeys=true 表示使用数据库自动增长的主键，该操作需要底层数据库的支持。keyProperty="id"表示将插入数据生成的主键设置到 user 对象的 id 当中。

程序清单：codes/11/ADMLTest/src/org/fkit/test/SelectTest.java

```
import org.apache.ibatis.session.SqlSession;
import org.fkit.domain.User;
import org.fkit.factory.FKSqlSessionFactory;
import org.fkit.mapper.UserMapper;
public class SelectTest {

    public static void main(String[] args) {
        // 定义 SqlSession 变量
        SqlSession sqlSession = null;
        try {
            // 创建 SqlSession 实例
```

```
            sqlSession = FKSqlSessionFactory.getSqlSession();
            // 创建 UserMapper 实例
            UserMapper um = sqlSession.getMapper(UserMapper.class);
            // 根据 id 查询 User 对象
            User user = um.selectUserById(1);
            // 查看查询到的 User 对象
            System.out.println(user);
            // 提交事务
            sqlSession.commit();
        } catch (Exception e) {
            // 回滚事务
            sqlSession.rollback();
            e.printStackTrace();
        }finally {
            // 关闭 SqlSession
            if(sqlSession != null)
                sqlSession.close();
        }
    }

}
```

运行 SelectTest 类的 main 方法，测试 selectUserById ()方法，控制台显示如下：

```
DEBUG [main] ==>  Preparing: SELECT * FROM TB_USER WHERE id = ?
DEBUG [main] ==> Parameters: 1(Integer)
DEBUG [main] <==      Total: 1
User [id=1, name=jack, sex=男, age=22]
```

selectUserById 方法的声明如下：

```
@Select("SELECT * FROM TB_USER WHERE id = #{id}")
@Results({
    @Result(id=true,column="id",property="id"),
    @Result(column="name",property="name"),
    @Result(column="sex",property="sex"),
    @Result(column="age",property="age")
})
User selectUserById(Integer id);
```

调用 selectUserById 方法，会执行@ Select 注解中的 SQL 语句。@Result 注解用于列和属性之间的结果映射，如果列和属性名称相同，则可以省略@Result 注解，MyBatis 会自动进行映射。

程序清单：codes/11/ADMLTest/src/org/fkit/test/UpdateTest.java

```
import org.apache.ibatis.session.SqlSession;
import org.fkit.domain.User;
import org.fkit.factory.FKSqlSessionFactory;
import org.fkit.mapper.UserMapper;
public class UpdateTest {

    public static void main(String[] args) {
        // 定义 SqlSession 变量
        SqlSession sqlSession = null;
        try {
            // 创建 SqlSession 实例
            sqlSession = FKSqlSessionFactory.getSqlSession();
            // 创建 UserMapper 实例
            UserMapper um = sqlSession.getMapper(UserMapper.class);
            // 查询 id=5 的 User 对象
            User user = um.selectUserById(5);
            // 修改查询到的 User 对象
            user.setName("update");
```

```
            user.setSex("女");
            user.setAge(19);
            // 修改 User 对象
            um.modifyUser(user);
            // 提交事务
            sqlSession.commit();
        } catch (Exception e) {
            // 回滚事务
            sqlSession.rollback();
            e.printStackTrace();
        }finally {
            // 关闭 SqlSession
            if(sqlSession != null)
                sqlSession.close();
        }
    }

}
```

运行 UpdateTest 类的 main 方法，测试 modifyUser ()方法，控制台显示如下：

```
    DEBUG [main] ==>  Preparing: SELECT * FROM TB_USER WHERE id = ?
    DEBUG [main] ==> Parameters: 5(Integer)
    DEBUG [main] <==      Total: 1
    DEBUG [main] ==>  Preparing: UPDATE TB_USER SET name = ?,sex = ?,age = ? WHERE id
= ?
    DEBUG [main] ==> Parameters: update(String), 女(String), 19(Integer), 5(Integer)
    DEBUG [main] <==    Updates: 1
```

modifyUser 方法的声明如下：

```
    @Update("UPDATE TB_USER SET name = #{name},sex = #{sex},age = #{age} WHERE id =
#{id}")
    void modifyUser(User user);
```

调用 modifyUser 方法，会执行@ Update 注解中的 SQL 语句。

程序清单：codes/11/ADMLTest /src/org/fkit/test/DeleteTest.java

```
import org.apache.ibatis.session.SqlSession;
import org.fkit.factory.FKSqlSessionFactory;
import org.fkit.mapper.UserMapper;
public class DeleteTest {

    public static void main(String[] args) {
        // 定义 SqlSession 变量
        SqlSession sqlSession = null;
        try {
            // 创建 SqlSession 实例
            sqlSession = FKSqlSessionFactory.getSqlSession();
            // 创建 UserMapper 实例
            UserMapper um = sqlSession.getMapper(UserMapper.class);
            // 删除 User 对象
            um.removeUser(5);
            // 提交事务
            sqlSession.commit();
        } catch (Exception e) {
            // 回滚事务
            sqlSession.rollback();
            e.printStackTrace();
        }finally {
            // 关闭 SqlSession
            if(sqlSession != null)
                sqlSession.close();
        }
```

```
    }
}
```

运行 DeleteTest 类的 main 方法，测试 removeUser ()方法，控制台显示如下：

```
DEBUG [main] ==>  Preparing: DELETE FROM TB_USER WHERE id = ?
DEBUG [main] ==> Parameters: 5(Integer)
DEBUG [main] <==    Updates: 1
```

removeUser 方法的声明如下：

```
@Delete("DELETE FROM TB_USER WHERE id = #{id}")
int removeUser(@Param("id") Integer id);
```

调用 removeUser 方法，会执行@ Delete 注解中的 SQL 语句。@Param("id")注解表示给该注解后面的变量取一个参数名称，对应@ Delete 注解中的#{id}。如果没有使用@Param 注解，则参数将会以它们的顺序位置来和 SQL 语句中的表达式进行映射。

程序清单：codes/11/ADMLTest/src/org/fkit/test/SelectAllTest.java

```
import java.util.List;
import org.apache.ibatis.session.SqlSession;
import org.fkit.domain.User;
import org.fkit.factory.FKSqlSessionFactory;
import org.fkit.mapper.UserMapper;
public class SelectAllTest {

    public static void main(String[] args) {
        // 定义 SqlSession 变量
        SqlSession sqlSession = null;
        try {
            // 创建 SqlSession 实例
            sqlSession = FKSqlSessionFactory.getSqlSession();
            // 创建 UserMapper 实例
            UserMapper um = sqlSession.getMapper(UserMapper.class);
            // 根据 id 查询 User 对象
            List<User> list = um.selectAllUser();
            // 查看查询到的 User 对象
            list.forEach(user->System.out.println(user));
            // 提交事务
            sqlSession.commit();
        } catch (Exception e) {
            // 回滚事务
            sqlSession.rollback();
            e.printStackTrace();
        }finally {
            // 关闭 SqlSession
            if(sqlSession != null)
                sqlSession.close();
        }
    }

}
```

运行 SelectAllTest 类的 main 方法，测试 selectAllUser ()方法，控制台显示如下：

```
DEBUG [main] ==>  Preparing: SELECT * FROM TB_USER
DEBUG [main] ==> Parameters:
DEBUG [main] <==      Total: 4
User [id=1, name=jack, sex=男, age=22]
User [id=2, name=rose, sex=女, age=19]
User [id=3, name=tom, sex=男, age=25]
```

```
User [id=4, name=mary, sex=女, age=20]
```

selectAllUser 方法的声明如下：

```
@Select("SELECT * FROM TB_USER ")
List<User> selectAllUser();
```

调用 selectAllUser 方法，会执行@ Select 注解中的 SQL 语句。

示例：测试一对一关联

本示例使用了 10.1.1 节中的一对一关系的 tb_crad 表和 tb_person 表，数据库脚本、Card.java、Person.java、FKSqlSessionFactory.java、mybatis-config.xml 和 log4j.xml 文件请参考第 10 章内容，此处不再赘述。

程序清单：codes/11/AOneToOneTest/src/org/fkit/mapper/CardMapper.java

```
import org.apache.ibatis.annotations.Select;
import org.fkit.domain.Card;
public interface CardMapper {
    @Select("SELECT * FROM TB_CARD WHERE ID = #{id} ")
    Card selectCardById(Integer id);

}
```

程序清单：codes/11/AOneToOneTest/src/org/fkit/mapper/PersonMapper.java

```
import org.apache.ibatis.annotations.One;
import org.apache.ibatis.annotations.Result;
import org.apache.ibatis.annotations.Results;
import org.apache.ibatis.annotations.Select;
import org.apache.ibatis.mapping.FetchType;
import org.fkit.domain.Person;
public interface PersonMapper {
    @Select("SELECT * FROM TB_PERSON WHERE ID = #{id}")
    @Results({
        @Result(id=true,column="id",property="id"),
        @Result(column="name",property="name"),
        @Result(column="sex",property="sex"),
        @Result(column="age",property="age"),
        @Result(column="card_id",property="card",
        one=@One(
                select="org.fkit.mapper.CardMapper.selectCardById",
                fetchType=FetchType.EAGER))
    })
    Person selectPersonById(Integer id);
}
```

selectPersonById 方法使用了@ Select 注解，其根据 id 查询对应的 Person 数据。因为需要将 Person 对应的 Card 数据也查询出来，所以 Person 的 Card 属性使用了一个@Result 结果映射。column="card_id",property="card"表示 Person 的 Card 属性对应 tb_person 表的 card_id 列，one 属性表示是一个一对一关联关系，@One 注解的 select 属性表示需要关联执行的 SQL 语句，fetchType 表示查询的类型是立即加载（EAGER）还是懒加载（LAZY）。

程序清单：codes/11/AOneToOneTest/src/org/fkit/test/OneToOneTest.java

```
import org.apache.ibatis.session.SqlSession;
import org.fkit.domain.Person;
import org.fkit.factory.FKSqlSessionFactory;
import org.fkit.mapper.PersonMapper;
public class OneToOneTest {

   public static void main(String[] args) {
```

```
        // 定义 SqlSession 变量
        SqlSession sqlSession = null;
        try {
            // 创建 SqlSession 实例
            sqlSession = FKSqlSessionFactory.getSqlSession();
            // 获取 PersonMapper 实例
            PersonMapper pm = sqlSession.getMapper(PersonMapper.class);
            // 根据 id 查询 Person 对象，同时需要获得关联的 Card 对象
            Person p = pm.selectPersonById(1);
            // 查看查询到的 Person 对象
            System.out.println(p);
            // 查看查询到的关联的 Card 对象
            System.out.println(p.getCard());
            // 提交事务
            sqlSession.commit();
        } catch (Exception e) {
            // 回滚事务
            sqlSession.rollback();
            e.printStackTrace();
        }finally {
            // 关闭 SqlSession
            if(sqlSession != null)
                sqlSession.close();
        }
    }
}
```

运行 OneToOneTest 类的 main 方法，该方法通过 SqlSession 的 getMapper(Class<T> type) 方法获得 mapper 接口的代理对象 PersonMapper。调用 selectPersonById 方法时会执行该方法上的注解。需要注意的是，Person 的一对一关联使用的注解@one 的 select 属性，要执行的 SQL 语句在 CardMapper 类的 selectCardById 方法的注解中。控制台显示如下：

```
DEBUG [main] ==>  Preparing: SELECT * FROM TB_PERSON WHERE ID = ?
DEBUG [main] ==> Parameters: 1(Integer)
DEBUG [main] ====>  Preparing: SELECT * FROM TB_CARD WHERE ID = ?
DEBUG [main] ====> Parameters: 1(Integer)
DEBUG [main] <====      Total: 1
DEBUG [main] <==      Total: 1
Person [id=1, name=jack, sex=男, age=23]
Card [id=1, code=432801198009191038]
```

可以看到，查询 Peson 信息时 Person 对应的 Card 对象也被查询出来了。

示例：测试一对多关联

本示例使用了 10.1.2 节中的一对多关系的 tb_calzz 表和 tb_student 表，数据库脚本、Clazz.java、Student.java、FKSqlSessionFactory.java、mybatis-config.xml 和 log4j.xml 文件请参考第 10 章内容，此处不再赘述。

程序清单：codes/11/AOneToManyTest/src/org/fkit/mapper/StudentMapper.java

```
import java.util.List;
import org.apache.ibatis.annotations.Result;
import org.apache.ibatis.annotations.Results;
import org.apache.ibatis.annotations.Select;
import org.fkit.domain.Student;
public interface StudentMapper {
    // 根据班级 id 查询班级所有学生
    @Select("SELECT * FROM TB_STUDENT WHERE CLAZZ_ID = #{id}")
    @Results({
        @Result(id=true,column="id",property="id"),
```

```
        @Result(column="name",property="name"),
        @Result(column="sex",property="sex"),
        @Result(column="age",property="age")
    })
    List<Student> selectByClazzId(Integer clazz_id);
}
```

程序清单：codes/11/AOneToManyTest/src/org/fkit/mapper/ClazzMapper.java

```
import org.apache.ibatis.annotations.Many;
import org.apache.ibatis.annotations.Result;
import org.apache.ibatis.annotations.Results;
import org.apache.ibatis.annotations.Select;
import org.apache.ibatis.mapping.FetchType;
import org.fkit.domain.Clazz;
public interface ClazzMapper {
    // 根据 id 查询班级信息
    @Select("SELECT * FROM TB_CLAZZ  WHERE ID = #{id}")
    @Results({
        @Result(id=true,column="id",property="id"),
        @Result(column="code",property="code"),
        @Result(column="name",property="name"),
        @Result(column="id",property="students",
        many=@Many(
                select="org.fkit.mapper.StudentMapper.selectByClazzId",
                fetchType=FetchType.LAZY))
    })
    Clazz selectById(Integer id);
}
```

selectById 方法使用了@ Select 注解，其根据 id 查询对应的 Clazz 数据。因为需要将 Clazz 所关联的所有 Student 查询出来，所以 Clazz 的 students 属性使用了一个@Result 结果映射。column="id"表示会使用 id 作为查询条件，property="students"表示 Clazz 的 students 属性（该属性是个集合），many 属性表示是一个一对多关联关系，@Many 注解的 select 属性表示需要关联执行的 SQL 语句，fetchType 表示查询的类型是立即加载（EAGER）还是懒加载（LAZY）。

程序清单：codes/11/AOneToManyTest/src/org/fkit/test/OneToManyTest.java

```
import org.apache.ibatis.session.SqlSession;
import org.fkit.domain.Clazz;
import org.fkit.factory.FKSqlSessionFactory;
import org.fkit.mapper.ClazzMapper;
public class OneToManyTest {

    public static void main(String[] args) {
        // 定义 SqlSession 变量
        SqlSession sqlSession = null;
        try {
            // 创建 SqlSession 实例
            sqlSession = FKSqlSessionFactory.getSqlSession();
            // 获取 ClazzMapper 实例
            ClazzMapper cm = sqlSession.getMapper(ClazzMapper.class);
            // 根据 id 查询 Clazz 对象
            Clazz clazz = cm.selectById(1);
            // 查看查询到的 Clazz 对象
            System.out.println(clazz.getId() + " "+ clazz.getCode() + " "+clazz.getName());
            // 查看关联的学生集合，因为配置使用的是 LAZY 懒加载，所以当使用时才执行 SQL 语句
            clazz.getStudents().forEach(student->System.out.println(student));
            // 提交事务
            sqlSession.commit();
        } catch (Exception e) {
            // 回滚事务
```

```
            sqlSession.rollback();
            e.printStackTrace();
        }finally {
            // 关闭 SqlSession
            if(sqlSession != null)
                sqlSession.close();
        }
    }
}
```

运行 OneToManyTest 类的 main 方法，该方法通过 SqlSession 的 getMapper(Class<T> type)方法获得 mapper 接口的代理对象 ClazzMapper。调用 selectById 方法时会执行该方法上的注解。需要注意的是，Clazz 的一对多关联使用的注解@Many 的 select 属性，要执行的 SQL 语句在 StudentMapper 类的 selectByClazzId 方法的注解中。由于使用的是懒加载，故需要用到关联的学生对象时才会执行 SQL 语句。控制台显示如下：

```
    DEBUG [main] ==>  Preparing: SELECT * FROM TB_CLAZZ WHERE ID = ?
    DEBUG [main] ==> Parameters: 1(Integer)
    DEBUG [main] <==      Total: 1
    1 j1601 Java 就业班
    DEBUG [main] ==>  Preparing: SELECT * FROM TB_STUDENT WHERE CLAZZ_ID = ?
    DEBUG [main] ==> Parameters: 1(Integer)
    DEBUG [main] <==      Total: 4
    Student [id=1, name=jack, sex=男, age=23]
    Student [id=2, name=rose, sex=女, age=18]
    Student [id=3, name=tom, sex=男, age=21]
    Student [id=4, name=alice, sex=女, age=20]
```

可以看到，首先查询了 id 为 1 的 Clazz 的数据，当使用 Clazz 关联的 Student 数据时才执行了根据班级 id 查询学生的 SQL 语句。

示例：测试多对多关联

本示例使用了 10.1.3 节中的多对多关系的 tb_user 表、tb_order 表和 tb_article 表，数据库脚本，User.java、Order.java、Article.java、FKSqlSessionFactory .java、mybatis-config.xml 和 log4j.xml 文件请参考第 10 章内容，此处不再赘述。

程序清单：codes/11/AManyToManyTest/src/org/fkit/mapper/UserMapper.java

```
    import org.apache.ibatis.annotations.Select;
    import org.fkit.domain.User;
    public interface UserMapper {
        @Select("SELECT * FROM TB_USER WHERE ID = #{id} ")
        User selectById(Integer id);
    }
```

程序清单：codes/11/AManyToManyTest/src/org/fkit/mapper/ArticleMapper.java

```
    import java.io.Serializable;
    import java.util.List;
    public interface ArticleMapper {
        @Select("SELECT * FROM tb_article WHERE id IN (SELECT article_id FROM tb_item
WHERE order_id = #{id} ) ")
        List<Article> selectByOrderId(Integer order_id);
    }
```

程序清单：codes/11/AManyToManyTest/src/org/fkit/mapper/OrdreMapper.java

```
    import java.io.Serializable;
    import java.util.List;
    public interface OrderMapper {
        @Select("SELECT * FROM TB_ORDER WHERE ID = #{id}")
```

```
    @Results({
        @Result(id=true,column="id",property="id"),
        @Result(column="code",property="code"),
        @Result(column="total",property="total"),
        @Result(column="user_id",property="user",
            one=@One(select="org.fkit.mapper.UserMapper.selectById",
        fetchType=FetchType.EAGER)),
        @Result(column="id",property="articles",
            many=@Many(select="org.fkit.mapper.ArticleMapper.selectByOrderId",
        fetchType=FetchType.LAZY))
    })
    Order selectById(Integer id);
}
```

selectOrderById 方法使用了@ Select 注解，其根据 id 查询对应的 Order 数据。Order 的 user 属性使用了一个@Result 结果映射。@One 注解的 select 属性表示需要关联执行的 SQL 语句，fetchType 表示查询的类型是立即加载（EAGER）还是懒加载（LAZY）。Order 的一对多关联使用的注解@Many 的 select 属性，要执行的 SQL 语句在 ArticleMapper 类的 selectByOrderId 方法的注解中。由于使用的是懒加载，故需要用到关联的 Article 对象时才会执行 SQL 语句。

程序清单：codes/11/AManyToManyTest/src/org/fkit/test/ManyToManyTest.java

```
import org.apache.ibatis.session.SqlSession;
import org.fkit.domain.Order;
import org.fkit.factory.FKSqlSessionFactory;
import org.fkit.mapper.OrderMapper;
public class ManyToManyTest {

    public static void main(String[] args) {
        // 定义 SqlSession 变量
        SqlSession sqlSession = null;
        try {
            // 创建 SqlSession 实例
            sqlSession = FKSqlSessionFactory.getSqlSession();
            // 获取 OrderMapper 实例
            OrderMapper om = sqlSession.getMapper(OrderMapper.class);
            // 根据 id 查询 Order 对象
            Order order = om.selectById(1);
            // 查看查询到的 Order 对象
            System.out.println(order.getId() + " " + order.getCode() + " " + order.getTotal());
            // 查看 Order 关联的 User 对象
            System.out.println(order.getUser());
            // 查看关联的 Article 集合，因为配置使用的是 LAZY 懒加载，所以当使用时才执行 SQL 语句
            order.getArticles().forEach(article -> System.out.println(article));
            // 提交事务
            sqlSession.commit();
        } catch (Exception e) {
            // 回滚事务
            sqlSession.rollback();
            e.printStackTrace();
        }finally {
            // 关闭 SqlSession
            if(sqlSession != null)
                sqlSession.close();
        }
    }
}
```

运行 ManyToManyTest 类的 main 方法，该方法通过 SqlSession 的 getMapper(Class<T> type) 方法获得 mapper 接口的代理对象 OrderMapper。调用 selectById 方法时会执行该方法上的注解。Order 和 User 之间是多对一关系，使用@One 关联；Order 和 Article 之间是多对多关系，使用

@Many 关联。控制台显示如下：

```
    DEBUG [main] ==>  Preparing: SELECT * FROM TB_ORDER WHERE ID = ?
    DEBUG [main] ==> Parameters: 1(Integer)
    DEBUG [main] ====>  Preparing: SELECT * FROM TB_USER WHERE ID = ?
    DEBUG [main] ====> Parameters: 1(Integer)
    DEBUG [main] <====      Total: 1
    DEBUG [main] <==      Total: 1
    1 6aa3fa359ff14619b77fab5990940a2d 388.6
    User [id=1, username=杰克, loginname=jack, password=123456, phone=13920001616,
address=广州]
    DEBUG [main] ==>  Preparing: SELECT * FROM tb_article WHERE id IN (SELECT article_id
FROM tb_item WHERE order_id = ? )
    DEBUG [main] ==> Parameters: 1(Integer)
    DEBUG [main] <==      Total: 3
    Article [id=1, name=疯狂 Java 讲义, price=108.9, remark=李刚老师经典著作]
    Article [id=2, name=疯狂 Android 讲义, price=99.9, remark=李刚老师经典著作]
    Article [id=3, name=疯狂 iOS 讲义, price=89.9, remark=李刚老师经典著作]
```

可以看到，程序首先查询了 id 为 1 的 Order 的数据，因为使用立即加载，所以立即执行了查询关联的 User 的 SQL 语句；而多对多关联使用懒加载，当使用到关联的 Article 的数据时才执行了根据 order_id 查询 Article 的 SQL 语句。

示例：测试动态 SQL

MyBatis 的注解也支持动态 SQL。MyBatis 提供了各种注解，如@InsertProvider、@UpdateProvider、@DeleteProvider 和@SelectProvider，来帮助构建动态 SQL 语句，然后 MyBatis 可以执行这些 SQL 语句。

以上 4 个 Provider 注解都有 type 属性，该属性指定了一个类。method 属性指定该类的方法，其用来提供需要执行的 SQL 语句。使用字符串拼接的方法构建 SQL 语句是非常困难的，并且容易出错。所以 MyBaits 提供了一个 SQL 工具类 org.apache.ibatis.jdbc.SQL，该类不使用字符串拼接的方式，并且会以合适的空格前缀和后缀来构造 SQL 语句。

SQL 类的常用方法如下：

- T SELECT(String columns)。启动或追加 SELECT 子句，参数通常是一个以逗号分隔的列的列表。
- T FROM(String table)。启动或追加 FROM 子句。调用可以超过一次，这些参数通常是一个表名。
- T JOIN(String join)。向 JOIN 子句添加一个新的查询条件，该参数通常是一个表，也可以包括一个标准的连接返回的结果集。
- T INNER_JOIN(String join)。同 JOIN 子句，连接方式是内连接（INNER_JOIN）。
- T LEFT_OUTER_JOIN(String join)。同 JOIN 子句，连接方式是左外连接（LEFT_OUTER_JOIN）。
- T RIGHT_OUTER_JOIN(String join)。同 JOIN 子句，连接方式是右外连接（RIGHT_OUTER_JOIN）。
- T WHERE(String conditions)。追加一个新的 WHERE 子句条件，可以多次调用。
- T OR()。使用 OR 拆分当前 WHERE 子句条件，可以不止一次调用。
- T AND()。使用 AND 拆分当前 WHERE 子句条件，可以不止一次调用。
- T GROUP_BY(String columns)。追加一个新的 GROUP BY 子句元素。

- T HAVING(String conditions)。追加一个新的 HAVING 子句条件。
- T ORDER_BY(String columns)。追加一个新的 ORDER BY 子句元素。
- T INSERT_INTO(String tableName)。启动 INSERT 语句插入到指定表，应遵循由一个或多个 VALUES（）调用。
- T VALUES(String columns,String values)。追加的 INSERT 语句。第一个参数是要插入的列，第二个参数是插入的值。
- T DELETE_FROM(String table)。启动 DELETE 语句，并指定删除表。
- T UPDATE(String table)。启动一个更新（UPDATE）语句，并指定更新表。
- T SET(String sets)。追加一个 SET 更新语句列表。

动态 SQL provider 方法可以接受以下参数：

- 无参数
- java 对象
- java.util.Map

本示例使用了 10.2 节中的动态 SQL 的 tb_employee 表，数据库脚本、Employee.java、FKSqlSessionFactory.java、mybatis-config.xml 和 log4j.xml 文件请参考第 10 章内容，此处不再赘述。

程序清单：codes/11/ADynamicSQLTest/src/org/fkit/mapper/EmployeeMapper.java

```
import java.util.List;
import java.util.Map;
import org.apache.ibatis.annotations.DeleteProvider;
import org.apache.ibatis.annotations.InsertProvider;
import org.apache.ibatis.annotations.Options;
import org.apache.ibatis.annotations.SelectProvider;
import org.apache.ibatis.annotations.UpdateProvider;
import org.fkit.domain.Employee;
public interface EmployeeMapper {
    // 动态查询
@SelectProvider(type=EmployeeDynaSqlProvider.class,method="selectWhitParam")
    List<Employee> selectWhitParam(Map<String, Object> param);
}
```

这里我们使用了@SelectProvider 指定了一个类 EmployeeDynaSqlProvider 及其内部方法 selectWhitParam，用来提供需要执行的 SQL 语句。

程序清单：codes/11/ADynamicSQLTest/src/org/fkit/mapper/EmployeeDynaSqlProvider.java

```
import java.util.Map;
import org.apache.ibatis.jdbc.SQL;
import org.fkit.domain.Employee;
public class EmployeeDynaSqlProvider {
    public String selectWhitParam(Map<String, Object> param){
        return new SQL(){
            {
                SELECT("*");
                FROM("tb_employee");
                if(param.get("id") != null){
                    WHERE(" id = #{id} ");
                }
                if(param.get("loginname") != null){
                    WHERE(" loginname = #{loginname} ");
                }
                if(param.get("password") != null){
                    WHERE("password = #{password}");
```

```
                }
                if(param.get("name")!= null){
                    WHERE("name = #{name}");
                }
                if(param.get("sex")!= null){
                    WHERE("sex = #{sex}");
                }
                if(param.get("age")!= null){
                    WHERE("age = #{age}");
                }
                if(param.get("phone")!= null){
                    WHERE("phone = #{phone}");
                }
                if(param.get("sal")!= null){
                    WHERE("sal = #{sal}");
                }
                if(param.get("state")!= null){
                    WHERE("state = #{state}");
                }

            }
        }.toString();
    }
}
```

selectWhitParam 方法会根据参数 Map 中的内容构建动态的 SELECT 语句。

程序清单：codes/11/ADynamicSQLTest/src/org/fkit/test/DynamicSQLTest.java

```
import java.util.HashMap;
import java.util.List;
import java.util.Map;
import org.apache.ibatis.session.SqlSession;
import org.fkit.domain.Employee;
import org.fkit.factory.FKSqlSessionFactory;
import org.fkit.mapper.EmployeeMapper;
public class DynamicSQLTest {

    public static void main(String[] args) {
        // 定义 SqlSession 变量
        SqlSession sqlSession = null;
        try {
            // 创建 SqlSession 实例
            sqlSession = FKSqlSessionFactory.getSqlSession();
            DynamicSQLTest t = new DynamicSQLTest();
            // 获取 EmployeeMapper 对象
            EmployeeMapper em = sqlSession.getMapper(EmployeeMapper.class);
            // 测试 testSelectWhitParam 方法
            t.testSelectWhitParam(em);
            // 提交事务
            sqlSession.commit();
        } catch (Exception e) {
            // 回滚事务
            sqlSession.rollback();
            e.printStackTrace();
        }finally {
            // 关闭 SqlSession
            if(sqlSession != null)
                sqlSession.close();
        }
    }

    // 根据动态参数查询员工数据
    public void testSelectWhitParam(EmployeeMapper em){
```

```
        // 使用 Map 装载参数
        Map<String, Object> param = new HashMap<String, Object>();
        param.put("loginname", "jack");
        param.put("password", "123456");
        // 调用 selectWhitParam 方法
        List<Employee> list = em.selectWhitParam(param);
        // 查看返回结果
        System.out.println(list);
    }
}
```

运行 DynamicSQLTest 类的 main 方法，执行 testSelectWhitParam 方法，控制台显示如下：

```
    DEBUG [main] ==>  Preparing: SELECT * FROM tb_employee WHERE ( loginname = ? AND
password = ?)
    DEBUG [main] ==> Parameters: jack(String), 123456(String)
    DEBUG [main] <==      Total: 1
    [Employee [id=1, loginname=jack, password=123456, name=杰克, sex=男, age=26,
phone=13902019999, sal=9800.0, state=ACTIVE]]
```

查看 MyBatis 执行的 SQL 语句，因为 Map 中只是设置了 loginname 和 password 参数，所以执行的 SQL 语句是"WHERE (loginname = ? AND password = ?)"。

selectWhitParam 方法也可以传递 Employee 对象作为参数，示例代码如下：

```
    public String selectWhitParam(Employee employee){
        return new SQL(){
            {
                SELECT("*");
                FROM("tb_employee");
                if(employee.getId() != null){
                    WHERE(" id = #{id} ");
                }
                if(employee.getLoginname() != null){
                    WHERE(" loginname = #{loginname} ");
                }
                // 省略部分动态参数设置代码……
            }
        }.toString();
    }
```

读者可以多使用几种不同的参数组合测试 selectWhitParam 方法，观察控制台 SQL 语句，以便更好地理解动态 SQL 查询。

程序清单：codes/11/ADynamicSQLTest/src/org/fkit/mapper/EmployeeMapper.java

```
    // 动态插入
    @InsertProvider(type=EmployeeDynaSqlProvider.class,method="insertEmployee")
    @Options(useGeneratedKeys = true, keyProperty = "id")
    int insertEmployee(Employee employee);
```

程序清单：codes/11/ADynamicSQLTest/src/org/fkit/mapper/EmployeeDynaSqlProvider.java

```
    public String insertEmployee(Employee employee){
        return new SQL(){
            {
                INSERT_INTO("tb_employee");
                if(employee.getLoginname() != null){
                    VALUES("loginname", "#{loginname}");
                }
                if(employee.getPassword() != null){
                    VALUES("password", "#{password}");
                }
                if(employee.getName()!= null){
                    VALUES("name", "#{name}");
                }
```

```
            if(employee.getSex()!= null){
                VALUES("sex", "#{sex}");
            }
            if(employee.getAge()!= null){
                VALUES("age", "#{age}");
            }
            if(employee.getPhone()!= null){
                VALUES("phone", "#{phone}");
            }
            if(employee.getSal()!= null){
                VALUES("sal", "#{sal}");
            }
            if(employee.getState()!= null){
                VALUES("state", "#{state}");
            }
        }
    }.toString();
}
```

insertEmployee 方法会根据参数 Employee 对象的属性动态构建 INSERT 语句。

程序清单：codes/11/ADynamicSQLTest/src/org/fkit/test/DynamicSQLTest.java

```
    // 根据设置的属性动态插入数据
    public void testInsertEmployee(EmployeeMapper em){
        // 创建员工对象并设置属性
        Enployee e = new Employee();
        e.setLoginname("mary");
        e.setPassword("123456");
        e.setName("玛丽");
        e.setSex("女");
        e.setAge(20);
        e.setPhone("13902019999");
        e.setSal(9800.99);
        // 注意：没有设置 state 属性，则 insert 语句中不会包含 state 列
        // e.setState("ACTIVE");
        em.insertEmployee(e);
        System.out.println("插入成功，返回 id：" + e.getId());
    }
```

运行 DynamicSQLTest 类的 main 方法，执行 testInsertEmployee 方法，控制台显示如下：

```
    DEBUG [main] ==>  Preparing: INSERT INTO tb_employee (loginname, password, name,
sex, age, phone, sal) VALUES (?, ?, ?, ?, ?, ?, ?)
    DEBUG [main] ==> Parameters: mary(String), 123456(String), 玛丽(String), 女(String),
20(Integer), 13902019999(String), 9800.99(Double)
    DEBUG [main] <==    Updates: 1
    插入成功，返回 id：5
```

读者可以多使用几种不同的属性组合测试 insertEmployee 方法，观察控制台 SQL 语句，以便更好地理解动态 SQL 插入。

程序清单：codes/11/ADynamicSQLTest/src/org/fkit/mapper/EmployeeMapper.java

```
    // 根据 id 查询
    @SelectProvider(type=EmployeeDynaSqlProvider.class,method="selectWhitParam")
    Employee selectEmployeeWithId(Map<String, Object> param);
    // 动态更新
    @UpdateProvider(type=EmployeeDynaSqlProvider.class,method="updateEmployee")
    void updateEmployee(Employee employee);
```

程序清单：codes/11/ADynamicSQLTest/src/org/fkit/mapper/EmployeeDynaSqlProvider.java

```
    public String updateEmployee(Employee employee){
        return new SQL(){
```

```
        {
            UPDATE("tb_employee");
            if(employee.getLoginname() != null){
                SET("loginname = #{loginname}");
            }
            if(employee.getPassword() != null){
                SET("password = #{password}");
            }
            if(employee.getName()!= null){
                SET("name = #{name}");
            }
            if(employee.getSex()!= null){
                SET("sex = #{sex}");
            }
            if(employee.getAge()!= null){
                SET("age = #{age}");
            }
            if(employee.getPhone()!= null){
                SET("phone = #{phone}");
            }
            if(employee.getSal()!= null){
                SET("sal = #{sal}");
            }
            if(employee.getState()!= null){
                SET("state = #{state}");
            }
            WHERE(" id = #{id} ");
        }
    }.toString();
}
```

updateEmployee 方法会根据参数 Employee 对象的属性动态构建 update 语句。

程序清单：codes/11/ADynamicSQLTest/src/org/fkit/test/DynamicSQLTest.java

```
// 根据设置的属性动态更新数据
public void testUpdateEmployee(EmployeeMapper em){
    // 使用 Map 装载参数
    Map<String, Object> param = new HashMap<String, Object>();
    param.put("id", 5);
    // 查询 id 为 1 的员工
    Employee e = em.selectEmployeeWithId(param);
    // 修改员工对象的三个属性
    e.setLoginname("update");
    e.setPassword("fkjava");
    e.setName("测试");
    // 动态更新
    em.updateEmployee(e);
}
```

运行 DynamicSQLTest 类的 main 方法，执行 testUpdateEmployee 方法，控制台显示如下：

```
DEBUG [main] ==>  Preparing: SELECT * FROM tb_employee WHERE ( id = ? )
DEBUG [main] ==> Parameters: 5(Integer)
DEBUG [main] <==      Total: 1
DEBUG [main] ==>  Preparing: UPDATE tb_employee SET loginname = ?, password = ?, name = ?, sex = ?, age = ?, phone = ?, sal = ? WHERE ( id = ? )
DEBUG [main] ==> Parameters: update(String), fkjava(String), 测试(String), 女(String), 20(Integer), 13902019999(String), 9800.99(Double), 5(Integer)
DEBUG [main] <==    Updates: 1
```

读者可以多使用几种不同的属性组合测试 updateEmployee 方法，观察控制台 SQL 语句，以便更好地理解动态 SQL 更新。

程序清单：codes/11/ADynamicSQLTest/src/org/fkit/mapper/EmployeeMapper.java

```
// 动态删除
@DeleteProvider(type=EmployeeDynaSqlProvider.class,method="deleteEmployee")
void deleteEmployee(Map<String, Object> param);
```

程序清单：codes/11/ADynamicSQLTest/src/org/fkit/mapper/ EmployeeDynaSqlProvider.java

```
public String deleteEmployee(Map<String, Object> param){
    return new SQL(){
        {
            DELETE_FROM("tb_employee");
            if(param.get("id") != null){
                WHERE(" id = #{id} ");
            }
            if(param.get("loginname") != null){
                WHERE(" loginname = #{loginname} ");
            }
            if(param.get("password") != null){
                WHERE("password = #{password}");
            }
            if(param.get("name")!= null){
                WHERE("name = #{name}");
            }
            if(param.get("sex")!= null){
                WHERE("sex = #{sex}");
            }
            if(param.get("age")!= null){
                WHERE("age = #{age}");
            }
            if(param.get("phone")!= null){
                WHERE("phone = #{phone}");
            }
            if(param.get("sal")!= null){
                WHERE("sal = #{sal}");
            }
            if(param.get("state")!= null){
                WHERE("state = #{state}");
            }
        }
    }.toString();
}
```

deleteEmployee 方法会根据参数 Map 中的内容动态构建 DELETE 语句。

程序清单：codes/11/ADynamicSQLTest/src/org/fkit/test/DynamicSQLTest.java

```
// 根据设置的属性动态删除数据
public void testDeleteEmployee (EmployeeMapper em){
    // 使用 Map 装载参数
    Map<String, Object> param = new HashMap<String, Object>();
    param.put("loginname", "jack");
    param.put("password", "123456");
    // 动态删除
    em.deleteEmployee(param);
}
```

运行 DynamicSQLTest 类的 main 方法，执行 testDeleteEmployee 方法，控制台显示如下：

```
    DEBUG [main] ==>  Preparing: DELETE FROM tb_employee WHERE ( loginname = ? AND
password = ?)
    DEBUG [main] ==> Parameters: jack(String), 123456(String)
    DEBUG [main] <==    Updates: 1
```

读者可以多使用几种不同的属性组合测试 deleteEmployee 方法，观察控制台 SQL 语句，以便更好地理解动态 SQL 删除。

示例：测试调用存储过程

本示例使用了 10.3 节中调用存储过程示例的 tb_user 表，数据库脚本、User.java、FKSqlSessionFactory.java、mybatis-config.xml 和 log4j.xml 文件请参考第 10 章内容，此处不再赘述。

程序清单：codes/11/AProcedureTest/src/org/fkit/mapper/UserMapper.java

```
import java.util.List;
import org.apache.ibatis.annotations.Delete;
import org.apache.ibatis.annotations.Insert;
import org.apache.ibatis.annotations.Options;
import org.apache.ibatis.annotations.Select;
import org.apache.ibatis.annotations.Update;
import org.apache.ibatis.mapping.StatementType;
import org.fkit.domain.User;
public interface UserMapper {

    /**
     * 插入数据
     * @param user 包含要查询的 User 信息
     */
    @Insert("{call insert_user(#{id,mode=OUT,jdbcType=INTEGER},"
          + "#{name,mode=IN},#{sex,mode=IN},#{age,mode=IN})}")
    @Options(statementType=StatementType.CALLABLE)
    void saveUser(User user);

    /**
     * 查询所有数据
     * @return List<User> 包含所有 User 数据的 List 集合
     */
    @Select("{call select_user()}")
    @Options(statementType=StatementType.CALLABLE)
    List<User> selectUser();

    /**
     * 根据 id 查询对应的数据
     * @param id 用户的 id
     * @return User 包含用户信息的 User 对象
     */
    @Select("{call select_user_by_id(#{id,mode=IN})}")
    @Options(statementType=StatementType.CALLABLE)
    User selectUserById(Integer id);

    /**
     * 修改数据
     * @param user 包含要修改的 User 信息
     */
    @Update("{call update_user(#{id,mode=IN},#{name,mode=IN},#{sex,mode=IN},
#{age,mode=IN})}")
    @Options(statementType=StatementType.CALLABLE)
    void modifyUser(User user);

    /**
     * 删除数据
     * @param user 包含要修改的 User 信息
     */
    @Delete("{call delete_user_by_id(#{id,mode=IN})}")
    @Options(statementType=StatementType.CALLABLE)
    void removeUser(User user);
```

```
}
```

UserMapper.java 只是将之前写在 XML 文件当中调用存储过程的 SQL 语句写在了注解当中，注意需要使用@Options(statementType=StatementType.CALLABLE)提供调用存储过程的 CALLBALE 选项，其他并无不同。

程序清单：codes/11/AProcedureTest/src/org/fkit/mapper/AProcedureTest.java

```
import java.util.List;
import org.apache.ibatis.session.SqlSession;
import org.fkit.domain.User;
import org.fkit.factory.FKSqlSessionFactory;
import org.fkit.mapper.UserMapper;
public class AProcedureTest {

    public static void main(String[] args) throws Exception {
        // 定义 SqlSession 变量
        SqlSession sqlSession = null;
        try {
            // 创建 SqlSession 实例
            sqlSession = FKSqlSessionFactory.getSqlSession();
            AProcedureTest t = new AProcedureTest();
            // 测试插入数据
            t.testSaveUser(sqlSession);
            // 测试查询所有数据
            // t.testSelectUser(sqlSession);
            // 测试根据 id 查询对应的数据
            // t.testSelectUserById(sqlSession);
            // 测试修改数据
            // t.testModifyUser(sqlSession);
            // 测试删除数据
            t.testRemoveUser(sqlSession);
            // 提交事务
            sqlSession.commit();
        } catch (Exception e) {
            // 回滚事务
            sqlSession.rollback();
            e.printStackTrace();
        }finally {
            // 关闭 SqlSession
            if(sqlSession != null)
                sqlSession.close();
        }
    }

    // 测试插入数据
    public void testSaveUser(SqlSession sqlSession){
        // 获得 UserMapper 接口的代理对象
        UserMapper um = sqlSession.getMapper(UserMapper.class);
        User user = new User("test","男",23);
        // User user = new User("jack","男",20);
        // User user = new User("rose","女",18);
        // User user = new User("tom","男",21);
        // User user = new User("mary","女",19);
        // User user = new User("alice","女",18);
        // User user = new User("scott","男",25);
        // 调用保存 User 的方法
        um.saveUser(user);
        System.out.println("插入成功，返回的 id = " + user.getId());
    }
```

```
    // 测试查询所有数据
    public void testSelectUser(SqlSession sqlSession){
        // 获得 UserMapper 接口的代理对象
        UserMapper um = sqlSession.getMapper(UserMapper.class);
        // 调用查询所有 User 数据的方法
        List<User> users = um.selectUser();
        // 遍历打印每一个 User 对象
        users.forEach(user -> System.out.println(user));
    }

    // 测试根据 id 查询对应的数据
    public void testSelectUserById(SqlSession sqlSession){
        // 获得 UserMapper 接口的代理对象
        UserMapper um = sqlSession.getMapper(UserMapper.class);
        // 调用根据 id 查询 User 数据的方法
        User user = um.selectUserById(1);
        System.out.println(user);
    }

    // 测试修改数据
    public void testModifyUser(SqlSession sqlSession){
        // 获得 UserMapper 接口的代理对象
        UserMapper um = sqlSession.getMapper(UserMapper.class);
        // 查询出 id 为 1 的 User 对象
        User user = um.selectUserById(1);
        // 修改对象的属性
        user.setName("fkit");
        user.setSex("男");
        user.setAge(19);
        // 调用修改数据的方法
        um.modifyUser(user);
    }

    // 测试删除数据
    public void testRemoveUser(SqlSession sqlSession){
        // 获得 UserMapper 接口的代理对象
        UserMapper um = sqlSession.getMapper(UserMapper.class);
        // 查询出 id 为 1 的 User 对象
        User user = um.selectUserById(1);
        // 调用删除数据的方法
        um.removeUser(user);
    }
}
```

所有测试结果和 10.3 节中的测试结果完全一致，读者可以自行测试，此处不再赘述。

示例：测试二级缓存

本示例使用了 10.5.2 节中二级缓存示例的 tb_user 表，数据库脚本、User.java、FKSqlSessionFactory.java、mybatis-config.xml 和 log4j.xml 文件请参考第 10 章内容，此处不再赘述。

程序清单：codes/11/ATwoLevelCacheTest/src/org/fkit/mapper/UserMapper.java

```
import org.apache.ibatis.annotations.CacheNamespace;
import org.apache.ibatis.annotations.Delete;
import org.apache.ibatis.annotations.Options;
import org.apache.ibatis.annotations.Select;
import org.apache.ibatis.cache.decorators.LruCache;
import org.fkit.domain.User;
```

```
@CacheNamespace(eviction=LruCache.class , flushInterval=60000, size=512,
readWrite=true)
public interface UserMapper {

    // 根据 id 查询 User
    @Select("SELECT * FROM TB_USER WHERE id = #{id}")
    @Options(useCache=true)
    User selectUserById(Integer id);

    // 根据 id 删除 User
    @Delete("DELETE FROM TB_USER WHERE id = #{id}")
    void deleteUserById(Integer id);

}
```

UserMapper.java 只是将之前写在 XML 文件当中的二级缓存配置写在了注解当中，其他并无不同。

@CacheNamespace 注解用来配置二级缓存，该注解同样可以配置各项属性。

eviction=LruCache.class 表示使用回收策略的 class，所有回收策略的类型都位于 org.apache.ibatis.cache.decorators 包下。

flushInterval=60000 表示刷新间隔。

size=512 表示缓存数目。

readWrite=true 表示只读，和之前 XML 中的 readOnly 属性一样。

程序清单：codes/11/ATwoLevelCacheTest/src/org/fkit/mapper/ATwoLevelCacheTest.java

```
import org.apache.ibatis.session.SqlSession;
import org.fkit.domain.User;
import org.fkit.factory.FKSqlSessionFactory;
import org.fkit.mapper.UserMapper;

public class ATwoLevelCacheTest {

    public static void main(String[] args)  {
        // 使用工厂类获得 SqlSession 对象
        SqlSession sqlSession1 = FKSqlSessionFactory.getSqlSession();
        // 获得 UserMapping 对象
        UserMapper um = sqlSession1.getMapper(UserMapper.class);
        // 查询 id 为 1 的 User 对象，会执行 select 语句
        User user = um.selectUserById(1);
        System.out.println(user);
        // 关闭一级缓存
        sqlSession1.close();
        // 重新获取一级缓存
        SqlSession sqlSession2 = FKSqlSessionFactory.getSqlSession();
        // 再次获得 UserMapping 对象
        um = sqlSession2.getMapper(UserMapper.class);
        // 再次查询 id 为 1 的 User 对象，虽然现在使用的是一个新的 SqlSession 对象，但是二级缓存
        // 中缓存了数据，所以不会再次执行 select 语句
        User user2 = um.selectUserById(1);
        System.out.println(user2);
        // 关闭 SqlSession 对象
        sqlSession2.close();
    }

}
```

所有测试结果和 10.5.2 节中的测试结果完全一致，读者可以自行测试，此处不再赘述。

11.3 本章小结

本章详细介绍了 MyBatis 的常用注解，包括增删改查和一对一关联、一对多关联、多对多关联、动态 SQL、调用存储过程和使用二级缓存的操作。

第 12 章将重点介绍 Spring MVC 和 MyBatis 的整合。

CHAPTER

12

第12章 Spring 5 整合 MyBatis 3

本章要点

- Spring 5 整合 MyBatis 3 配置
- Spring 5 整合 MyBatis 3 的优势

前面章节详细介绍了 Spring MVC 的常用功能和 MyBatis 的基本用法。在实际的项目开发中，我们需要将 Spring 和 MyBatis 进行整合，从而使用 Spring 依赖注入以减少代码的耦合，使用 Spring MVC 处理请求并作出响应，使用 MyBatis 更加简捷地完成数据库操作。本章将重点介绍 Spring 和 MyBatis 的整合过程。

12.1 开发环境搭建

- 安装 JDK 8，配置环境变量 JAVA_HOME、PATH 和 CLASSPATH。
- 安装 Eclipse 并配置 JRE 环境。
- 安装 Tomcat 8，并在 Eclipse 中集成 Tomcat 8。
- 安装 MySQL 5 以上数据库。

提示

关于开发环境的内容请参考《疯狂 Java 讲义》。

12.2 准备所需的 jar 包

- **Spring 框架 jar 包**。spring-framework-5.0.1.RELEASE 文件夹下 libs 目录下所有模块 class 文件的 jar 包和 Spring 的核心容器必须依赖的 common-logging 的 jar 包（本书示例是 commons-logging-1.2.jar），共 22 个。
- **MyBatis 框架 jar 包**。mybatis-3.4.5.jar 和 mybatis-3.4.5 文件夹下的 lib 目录下所有 jar 包，共 13 个。
- **MyBatis 整合 Spring 中间件 jar 包**。根据 MyBatis 官方的说法，在 MyBatis 3 问世之前，Spring 3 的开发工作就已经完成了，所以 Spring 3 中没有提供对 MyBatis 3 的支持。因此由 MyBatis 社区自己开发了一个 MyBatis-Spring 中间件用来满足 MyBatis 用户整合 Spring 的需求，该中间件有如下两个作用：
  - 在 Spring 中配置 MyBatis 工厂类。
  - 在 DAO 层使用 Spring 注入的工具 Bean 对数据进行操作。

  本书成书时该中间件最高版本是 mybatis-spring-1.3.1.jar。
- **aspectj 框架 jar 包**。aspectjrt.jar、aspectjtools.jar、aspectjweaver.jar 和 org.aspectj.matcher.jar。本书成书时 aspectj 框架最高版本是 aspectj-1.8.13。
- **数据库驱动 jar 包**。本书成书时最高版本是 mysql-connector-java-5.1.44.jar。
- **数据源 C3P0 所需 jar 包**。本书成书时最高版本是 c3p0-0.9.5.2.jar、hibernate-c3p0-5.2.10.Final.jar 和 mchange- commons-java-0.2.11.jar。
- **JSTL 标签库 jar 包**。javax.servlet.jsp.jstl-1.2.1.jar 和 javax.servlet.jsp.jstl-api-1.2.1.jar。

12.3 准备数据库资源

在 mybatis 数据库中创建两个表 tb_user 和 tb_book，并插入测试数据，SQL 脚本如下：

```
# 如果存在 tb_user 表则删除
DROP TABLE IF EXISTS tb_user;
```

```
#创建用户表
CREATE TABLE tb_user(
id INT PRIMARY KEY AUTO_INCREMENT,  #id
loginname VARCHAR(50) UNIQUE,       #登录名 邮箱
PASSWORD VARCHAR(18),               #密码
username VARCHAR(18),               #用户名
phone VARCHAR(18),                  #电话
address VARCHAR(255)                #地址
);
INSERT INTO tb_user(loginname,PASSWORD,username,phone,address)
VALUES('jack','123456','杰克','13920001234','广州市天河区');
# 如果存在 tb_book 表则删除
DROP TABLE IF EXISTS tb_book;
CREATE TABLE tb_book (
   id INT (11) PRIMARY KEY AUTO_INCREMENT,
   name VARCHAR (54),
   author VARCHAR (54),
   publicationdate DATE ,
   publication VARCHAR (150),
   price DOUBLE ,
   image VARCHAR (54),
   remark VARCHAR (600)
);
INSERT INTO tb_book (id, name, author, publicationdate, publication, price, image,
remark) VALUES('1','疯狂 Java 讲义(附光盘)','李刚 编著','2008-10-01','电子工业出版社
','74.2','java.jpg','疯狂源自梦想,技术成就辉煌 本书来自作者 3 年的 Java 培训经历,凝结了作者近
3000 个小时的授课经验,总结了几百个 Java 学员学习过程中的典型错误.');
INSERT INTO tb_book (id, name, author, publicationdate, publication, price, image,
remark) VALUES('2','轻量级 Java EE 企业应用实战','李刚 编著','2008-11-01','电子工业出版社
','59.2','ee.jpg','本书主要介绍以 Spring+Hibernate 为基础的 Java EE 应用.');
INSERT INTO tb_book (id, name, author, publicationdate, publication, price, image,
remark) VALUES('3','疯狂 Android 讲义(含 CD 光盘 1','李刚 编著','2011-07-01','电子工业出版社
','60.6','android.jpg','计算机便携化是未来的发展趋势，而 Android 作为最受欢迎的手机、平板电脑操
作系统之一，其发展的上升势头是势不可当的。而 Android 应用选择了 Java 作为其开发语言，对于 Java 来说也
是一次极好的机会。');
INSERT INTO tb_book (id, name, author, publicationdate, publication, price, image,
remark) VALUES('4','疯狂 Ajax 讲义(含 CD 光盘 1 张)','李刚 编著','2011-07-01','电子工业出版社
','66.6','ajax.jpg','异步访问技术,现在所有网站都在用.');
```

12.4　完成配置文件

新建一个项目 fkbookapp，加入在 12.2 节中准备的 jar 包。

程序清单：codes/12/fkbookapp/src/db.properties

```
dataSource.driverClass=com.mysql.jdbc.Driver
dataSource.jdbcUrl=jdbc:mysql://127.0.0.1:3306/mybatis
dataSource.user=root
dataSource.password=root
dataSource.maxPoolSize=20
dataSource.maxIdleTime = 1000
dataSource.minPoolSize=6
dataSource.initialPoolSize=5
```

程序清单：codes/12/fkbookapp/WebContent/WEB-INF/applicationContext.xml

```
<?xml version="1.0" encoding="UTF-8"?>
<beans xmlns="http://www.springframework.org/schema/beans"
   xmlns:mybatis="http://mybatis.org/schema/mybatis-spring"
   xmlns:xsi="http://www.w3.org/2001/XMLSchema-instance"
   xmlns:p="http://www.springframework.org/schema/p"
```

```
    xmlns:context="http://www.springframework.org/schema/context"
    xmlns:mvc="http://www.springframework.org/schema/mvc"
    xmlns:tx="http://www.springframework.org/schema/tx"
    xsi:schemaLocation="http://www.springframework.org/schema/beans

http://www.springframework.org/schema/beans/spring-beans.xsd
                        http://www.springframework.org/schema/context
                        http://www.springframework.org/schema/context/spring-context.xsd
                        http://www.springframework.org/schema/mvc
                        http://www.springframework.org/schema/mvc/spring-mvc.xsd
                        http://www.springframework.org/schema/tx
                        http://www.springframework.org/schema/tx/spring-tx.xsd
                        http://mybatis.org/schema/mybatis-spring
http://mybatis.org/schema/mybatis-spring.xsd ">
     <!-- mybatis:scan 会将 org.fkit.mapper 包里的所有接口当作 mapper 配置，之后可以自动引
      入 mapper 类-->
    <mybatis:scan base-package="org.fkit.mapper"/>
     <!-- 扫描 org.fkit 包下面的 java 文件,若有 Spring 的相关注解的类,则把这些类注册为 Spring
      的 bean -->
    <context:component-scan base-package="org.fkit"/>
    <!-- 使用 PropertyOverrideConfigurer 后处理器加载数据源参数 -->
    <context:property-override location="classpath:db.properties"/>
    <!-- 配置 c3p0 数据源 -->
    <bean id="dataSource" class="com.mchange.v2.c3p0.ComboPooledDataSource"/>
    <!-- 配置 SqlSessionFactory,org.mybatis.spring.SqlSessionFactoryBean 是 MyBatis
     社区开发用于整合 Spring 的 bean -->
    <bean id="sqlSessionFactory" class="org.mybatis.spring.SqlSessionFactoryBean"
        p:dataSource-ref="dataSource"/>
    <!-- JDBC 事务管理器 -->
    <bean id="transactionManager"
    class="org.springframework.jdbc.datasource.DataSourceTransactionManager"
        p:dataSource-ref="dataSource"/>
    <!-- 启用支持 annotation 注解方式事务管理 -->
    <tx:annotation-driven transaction-manager="transactionManager"/>
</beans>
```

程序清单：codes/12/fkbookapp/WebContent/WEB-INF/springmvc-config.xml

```
<?xml version="1.0" encoding="UTF-8"?>
<beans xmlns="http://www.springframework.org/schema/beans"
    xmlns:xsi="http://www.w3.org/2001/XMLSchema-instance"
    xmlns:mvc="http://www.springframework.org/schema/mvc"
    xmlns:context="http://www.springframework.org/schema/context"
    xsi:schemaLocation="
        http://www.springframework.org/schema/beans
        http://www.springframework.org/schema/beans/spring-beans.xsd
        http://www.springframework.org/schema/mvc
        http://www.springframework.org/schema/mvc/spring-mvc.xsd
        http://www.springframework.org/schema/context
        http://www.springframework.org/schema/context/spring-context.xsd">
    <!-- 自动扫描该包，Spring MVC 会将包下使用了@controller 注解的类注册为 Spring 的
     controller -->
    <context:component-scan base-package="org.fkit.controller"/>
    <!-- 设置默认配置方案 -->
    <mvc:annotation-driven/>
     <!-- 视图解析器  p:prefix 属性表示前缀  p:suffix 表示后缀  -->
       <bean id="viewResolver"
  class="org.springframework.web.servlet.view.InternalResourceViewResolver"
  p:prefix="/WEB-INF/content/" p:suffix=".jsp"/>
</beans>
```

程序清单：codes/12/fkbookapp/WebContent/WEB-INF/web.xml

```
<?xml version="1.0" encoding="UTF-8"?>
<web-app xmlns:xsi="http://www.w3.org/2001/XMLSchema-instance"
   xmlns="http://xmlns.jcp.org/xml/ns/javaee"
   xsi:schemaLocation="http://xmlns.jcp.org/xml/ns/javaee
   http://xmlns.jcp.org/xml/ns/javaee/web-app_3_1.xsd"
   id="WebApp_ID" version="3.1">
   <!-- 配置spring核心监听器，默认会以 /WEB-INF/applicationContext.xml 作为配置文件 -->
   <listener>
      <listener-class>org.springframework.web.context.ContextLoaderListener
       </listener-class>
   </listener>
   <!-- contextConfigLocation 参数用来指定 Spring 的配置文件 -->
   <context-param>
      <param-name>contextConfigLocation</param-name>
      <param-value>/WEB-INF/applicationContext*.xml</param-value>
   </context-param>
   <!-- 定义 Spring MVC 的前端控制器 -->
  <servlet>
   <servlet-name>springmvc</servlet-name>
   <servlet-class>
      org.springframework.web.servlet.DispatcherServlet
   </servlet-class>
   <init-param>
     <param-name>contextConfigLocation</param-name>
     <param-value>/WEB-INF/springmvc-config.xml</param-value>
   </init-param>
   <load-on-startup>1</load-on-startup>
  </servlet>
  <!-- 让Spring MVC的前端控制器拦截所有请求 -->
  <servlet-mapping>
   <servlet-name>springmvc</servlet-name>
   <url-pattern>/</url-pattern>
  </servlet-mapping>
  <!-- 编码过滤器 -->
  <filter>
      <filter-name>characterEncodingFilter</filter-name>
      <filter-class>org.springframework.web.filter.CharacterEncodingFilter
</filter-class>
      <init-param>
         <param-name>encoding</param-name>
         <param-value>UTF-8</param-value>
      </init-param>
 </filter>
   <filter-mapping>
      <filter-name>characterEncodingFilter</filter-name>
      <url-pattern>/*</url-pattern>
   </filter-mapping>
</web-app>
```

12.5 持久层功能实现

程序清单：codes/12/fkbookapp/src/org/fkit/domain/User.java

```
import java.io.Serializable;
public class User implements Serializable{
   private Integer id;            // id
   private String loginname;      // 登录名
   private String password;       // 密码
   private String username;       // 用户名
   private String phone;          // 电话
```

```
    private String address;              // 地址
    // 省略构造器和 set/get 方法......
}
```

程序清单：codes/12/fkbookapp/src/org/fkit/domain/Book.java

```
import java.io.Serializable;
import java.util.Date;
public class Book implements Serializable{
    private Integer id;                  // id
    private String name;                 // 书名
    private String author;               // 作者
    private String publication;          // 出版社
    private Date publicationdate;        // 出版日期
    private Double price;                // 价格
    private String image;                // 封面图片
    private String remark;               // 详细描述
    // 省略构造器和 set/get 方法......
}
```

程序清单：codes/12/fkbookapp/src/org/fkit/mapper/UserMapper.java

```
import org.apache.ibatis.annotations.Param;
import org.apache.ibatis.annotations.Select;
import org.fkit.domain.User;
public interface UserMapper {
    /**
     * 根据登录名和密码查询用户
     * @param String loginname
     * @param String password
     * @return 找到返回 User 对象，没有找到返回 null
     * */
    @Select("select * from tb_user where loginname = #{loginname} and password =
     #{password}")
    User findWithLoginnameAndPassword(@Param("loginname")String loginname,
            @Param("password") String password);
}
```

程序清单：codes/12/fkbookapp/src/org/fkit/mapper/BookMapper.java

```
import java.util.List;
import org.apache.ibatis.annotations.Select;
import org.fkit.domain.Book;
public interface BookMapper {
    /**
     * 查询所有图书
     * @return 图书对象集合
     * */
    @Select(" select * from tb_book ")
    List<Book> findAll();
}
```

持久层包括和数据库表映射的User.java和Book.java两个JavaBean对象，并使用了MyBatis的注解映射了对应的 SQL 语句。

12.6 服务层功能实现

程序清单：codes/12/fkbookapp/src/org/fkit/service/UserService.java

```
import org.fkit.domain.User;
/**
 * User 服务层接口
```

```
 * */
public interface UserService {
    /**
     * 判断用户登录
     * @param String loginname
     * @param String password
     * @return 找到返回User对象，没有找到返回null
     * */
    User login(String loginname,String password);
}
```

程序清单：codes/13/fkbookapp/src/org/fkit.service/impl/UserServiceImpl.java

```
import org.fkit.domain.User;
import org.fkit.mapper.UserMapper;
import org.fkit.service.UserService;
import org.springframework.beans.factory.annotation.Autowired;
import org.springframework.stereotype.Service;
import org.springframework.transaction.annotation.Isolation;
import org.springframework.transaction.annotation.Propagation;
import org.springframework.transaction.annotation.Transactional;
/**
 * User服务层接口实现类
 * @Service("userService")用于将当前类注释为一个Spring的bean，名为userService
 * */
@Service("userService")
public class UserServiceImpl implements UserService {
    /**
     * 自动注入UserMapper
     * */
    @Autowired
    private UserMapper userMapper;
    /**
     * UserService接口login方法实现
     * @see { UserService }
     * */
    @Override
    public User login(String loginname, String password) {
        return userMapper.findWithLoginnameAndPassword(loginname, password);
    }
}
```

程序清单：codes/13/fkbookapp/src/org/fkit/service/BookService.java

```
import java.util.List;
import org.fkit.domain.Book;
/**
 * Book服务层接口
 * */
public interface BookService {
    /**
     * 查找所有图书
     * @return Book对象集合
     * */
    List<Book> getAll();
}
```

程序清单：codes/13/fkbookapp/src/org/fkit/service/impl/BookServiceImpl.java

```
import java.util.List;
import org.fkit.domain.Book;
import org.fkit.mapper.BookMapper;
import org.fkit.service.BookService;
import org.springframework.beans.factory.annotation.Autowired;
import org.springframework.stereotype.Service;
```

```
import org.springframework.transaction.annotation.Isolation;
import org.springframework.transaction.annotation.Propagation;
import org.springframework.transaction.annotation.Transactional;
/**
 * Book 服务层接口实现类
 * @Service("bookService")用于将当前类注释为一个 Spring 的 bean，名为 bookService
 * */
@Service("bookService")
public class BookServiceImpl implements BookService {
    /**
     * 自动注入 BookMapper
     * */
    @Autowired
    private BookMapper bookMapper;
    /**
     * BookService 接口 getAll 方法实现
     * @see { BookService }
     * */
    @Override
    public List<Book> getAll() {
        return bookMapper.findAll();
    }
}
```

服务层使用了 Spring 的@Autowired 注解自动注入持久层的 Mapper 对象，并且使用了@Service 注解将类注释成为 Spring 的 Bean。

12.7 控制层功能实现

程序清单：codes/12/fkbookapp/src/org/fkit/controller/FormController.java

```
import org.springframework.stereotype.Controller;
import org.springframework.web.bind.annotation.RequestMapping;
/**
 * 页面跳转控制器
 * */
@Controller
public class FormController{
    @RequestMapping(value= "/loginForm")
     public String loginForm(){
        // 跳转页面
        return "/loginForm";
    }
}
```

程序清单：codes/12/fkbookapp/src/org/fkit/controller/UserController.java

```
import javax.servlet.http.HttpSession;
import org.fkit.domain.User;
import org.fkit.service.UserService;
import org.springframework.beans.factory.annotation.Autowired;
import org.springframework.beans.factory.annotation.Qualifier;
import org.springframework.stereotype.Controller;
import org.springframework.web.bind.annotation.PostMapping;
import org.springframework.web.servlet.ModelAndView;
import org.springframework.web.servlet.view.RedirectView;
/**
 * 处理用户请求控制器
 * */
@Controller
public class UserController {
    /**
```

```
     * 自动注入UserService
     * */
    @Autowired
    @Qualifier("userService")
    private UserService userService;
    /**
     * 处理/login请求
     * */
    @PostMapping(value="/login")
     public ModelAndView login(
            String loginname,String password,
            ModelAndView mv,
            HttpSession session){
        // 根据登录名和密码查找用户，判断用户登录
        User user = userService.login(loginname, password);
        if(user != null){
            // 登录成功，将user对象设置到HttpSession作用域
            session.setAttribute("user", user);
            // 转发到main请求
            mv.setView(new RedirectView("/fkbookapp/main"));
        }else{
            // 登录失败，设置失败提示信息，并跳转到登录页面
            mv.addObject("message", "登录名或密码错误，请重新输入!");
            mv.setViewName("loginForm");
        }
        return mv;
    }
}
```

程序清单：codes/12/fkbookapp/src/org/fkit/controller/BookController.java

```
import java.util.List;
import org.fkit.domain.Book;
import org.fkit.service.BookService;
import org.springframework.beans.factory.annotation.Autowired;
import org.springframework.beans.factory.annotation.Qualifier;
import org.springframework.stereotype.Controller;
import org.springframework.ui.Model;
import org.springframework.web.bind.annotation.RequestMapping;
/**
 * 处理图书请求控制器
 * */
@Controller
public class BookController {
    /**
     * 自动注入BookService
     * */
    @Autowired
    @Qualifier("bookService")
    private BookService bookService;
    /**
     * 处理/main请求
     * */
    @RequestMapping(value="/main")
     public String main(Model model){
        // 获得所有图书集合
        List<Book> book_list = bookService.getAll();
        // 将图书集合添加到model当中
        model.addAttribute("book_list", book_list);
        // 跳转到main页面
        return "main";
    }
}
```

控制层使用了 Spring 的@Autowired 注解自动注入服务层的 Service 对象，@Qualifier 注解用于指明需要注入的具体类型，并且使用@Controller 注解将类注释成为 Spring MVC 的 Controller。

12.8 JSP 页面

程序清单：codes/12/fkbookapp/WebContent/WEB-INF/content/loginForm.jsp

```
<html>
<head>
<meta http-equiv="Content-Type" content="text/html; charset=UTF-8">
<title>登录页面</title>
</head>
<body>
<h3>登录页面</h3>
<form action="login" method="post">
   <font color="red">${requestScope.message }</font>
    <table>
       <tr>
          <td><label>登录名：</label></td>
          <td><input type="text" id="loginname" name="loginname" ></td>
       </tr>
       <tr>
          <td><label>密码：</label></td>
          <td><input type="password" id="password" name="password" ></td>
       </tr>
       <tr>
          <td><input type="submit" value="登录"></td>
       </tr>
    </table>
</form>
</body>
</html>
```

程序清单：codes/12/fkbookapp/WebContent/WEB-INF/content/main.jsp

```
<html>
<head>
<meta http-equiv="Content-Type" content="text/html; charset=UTF-8">
<title>首页</title>
<style type="text/css">
   table{border-collapse:collapse;border-spacing:0;border-left:1px solid #888;
border-top:1px solid #888;background:#efefef;}
   th,td{border-right:1px solid #888;border-bottom:1px solid #888;padding:5px
15px;}
   th{font-weight:bold;background:#ccc;}
</style>
</head>
<body>
欢迎[${sessionScope.user.username }]访问
<br>
<table border="1">
   <tr>
      <th>封面</th><th>书名</th><th>作者</th><th>价格</th>
   </tr>
   <c:forEach items="${requestScope.book_list }" var="book">
      <tr>
         <td><img src="images/${book.image }" height="60"></td>
         <td>${book.name }</td>
         <td>${book.author }</td>
         <td>${book.price }</td>
```

```
        </tr>
    </c:forEach>
</table>
</body>
</html>
```

loginForm.jsp 是一个登录页面，可以在此输入登录名和密码进行登录。登录完成后 main.jsp 显示从数据库 tb_book 表读取出的书籍信息。

12.9 测试 Spring 5 整合 MyBatis 3

部署 fkbookapp 项目，在浏览器中输入如下 URL 来测试应用：

```
http://localhost:8080/fkbookapp/loginForm
```

会看到如图 12.1 所示的界面，表示 Spring MVC 访问成功。

图 12.1 登录页面

输入登录名“jack”，密码“123456”，单击“登录”按钮，请求将会被提交到 UserController 类的 login 方法进行登录验证。验证成功，请求将会被转发到 BookController 的 main 方法，获取所有书籍信息，之后跳转到 main.jsp 页面，如图 12.2 所示。

图 12.2 书籍页面

如果登录名和密码输入错误，会重新跳转到登录页面提示用户重新输入，如图 12.3 所示。

图 12.3　登录失败页面

12.10　本章小结

本章详细介绍了 Spring 5 整合 MyBatis 3 的过程，重点在于 MyBatis 的注解映射、Spring MVC 的配置和注解映射、Spring 整合 MyBatis 的配置。

第 13 章将重点介绍 Spring 5 整合 MyBatis 3 的人事管理系统项目。

CHAPTER

13

第13章 实战项目：人事管理系统

本章要点

- 人事管理系统功能模块设计
- 人事管理系统持久化组件设计
- 人事管理系统业务逻辑组件设计
- Spring MVC 整合 MyBatis 的优势

本章将会综合运用前面章节所介绍的知识来开发一个完整的人事管理（HRM）系统。该系统包括用户管理、部门管理、职位管理、员工管理、公告管理、下载中心等常用的人事管理系统功能。

本系统采用前面介绍的 Java EE 架构：Spring MVC 5 + MyBatis 3，该系统架构成熟，性能良好，运行稳定。Spring 的 IoC 容器负责管理业务逻辑组件、持久层组件及控制层组件，这样可以充分利用 Spring 的依赖注入的优势，进一步增强系统的解耦性，从而提高应用的可扩展性，降低系统重构的成本。

13.1 项目简介及系统结构

人事管理系统是办公信息自动化建设中最常见的项目，几乎每个 OA 系统中都需要人事管理系统。图 13.1 显示了我们开发的人事管理系统界面和功能菜单。

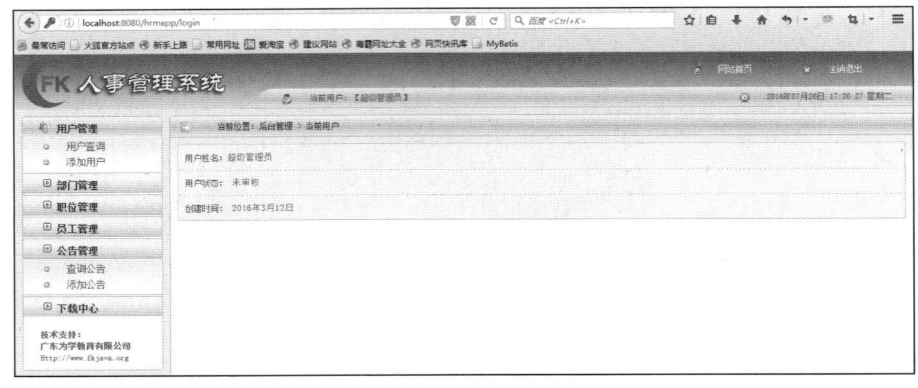

图 13.1　人事管理系统界面和功能菜单

该项目包含了用户管理、部门管理、职位管理、员工管理、公告管理、下载中心等多个模块，页面使用 jQuery 框架完成动态功能（关于 jQuery 的知识请参考“疯狂 Java 系列”之《疯狂前端开发讲义》），用户管理、部门管理等模块包含了实际项目开发中常用的增删改查动作，下载中心包含了 Spring MVC 的文件上传、下载等功能。

13.1.1　系统功能介绍

用户管理的功能包括：添加用户，用户可以为管理员或者普通用户；查询用户，可以查询所有用户或根据用户名和用户状态进行模糊查询；删除用户；修改用户。

部门管理的功能包括：添加部门；查询部门，可以查询所有部门或根据部门名称进行模糊查询；删除部门；修改部门。

职位管理的功能包括：添加职位；查询职位，可以查询所有职位或根据职位名称进行模糊查询；删除职位；修改职位。

员工管理的功能包括：添加员工；查询员工，可以查询所有员工或根据员工姓名、身份证号、手机号、性别、职位、部门进行模糊查询；删除员工；修改员工。

公告管理的功能包括：添加公告；查询公告，可以查询所有公告或根据公告名称、公告内容进行模糊查询；删除公告；修改公告。

下载中心的功能包括：上传文件；查询文件，可以查询所有文件或根据文件标题进行模糊查询；预览文件内容；删除文件；下载文件。

所有查询页面统一使用分页处理。

13.1.2 相关技术介绍

本系统主要涉及三个开源框架：MyBatis 3、Spring 5 和表现层框架 jQuery1.11。同时还使用了 JSP 作为表现层技术。本系统将这 4 种技术有机地结合在一起，从而构建出一个健壮的 Java EE 应用。

1. 传统表现层技术：JSP

本系统使用 JSP 作为表现层，负责收集用户请求数据以及业务数据的表示。

JSP 是最传统也最有效的表现层技术。本系统的 JSP 页面是单纯的表现层，所有的 JSP 页面不再使用 Java 脚本。结合 EL 表达式和 JSTL 标签库，JSP 可完成全部的表现层功能，如数据收集、数据展现。

2. MVC 框架

本系统使用 Spring MVC 作为 MVC 框架。Spring MVC 是一个设计优良的 MVC 框架，大有取代 Struts 2 之势。本应用的所有用户请求，包括系统的超链接和表单提交等，都不再直接发送到表现层 JSP 页面，而是必须发送给 Spring MVC 的 Controller，Spring MVC 控制所有请求的处理和转发。

通过 Controller 拦截所有请求有个好处：将所有的 JSP 页面放入 WEB-INF/路径下，可以避免用户直接访问 JSP 页面，从而提高系统的安全性。

本应用使用基于 Spring MVC 拦截器的权限控制，应用中控制器没有进行权限检查，但每个控制器都需要检查调用者是否有足够的访问权限，这种通用操作正是 Spring MVC 拦截器的优势所在。

3. Spring 框架的作用

Spring 框架是系统的核心部分，Spring 提供的 IoC 容器是业务逻辑组件和 DAO 组件的工厂，它负责生成并管理这些实例。

借助于 Spring 的依赖注入，各组件以松耦合的方式组合在一起，组件与组件之间的依赖正是通过 Spring 的依赖注入管理的。其 Service 组件和 DAO 对象都采用面向接口编程的方式，从而降低了系统重构的成本，极好地提高了系统的可维护性、可扩展性。

应用事务采用 Spring 的声明式事务框架。通过声明式事务，无须将事务策略以硬编码的方式与代码耦合在一起，而是放在配置文件中声明，使业务逻辑组件可以更加专注于业务的实现，从而简化开发。同时，声明式事务降低了不同事务策略的切换代价。

4. MyBatis 的作用

MyBatis 的 SQL Mapping 功能简化了数据库的访问，并在 JDBC 层提供了更好的封装。

MyBatis 可以优雅及灵活的方式操作数据库，由此开发者可以更灵活地编写 SQL 语句，满足更好的性能需求，为底层 DAO 对象的实现提供支持。

13.1.3 系统结构

本系统采用严格的 Java EE 应用结构，主要有如下几个分层：

- **表现层**。由 JSP 页面组成。
- **MVC 层**。使用 Spring MVC 技术。

- **业务逻辑层**。主要由 Spring IoC 容器管理的业务逻辑组件组成。
- **DAO 层**。由 6 个 DAO 组件组成。
- **领域对象层**。由 6 个 Domain Object 对象组成。
- **数据库服务层**。使用 MySQL 数据库存储持久化数据。

本应用中的领域对象实际上只是一些简单的 JavaBean 类，并未提供任何业务逻辑方法，所有的业务逻辑方法都由系统的业务逻辑组件来提供。这种模式简单、直接，系统分层清晰，比较适用于实际项目开发。

13.1.4 系统的功能模块

本系统可以大致分为 6 个模块：用户管理、部门管理、职位管理、员工管理、公告管理和文件下载，其主要业务逻辑通过 HrmService 业务逻辑组件实现，因此可以使用这个业务逻辑组件来封装 DAO 组件。

提示

通常建议按细粒度的模块来设计 Service 组件，让业务逻辑组件作为 DAO 组件的门面，这符合门面模式的设计。同时让 DAO 组件负责系统持久化逻辑，可以将系统在持久化技术这个维度上的变化独立出去，而业务逻辑组件负责业务逻辑这个维度的改变。

系统以业务逻辑组件作为 DAO 组件的门面，封装这些 DAO 组件，业务逻辑组件底层依赖于这些 DAO 组件，向上实现系统的业务逻辑功能。

本系统主要有如下 6 个 DAO 对象：

- UserDao。提供对 user_inf 表的基本操作。
- DeptDao。提供对 dept_inf 表的基本操作。
- JobDao。提供对 job_inf 表的基本操作。
- EmployeeDao。提供对 employee_inf 表的基本操作。
- NoticeDao。提供对 notice_inf 表的基本操作。
- DocumentDao。提供对 document_inf 表的基本操作。

本系统还提供一个业务逻辑组件：

- HrmService。提供所有的业务逻辑功能的实现。

13.2 数据表和持久化类

通过使用 MyBatis 持久层，可以避免使用传统的 JDBC 方式来操作数据库，如此可以使 SQL 语句更加灵活，并且可以直接通过面向对象的方式操作数据。

13.2.1 设计数据库表

首先，根据模块功能设计数据库表，具体 SQL 语句如下：

```
#创建数据库 hrm_db
CREATE DATABASE hrm_db;
#使用数据库 hrm_db
USE hrm_db;
#创建表 dept_inf
```

```
CREATE TABLE dept_inf (
  ID INT(11) NOT NULL AUTO_INCREMENT,
  NAME VARCHAR(50) NOT NULL,
  REMARK VARCHAR(300) DEFAULT NULL,
  PRIMARY KEY (ID)
) ENGINE=INNODB AUTO_INCREMENT=10 DEFAULT CHARSET=utf8;
INSERT  INTO dept_inf(ID,NAME,REMARK) VALUES (1,'技术部','技术部'),(2,'运营部','运营
 部'),(3,'财务部','财务部'),(5,'总公办','总公办'),(6,'市场部','市场部'),(7,'教学部','教
 学部');
#创建表job_inf
CREATE TABLE job_inf (
  ID INT(11) NOT NULL AUTO_INCREMENT,
  NAME VARCHAR(50) NOT NULL,
  REMARK VARCHAR(300) DEFAULT NULL,
  PRIMARY KEY (ID)
) ENGINE=INNODB AUTO_INCREMENT=10 DEFAULT CHARSET=utf8;
INSERT  INTO job_inf(ID,NAME,REMARK) VALUES (1,'职员','职员'),(2,'Java开发工程师
  ','Java开发工程师'),(3,'Java中级开发工程师','Java中级开发工程师'),(4,'Java高级开发工
  程师','Java高级开发工程师'),(5,'系统管理员','系统管理员'),(6,'架构师','架构师'),(7,'主
  管','主管'),(8,'经理','经理'),(9,'总经理','总经理');
#创建表user_inf
CREATE TABLE user_inf (
  ID INT(11) NOT NULL AUTO_INCREMENT,
  loginname VARCHAR(20) NOT NULL,
  PASSWORD VARCHAR(16) NOT NULL,
  USERSTATUS INT(11) NOT NULL DEFAULT '1',
  createdate TIMESTAMP NOT NULL DEFAULT CURRENT_TIMESTAMP,
  username VARCHAR(20) DEFAULT NULL,
  PRIMARY KEY (ID)
) ENGINE=INNODB AUTO_INCREMENT=2 DEFAULT CHARSET=utf8;
INSERT  INTO user_inf(ID,loginname,PASSWORD,USERSTATUS,createdate,username)
 VALUES (1,'admin','123456',2,'2016-03-12 09:34:28','超级管理员');
#创建表employee_inf
CREATE TABLE employee_inf (
  ID INT(11) NOT NULL AUTO_INCREMENT,
  DEPT_ID INT(11) NOT NULL,
  JOB_ID INT(11) NOT NULL,
  NAME VARCHAR(20) NOT NULL,
  CARD_ID VARCHAR(18) NOT NULL,
  ADDRESS VARCHAR(50) NOT NULL,
  POST_CODE VARCHAR(50) DEFAULT NULL,
  TEL VARCHAR(16) DEFAULT NULL,
  PHONE VARCHAR(11) NOT NULL,
  QQ_NUM VARCHAR(10) DEFAULT NULL,
  EMAIL VARCHAR(50) NOT NULL,
  SEX INT(11) NOT NULL DEFAULT '1',
  PARTY VARCHAR(10) DEFAULT NULL,
  BIRTHDAY DATETIME DEFAULT NULL,
  RACE VARCHAR(100) DEFAULT NULL,
  EDUCATION VARCHAR(10) DEFAULT NULL,
  SPECIALITY VARCHAR(20) DEFAULT NULL,
  HOBBY VARCHAR(100) DEFAULT NULL,
  REMARK VARCHAR(500) DEFAULT NULL,
  CREATE_DATE TIMESTAMP NOT NULL DEFAULT CURRENT_TIMESTAMP,
  PRIMARY KEY (ID),
  KEY FK_EMP_DEPT (DEPT_ID),
  KEY FK_EMP_JOB (JOB_ID),
  CONSTRAINT FK_EMP_DEPT FOREIGN KEY (DEPT_ID) REFERENCES dept_inf (ID),
  CONSTRAINT FK_EMP_JOB FOREIGN KEY (JOB_ID) REFERENCES job_inf (ID)
) ENGINE=INNODB AUTO_INCREMENT=21 DEFAULT CHARSET=utf8;
INSERT  INTO employee_inf(ID,DEPT_ID,JOB_ID,NAME,CARD_ID,ADDRESS,POST_CODE,TEL,
PHONE,QQ_NUM,EMAIL,SEX,PARTY,BIRTHDAY,RACE,EDUCATION,SPECIALITY,HOBBY,REMARK,CREATE
_DATE)
```

```
VALUES (1,1,8,'爱丽丝','4328011988','广州天河','510000','020-77777777',
    '13902001111','36750066','251425887@qq.com',0,'党员','1980-01-01 00:00:00',
    '满','本科','美声','唱歌','四大天王','2016-03-14 11:35:18'),
(2,2,1,'杰克','22623','43234','42427424','42242','4247242','42424',
    '251425887@qq.com',2,NULL,NULL,NULL,NULL,NULL,NULL,NULL,'2016-03-14 11:35:18'),
(3,1,2,'bb','432801197711251038','广州','510000','020-99999999','13907351532',
   '36750064','36750064@qq.com',1,'党员','1977-11-25 00:00:00','汉','本科','计算机
   ','爬山','无','2016-07-14 09:54:52');
#创建表 notice_inf
CREATE TABLE notice_inf (
  ID INT(11) NOT NULL AUTO_INCREMENT,
  TITLE VARCHAR(50) NOT NULL,
  CONTENT TEXT NOT NULL,
  CREATE_DATE TIMESTAMP NOT NULL DEFAULT CURRENT_TIMESTAMP,
  USER_ID INT(11) DEFAULT NULL,
  PRIMARY KEY (ID),
  KEY FK_NOTICE_USER (USER_ID),
  CONSTRAINT FK_NOTICE_USER FOREIGN KEY (USER_ID) REFERENCES user_inf (ID)
) ENGINE=INNODB AUTO_INCREMENT=19 DEFAULT CHARSET=utf8;
#创建表 document_inf
CREATE TABLE document_inf (
  ID INT(11) NOT NULL AUTO_INCREMENT,
  TITLE VARCHAR(50) NOT NULL,
  filename VARCHAR(300) NOT NULL,
  REMARK VARCHAR(300) DEFAULT NULL,
  CREATE_DATE TIMESTAMP NOT NULL DEFAULT CURRENT_TIMESTAMP,
  USER_ID INT(11) DEFAULT NULL,
  PRIMARY KEY (ID),
  KEY FK_DOCUMENT_USER (USER_ID),
  CONSTRAINT FK_DOCUMENT_USER FOREIGN KEY (USER_ID) REFERENCES user_inf (ID)
) ENGINE=INNODB AUTO_INCREMENT=7 DEFAULT CHARSET=utf8;
```

13.2.2 设计持久化实体

面向对象分析，是指根据系统需求提取应用中的对象，将这些对象抽象成类，再抽取出需要持久化保存的类，这些需要持久化保存的类就是持久化对象（PO）。本系统设计了 6 个持久化类。具体包括：

- User。对应用户，包括用户编号、用户名、登录名、密码、状态及建档日期等属性。
- Dept。对应部门，包括部门编号、部门名称及详细描述等属性。
- Job。对应职位，包括职位编号、职位名称及详细描述等属性。
- Employee。对应员工，包括员工编号、员工名称、身份证号码、地址、邮政编码、电话号码、手机号码、qq 号码、邮箱、性别、政治面貌、生日、民族、学历、所学专业、爱好、备注及建档日期等属性。
- Notice。对应公告，包括公告编号、公告标题、公告内容、发布日期等属性。
- Document。对应文件，包括文件编号、文件标题、文件名、文件描述、上传时间等属性。

在领域模式的设计中，这 6 个 PO 对象也应该包含系统的业务逻辑方法，也就是使用领域模型对象来为它们建模；但本应用不打算为它们提供任何业务逻辑方法，而是将所有的业务逻辑方法放到业务逻辑组件中实现。

将所有的业务逻辑方法放到业务逻辑组件中实现，这样系统中的领域对象显得十分简洁，它们都是单纯的数据类，不需要考虑到底应该包含哪些业务逻辑方法，因此开发起来非常便捷；而系统的所有业务逻辑都由业务逻辑组件负责实现，这样可以将业务逻辑的变化限制在业务逻

辑层内，从而避免扩散到其他两个层，因此降低了系统的开发难度。

客观世界中的对象不是孤立存在的，以上 6 个 PO 类也不是孤立存在的，它们之间存在复杂的关联关系，分析关联关系是面向对象分析的必要步骤。

这 6 个 PO 的关系如下：

Dept 和 Employee 之间存在 1 对 *N* 的关系，即一个 Dept 可以有多个 Employee。

Employee 和 Dept 之间存在 *N* 对 1 的关系，即一个 Employee 只属于一个 Dept。

Employee 和 Job 之间存在 *N* 对 1 的关系，即一个 Employee 只能有一个 Job。

User 和 Notice 之间存在 1 对 *N* 的关系，即一个 User 可以发布多个 Notice。

User 和 Document 之间存在 1 对 *N* 的关系，即一个 User 可以上传多个 Document。

这 6 个类之间的类关系如图 13.2 所示。

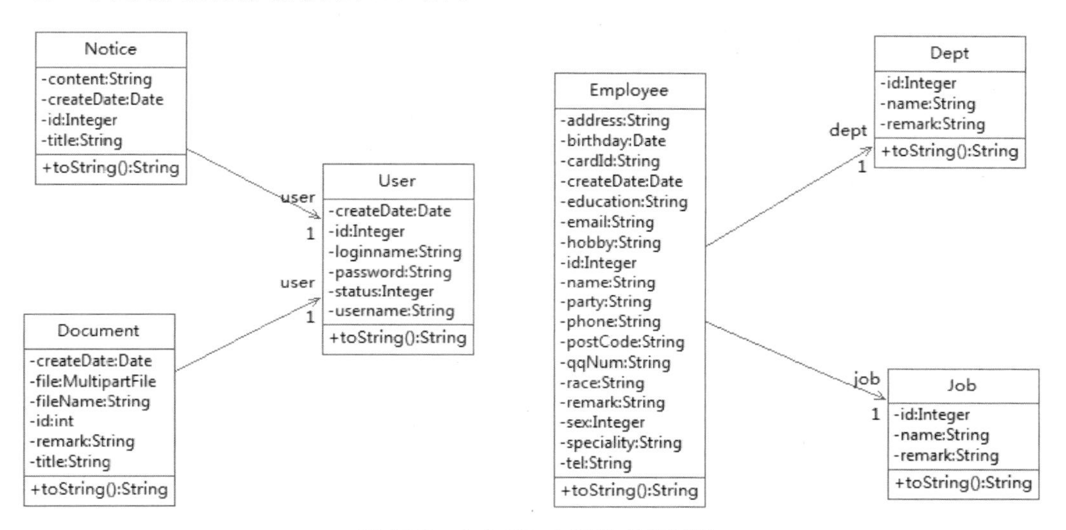

图 13.2　6 个 PO 之间的类关系图

13.2.3　创建持久化实体类

从图 13.2 可以看出，持久化对象之间的关联关系以成员变量的方式表现出来，当然，这些成员变量同样需要 setter 和 getter 方法的支持。持久化类之间的关联关系通常对应数据库里的主、外键约束。

除此之外，持久化对象还有自己的普通类型的成员变量，这些成员变量通常对应数据库的字段。

下面是 6 个持久化类的源代码。

程序清单：codes/13/hrmapp/src/org/fkit/hrm/domain/User.java

```
import java.io.Serializable;
import java.util.Date;
public class User implements Serializable {
   private Integer id;          // id
   private String username;    // 用户名
   private String loginname;    // 登录名
   private String password;    // 密码
   private Integer userstatus;        // 状态
   private Date createDate;    // 建档日期
   // 无参数构造器
   public User() {
```

```
        super();
    }
    // setter 和 getter 方法
    public Integer getId() {
        return id;
    }
    public void setId(Integer id) {
        this.id = id;
    }
    public String getUsername() {
        return username;
    }
    public void setUsername(String username) {
        this.username = username;
    }
    public String getLoginname() {
        return loginname;
    }
    public void setLoginname(String loginname) {
        this.loginname = loginname;
    }
    public String getPassword() {
        return password;
    }
    public void setPassword(String password) {
        this.password = password;
    }
    public Integer getUserstatus() {
    return userstatus;
    }
    public void setUserstatus(Integer userstatus) {
    this.userstatus = userstatus;
    }
    public Date getCreateDate() {
        return createDate;
    }
    public void setCreateDate(Date createDate) {
        this.createDate = createDate;
    }
}
```

程序清单：codes/13/hrmapp/src/org/fkit/hrm/domain/Dept.java

```
import java.io.Serializable;
public class Dept implements Serializable{
    private Integer id;         // id
    private String name;     // 部门名称
    private String remark;     // 详细描述
    // 无参数构造器
    public Dept() {
        super();
    }
    // setter 和 getter 方法
    public void setId(Integer id){
        this.id = id;
    }
    public Integer getId(){
        return this.id;
    }
    public void setName(String name){
        this.name = name;
    }
    public String getName(){
        return this.name;
    }
```

```
    public void setRemark(String remark){
        this.remark = remark;
    }
    public String getRemark(){
        return this.remark;
    }
}
```

程序清单：codes/13/hrmapp/src/org/fkit/hrm/domain/Job.java

```
import java.io.Serializable;
public class Job implements Serializable{
    private Integer id;           // id
    private String name;        // 职位名称
    private String remark;        // 详细描述
    // 无参数构造器
    public Job() {
        super();
    }
    // setter 和 getter 方法
    public void setId(Integer id){
        this.id = id;
    }
    public Integer getId(){
        return this.id;
    }
    public void setName(String name){
        this.name = name;
    }
    public String getName(){
        return this.name;
    }
    public void setRemark(String remark){
        this.remark = remark;
    }
    public String getRemark(){
        return this.remark;
    }
}
```

程序清单：codes/13/hrmapp/src/org/fkit/hrm/domain/Employee.java

```
import org.springframework.format.annotation.DateTimeFormat;
import java.io.Serializable;
public class Employee implements Serializable{
    private Integer id;           // id
    // 员工关联的部门对象
    private Dept dept;           // 部门
    // 员工关联的职位对象
    private Job job;           // 职位
    private String name;        // 名称
    private String cardId;        // 身份证
    private String address;        // 地址
    private String postCode;    // 邮政编码
    private String tel;           // 电话
    private String phone;        // 手机
    private String qqNum;        // qq
    private String email;        // 邮箱
    private Integer sex;        // 性别
    private String party;        // 政治面貌
    /**
     * 使用@ModelAttribute 接受参数
     * form 表单中有日期时,Spring 不知道该如何转换
```

```
 *  要在实体类的日期属性上加@DateTimeFormat(pattern="yyyy-MM-dd")注解
 */
@DateTimeFormat(pattern="yyyy-MM-dd")
private java.util.Date birthday;    //生日
private String race;               // 民族
private String education;           // 学历
private String speciality;           // 专业
private String hobby;              // 爱好
private String remark;              // 备注
private java.util.Date createDate;    // 建档日期
// 无参数构造器
public Employee() {
    super();
    // TODO Auto-generated constructor stub
}
// setter 和 getter 方法
public void setId(Integer id){
    this.id = id;
}
public Integer getId(){
    return this.id;
}

public Dept getDept() {
    return dept;
}
public void setDept(Dept dept) {
    this.dept = dept;
}
public Job getJob() {
    return job;
}
public void setJob(Job job) {
    this.job = job;
}
public void setName(String name){
    this.name = name;
}
public String getName(){
    return this.name;
}
public void setCardId(String cardId){
    this.cardId = cardId;
}
public String getCardId(){
    return this.cardId;
}
public void setAddress(String address){
    this.address = address;
}
public String getAddress(){
    return this.address;
}
public void setPostCode(String postCode){
    this.postCode = postCode;
}
public String getPostCode(){
    return this.postCode;
}
public void setTel(String tel){
    this.tel = tel;
}
public String getTel(){
```

```
        return this.tel;
    }
    public void setPhone(String phone){
        this.phone = phone;
    }
    public String getPhone(){
        return this.phone;
    }
    public void setQqNum(String qqNum){
        this.qqNum = qqNum;
    }
    public String getQqNum(){
        return this.qqNum;
    }
    public void setEmail(String email){
        this.email = email;
    }
    public String getEmail(){
        return this.email;
    }
    public void setSex(Integer sex){
        this.sex = sex;
    }
    public Integer getSex(){
        return this.sex;
    }
    public void setParty(String party){
        this.party = party;
    }
    public String getParty(){
        return this.party;
    }
    public void setBirthday(java.util.Date birthday){
        this.birthday = birthday;
    }
    public java.util.Date getBirthday(){
        return this.birthday;
    }
    public void setRace(String race){
        this.race = race;
    }
    public String getRace(){
        return this.race;
    }
    public void setEducation(String education){
        this.education = education;
    }
    public String getEducation(){
        return this.education;
    }
    public void setSpeciality(String speciality){
        this.speciality = speciality;
    }
    public String getSpeciality(){
        return this.speciality;
    }
    public void setHobby(String hobby){
        this.hobby = hobby;
    }
    public String getHobby(){
        return this.hobby;
    }
    public void setRemark(String remark){
        this.remark = remark;
    }
```

```
    public String getRemark(){
        return this.remark;
    }
    public void setCreateDate(java.util.Date createDate){
        this.createDate = createDate;
    }
    public java.util.Date getCreateDate(){
        return this.createDate;
    }


}
```

程序清单：codes/13/hrmapp/src/org/fkit/hrm/domain/Notice.java

```
import java.io.Serializable;
public class Notice implements Serializable{
    private Integer id;        // 编号
    private String title;   // 标题
    private String content; // 内容
    private java.util.Date createDate;  // 发布日期
    private User user;        // 发布人
    // 无参数构造器
    public Notice() {
        super();
    }
    // setter 和 getter 方法
    public void setId(Integer id){
        this.id = id;
    }
    public Integer getId(){
        return this.id;
    }
    public void setTitle(String title){
        this.title = title;
    }
    public String getTitle(){
        return this.title;
    }
    public void setContent(String content){
        this.content = content;
    }
    public String getContent(){
        return this.content;
    }
    public void setCreateDate(java.util.Date createDate){
        this.createDate = createDate;
    }
    public java.util.Date getCreateDate(){
        return this.createDate;
    }
    public User getUser() {
        return user;
    }
    public void setUser(User user) {
        this.user = user;
    }
}
```

程序清单：codes/13/hrmapp/src/org/fkit/hrm/domain/Document.java

```
import org.springframework.web.multipart.MultipartFile;
import java.io.Serializable;
public class Document implements Serializable{
    private int id;                    // 编号
```

```
    private String title;            // 标题
    private String fileName;         // 文件名
    private MultipartFile file;        // 文件
    private String remark;           // 描述
    private java.util.Date createDate;    // 上传时间
    private User user;               // 上传人
    // 无参数构造器
    public Document() {
       super();
    }
    // setter 和 getter 方法
    public void setId(int id){
       this.id = id;
    }
    public int getId(){
       return this.id;
    }
    public void setTitle(String title){
       this.title = title;
    }
    public String getTitle(){
       return this.title;
    }

    public String getFileName() {
       return fileName;
    }
    public void setFileName(String fileName) {
       this.fileName = fileName;
    }
    public MultipartFile getFile() {
       return file;
    }
    public void setFile(MultipartFile file) {
       this.file = file;
    }
    public void setRemark(String remark){
       this.remark = remark;
    }
    public String getRemark(){
       return this.remark;
    }
    public void setCreateDate(java.util.Date createDate){
       this.createDate = createDate;
    }
    public java.util.Date getCreateDate(){
       return this.createDate;
    }
    public User getUser() {
       return user;
    }
    public void setUser(User user) {
       this.user = user;
    }
}
```

13.3 实现 DAO 持久层

MyBatis 建议定义接口完成 SQL 语句的映射，该接口可以直接作为 DAO 组件使用。使用 DAO 模式，既能体现业务逻辑组件封装 DAO 组件的门面模式，也可分离业务逻辑组件和 DAO

组件的功能：业务逻辑组件负责业务逻辑的变化，而 DAO 组件负责持久化技术的变化，这正是桥接模式的应用。

引入 DAO 模式后，每个 DAO 组件包含了数据库的访问逻辑；每个 DAO 组件可对一个数据库表完成基本的 CRUD 等操作。

13.3.1 公共常量类

程序清单：codes/13/hrmapp/src/org/fkit/hrm/util/common/HrmConstants.java

```
public class HrmConstants {
    // 数据库表常量
    public static final String USERTABLE = "user_inf";
    public static final String DEPTTABLE = "dept_inf";
    public static final String JOBTABLE = "job_inf";
    public static final String EMPLOYEETABLE = "employee_inf";
    public static final String NOTICETABLE = "notice_inf";
    public static final String DOCUMENTTABLE = "document_inf";
    // 登录
    public static final String LOGIN = "loginForm";
    // 用户的 session 对象
    public static final String USER_SESSION = "user_session";
    // 默认每页 4 条数据
    public static int PAGE_DEFAULT_SIZE = 4;
}
```

在 HrmConstants 类中定义了本系统中使用的常量。

13.3.2 定义 DAO 接口

下面是 UserDao 接口的源代码。

程序清单：codes/13/hrmapp/src/org/fkit/hrm/dao/UserDao.java

```
import java.util.List;
import java.util.Map;
import org.apache.ibatis.annotations.Delete;
import org.apache.ibatis.annotations.Param;
import org.apache.ibatis.annotations.Select;
import org.apache.ibatis.annotations.SelectProvider;
import org.fkit.hrm.dao.provider.UserDynaSqlProvider;
import org.fkit.hrm.domain.User;
import static org.fkit.hrm.util.common.HrmConstants.USERTABLE;
public interface UserDao {
    // 根据登录名和密码查询员工
    @Select("select * from "+USERTABLE+" where loginname = #{loginname} and password
      = #{password}")
    User selectByLoginnameAndPassword(
            @Param("loginname") String loginname,
            @Param("password") String password);
    // 根据 id 查询用户
    @Select("select * from "+USERTABLE+" where ID = #{id}")
    User selectById(Integer id);
    // 根据 id 删除用户
    @Delete(" delete from "+USERTABLE+" where id = #{id} ")
    void deleteById(Integer id);
    // 动态修改用户
    @SelectProvider(type=UserDynaSqlProvider.class,method="updateUser")
    void update(User user);
    // 动态查询
    @SelectProvider(type=UserDynaSqlProvider.class,method="selectWhitParam")
    List<User> selectByPage(Map<String, Object> params);
```

```
    // 根据参数查询用户总数
    @SelectProvider(type=UserDynaSqlProvider.class,method="count")
    Integer count(Map<String, Object> params);
    // 动态插入用户
    @SelectProvider(type=UserDynaSqlProvider.class,method="insertUser")
    void save(User user);
}
```

在 UserDao 接口中使用了动态 SQL 提供类 UserDynaSqlProvider。

程序清单：codes/13/hrmapp/src/org/fkit/hrm/dao/provider/UserDynaSqlProvider.java

```
import java.util.Map;
import org.apache.ibatis.jdbc.SQL;
import org.fkit.hrm.domain.User;
import static org.fkit.hrm.util.common.HrmConstants.USERTABLE;
public class UserDynaSqlProvider {
    // 分页动态查询
    public String selectWhitParam(Map<String, Object> params){
        String sql =  new SQL(){
            {
                SELECT("*");
                FROM(USERTABLE);
                if(params.get("user") != null){
                    User user = (User)params.get("user");
                    if(user.getUsername() != null && !user.getUsername().equals("")){
                        WHERE("  username LIKE CONCAT ('%',#{user.username},'%') ");
                    }
                    if(user. getUserstatus() != null && !user.getUserstatus ().equals ("")){
                        WHERE("userstatus LIKE CONCAT ('%',#{user.userstatus},'%') ");
                    }
                }
            }
        }.toString();
        if(params.get("pageModel") != null){
            sql += " limit #{pageModel.firstLimitParam} , #{pageModel.pageSize}  ";
        }
        return sql;
    }
    // 动态查询总数量
    public String count(Map<String, Object> params){
        return new SQL(){
            {
                SELECT("count(*)");
                FROM(USERTABLE);
                if(params.get("user") != null){
                    User user = (User)params.get("user");
                    if(user.getUsername() != null && !user.getUsername().equals("")){
                        WHERE(" username LIKE CONCAT ('%',#{user.username},'%') ");
                    }
                    if(user.getUserstatus() != null && !user.getUserstatus ().equals ("")){
                        WHERE("userstatus LIKE CONCAT ('%',#{user.userstatus},'%') ");
                    }
                }
            }
        }.toString();
    }
    // 动态插入
    public String insertUser(User user){
        return new SQL(){
            {
                INSERT_INTO(USERTABLE);
                if(user.getUsername() != null && !user.getUsername().equals("")){
```

```
                VALUES("username", "#{username}");
            }
            if(user.getUserstatus() != null && !user.getUserstatus().equals("")){
                VALUES("userstatus", "#{userstatus}");
            }
            if(user.getLoginname() != null && !user.getLoginname().equals("")){
                VALUES("loginname", "#{loginname}");
            }
            if(user.getPassword() != null && !user.getPassword().equals("")){
                VALUES("password", "#{password}");
            }
        }
    }.toString();
}
// 动态更新
public String updateUser(User user){
    return new SQL(){
        {
            UPDATE(USERTABLE);
            if(user.getUsername() != null){
                SET(" username = #{username} ");
            }
            if(user.getLoginname() != null){
                SET(" loginname = #{loginname} ");
            }
            if(user.getPassword()!= null){
                SET(" password = #{password} ");
            }
            if(user.getUserstatus ()!= null){
                SET("userstatus = #{ userstatus } ");
            }
            if(user.getCreateDate()!= null){
                SET(" create_date = #{createDate} ");
            }
            WHERE(" id = #{id} ");
        }
    }.toString();
}
}
```

通过上面的 DAO 接口完成数据库的操作，这种简单的实现较之传统的 JDBC 持久化访问，简直不可同日而语。

提示

在学习框架的过程中也许会有少许的坎坷，但一旦掌握了框架的作用，将大幅提高开发效率，而且通过了解好的框架所倡导的软件架构理念还会提高架构设计水平。

下面是 DeptDao 接口的源代码。

程序清单：codes/13/hrmapp/src/org/fkit/hrm/dao/DeptDao.java

```
import java.util.List;
import java.util.Map;
import org.apache.ibatis.annotations.Delete;
import org.apache.ibatis.annotations.Select;
import org.apache.ibatis.annotations.SelectProvider;
import org.fkit.hrm.dao.provider.DeptDynaSqlProvider;
import org.fkit.hrm.domain.Dept;
import static org.fkit.hrm.util.common.HrmConstants.DEPTTABLE;
public interface DeptDao {
    // 动态查询
    @SelectProvider(type=DeptDynaSqlProvider.class,method="selectWhitParam")
    List<Dept> selectByPage(Map<String, Object> params);
```

```
    @SelectProvider(type=DeptDynaSqlProvider.class,method="count")
    Integer count(Map<String, Object> params);
    @Select("select * from "+DEPTTABLE+" ")
    List<Dept> selectAllDept();
    @Select("select * from "+DEPTTABLE+" where ID = #{id}")
    Dept selectById(int id);
    // 根据 id 删除部门
    @Delete(" delete from "+DEPTTABLE+" where id = #{id} ")
    void deleteById(Integer id);
    // 动态插入部门
    @SelectProvider(type=DeptDynaSqlProvider.class,method="insertDept")
    void save(Dept dept);
    // 动态修改用户
    @SelectProvider(type=DeptDynaSqlProvider.class,method="updateDept")
    void update(Dept dept);
}
```

在 DeptDao 接口中使用了动态 SQL 提供类 DeptDynaSqlProvider。

程序清单：codes/13/hrmapp/src/org/fkit/hrm/dao/ provider/DeptDynaSqlProvider.java

```
import java.util.Map;
import org.apache.ibatis.jdbc.SQL;
import org.fkit.hrm.domain.Dept;
import static org.fkit.hrm.util.common.HrmConstants.DEPTTABLE;
public class DeptDynaSqlProvider {
    // 分页动态查询
    public String selectWhitParam(Map<String, Object> params){
        String sql = new SQL(){
            {
                SELECT("*");
                FROM(DEPTTABLE);
                if(params.get("dept") != null){
                    Dept dept = (Dept) params.get("dept");
                    if(dept.getName() != null && !dept.getName().equals("")){
                        WHERE(" name LIKE CONCAT ('%',#{dept.name},'%') ");
                    }
                }
            }
        }.toString();

        if(params.get("pageModel") != null){
            sql += " limit #{pageModel.firstLimitParam} , #{pageModel.pageSize}  ";
        }
        return sql;
    }
    // 动态查询总数量
    public String count(Map<String, Object> params){
        return new SQL(){
            {
                SELECT("count(*)");
                FROM(DEPTTABLE);
                if(params.get("dept") != null){
                    Dept dept = (Dept) params.get("dept");
                    if(dept.getName() != null && !dept.getName().equals("")){
                        WHERE(" name LIKE CONCAT ('%',#{dept.name},'%') ");
                    }
                }
            }
        }.toString();
    }
    // 动态插入
    public String insertDept(Dept dept){
        return new SQL(){
            {
```

```
                INSERT_INTO(DEPTTABLE);
                if(dept.getName() != null && !dept.getName().equals("")){
                    VALUES("name", "#{name}");
                }
                if(dept.getRemark() != null && !dept.getRemark().equals("")){
                    VALUES("remark", "#{remark}");
                }
            }
        }.toString();
    }
    // 动态更新
    public String updateDept(Dept dept){

        return new SQL(){
            {
                UPDATE(DEPTTABLE);
                if(dept.getName() != null){
                    SET(" name = #{name} ");
                }
                if(dept.getRemark() != null){
                    SET(" remark = #{remark} ");
                }
                WHERE(" id = #{id} ");
            }
        }.toString();
    }
}
```

下面是 JobDao 接口的源代码。

程序清单：codes/13/hrmapp/src/org/fkit/hrm/dao/JobDao.java

```
import java.util.List;
import java.util.Map;
import org.apache.ibatis.annotations.Delete;
import org.apache.ibatis.annotations.Select;
import org.apache.ibatis.annotations.SelectProvider;
import org.fkit.hrm.dao.provider.JobDynaSqlProvider;
import org.fkit.hrm.domain.Job;
import static org.fkit.hrm.util.common.HrmConstants.JOBTABLE;
public interface JobDao {
    @Select("select * from "+JOBTABLE+" where ID = #{id}")
    Job selectById(int id);
    @Select("select * from "+JOBTABLE+" ")
    List<Job> selectAllJob();
    // 动态查询
    @SelectProvider(type=JobDynaSqlProvider.class,method="selectWhitParam")
    List<Job> selectByPage(Map<String, Object> params);
    @SelectProvider(type=JobDynaSqlProvider.class,method="count")
    Integer count(Map<String, Object> params);
    // 根据 id 删除部门
    @Delete(" delete from "+JOBTABLE+" where id = #{id} ")
    void deleteById(Integer id);
    // 动态插入部门
    @SelectProvider(type=JobDynaSqlProvider.class,method="insertJob")
    void save(Job job);
    // 动态修改用户
    @SelectProvider(type=JobDynaSqlProvider.class,method="updateJob")
    void update(Job job);
}
```

在 JobDao 接口中使用了动态 SQL 提供类 JobDynaSqlProvider。

程序清单：codes/13/hrmapp/src/org/fkit/hrm/dao/ provider/JobDynaSqlProvider.java

```
import java.util.Map;
import org.apache.ibatis.jdbc.SQL;
import org.fkit.hrm.domain.Job;
import static org.fkit.hrm.util.common.HrmConstants.JOBTABLE;
public class JobDynaSqlProvider {
    // 分页动态查询
    public String selectWhitParam(Map<String, Object> params){
        String sql =  new SQL(){
            {
                SELECT("*");
                FROM(JOBTABLE);
                if(params.get("job") != null){
                    Job job = (Job) params.get("job");
                    if(job.getName() != null && !job.getName().equals("")){
                        WHERE("  name LIKE CONCAT ('%',#{job.name},'%') ");
                    }
                }
            }
        }.toString();
        if(params.get("pageModel") != null){
            sql += " limit #{pageModel.firstLimitParam} , #{pageModel.pageSize}  ";
        }
        return sql;
    }
    // 动态查询总数量
    public String count(Map<String, Object> params){
        return new SQL(){
            {
                SELECT("count(*)");
                FROM(JOBTABLE);
                if(params.get("job") != null){
                    Job job = (Job) params.get("job");
                    if(job.getName() != null && !job.getName().equals("")){
                        WHERE("  name LIKE CONCAT ('%',#{job.name},'%') ");
                    }
                }
            }
        }.toString();
    }
    // 动态插入
    public String insertJob(Job job){
        return new SQL(){
            {
                INSERT_INTO(JOBTABLE);
                if(job.getName() != null && !job.getName().equals("")){
                    VALUES("name", "#{name}");
                }
                if(job.getRemark() != null && !job.getRemark().equals("")){
                    VALUES("remark", "#{remark}");
                }
            }
        }.toString();
    }
    // 动态更新
    public String updateJob(Job job){
        return new SQL(){
            {
                UPDATE(JOBTABLE);
                if(job.getName() != null){
                    SET(" name = #{name} ");
                }
                if(job.getRemark() != null){
```

```
                    SET(" remark = #{remark} ");
                }
                WHERE(" id = #{id} ");
            }
        }.toString();
    }
}
```

下面是 EmployeeDao 接口的源代码。

程序清单：codes/13/hrmapp/src/org/fkit/hrm/dao/EmployeeDao.java

```
import java.util.List;
import java.util.Map;
import org.apache.ibatis.annotations.Delete;
import org.apache.ibatis.annotations.One;
import org.apache.ibatis.annotations.Result;
import org.apache.ibatis.annotations.Results;
import org.apache.ibatis.annotations.Select;
import org.apache.ibatis.annotations.SelectProvider;
import org.apache.ibatis.mapping.FetchType;
import org.fkit.hrm.dao.provider.EmployeeDynaSqlProvider;
import org.fkit.hrm.domain.Employee;
import static org.fkit.hrm.util.common.HrmConstants.EMPLOYEETABLE;
public interface EmployeeDao {
    // 根据参数查询员工总数
    @SelectProvider(type=EmployeeDynaSqlProvider.class,method="count")
    Integer count(Map<String, Object> params);
    // 根据参数动态查询员工

@SelectProvider(type=EmployeeDynaSqlProvider.class,method="selectWhitParam")
    @Results({
        @Result(id=true,column="id",property="id"),
        @Result(column="CARD_ID",property="cardId"),
        @Result(column="POST_CODE",property="postCode"),
        @Result(column="QQ_NUM",property="qqNum"),
        @Result(column="BIRTHDAY",property="birthday",javaType=java.util.Date.class),
        @Result(column="CREATE_DATE",property="createDate",javaType=java.util.
         Date.class),
        @Result(column="DEPT_ID",property="dept",
            one=@One(select="org.fkit.hrm.dao.DeptDao.selectById",
        fetchType=FetchType.EAGER)),
        @Result(column="JOB_ID",property="job",
            one=@One(select="org.fkit.hrm.dao.JobDao.selectById",
        fetchType=FetchType.EAGER))
    })
    List<Employee> selectByPage(Map<String, Object> params);
    // 动态插入员工
    @SelectProvider(type=EmployeeDynaSqlProvider.class,method="insertEmployee")
    void save(Employee employee);
    // 根据id删除员工
    @Delete(" delete from "+EMPLOYEETABLE+" where id = #{id} ")
    void deleteById(Integer id);
    // 根据id查询员工
    @Select("select * from "+EMPLOYEETABLE+" where ID = #{id}")
    @Results({
        @Result(id=true,column="id",property="id"),
        @Result(column="CARD_ID",property="cardId"),
        @Result(column="POST_CODE",property="postCode"),
        @Result(column="QQ_NUM",property="qqNum"),
        @Result(column="BIRTHDAY",property="birthday",javaType=java.util.Date.
         class),
        @Result(column="CREATE_DATE",property="createDate",javaType=java.util.
         Date.class),
        @Result(column="DEPT_ID",property="dept",
```

```
            one=@One(select="org.fkit.hrm.dao.DeptDao.selectById",
        fetchType=FetchType.EAGER)),
        @Result(column="JOB_ID",property="job",
            one=@One(select="org.fkit.hrm.dao.JobDao.selectById",
        fetchType=FetchType.EAGER))
    })
    Employee selectById(Integer id);
    // 动态修改员工
    @SelectProvider(type=EmployeeDynaSqlProvider.class,method="updateEmployee")
    void update(Employee employee);
}
```

在 EmployeeDao 接口中使用了动态 SQL 提供类 EmployeeDynaSqlProvider。

程序清单：codes/13/hrmapp/src/org/fkit/hrm/dao/provider/EmployeeDynaSqlProvider.java

```
import java.util.Map;
import org.apache.ibatis.jdbc.SQL;
import org.fkit.hrm.domain.Employee;
import static org.fkit.hrm.util.common.HrmConstants.EMPLOYEETABLE;
public class EmployeeDynaSqlProvider {
    // 分页动态查询
    public String selectWhitParam(Map<String, Object> params){
        String sql =  new SQL(){
            {
                SELECT("*");
                FROM(EMPLOYEETABLE);
                if(params.get("employee") != null){
                    Employee employee = (Employee)params.get("employee");
                    if(employee.getDept() != null && employee.getDept().getId() !=
                        null && employee.getDept().getId() != 0){
                        WHERE(" DEPT_ID = #{employee.dept.id} ");
                    }
                    if(employee.getJob() != null && employee.getJob().getId() != null
                        && employee.getJob().getId() != 0){
                        WHERE(" JOB_ID = #{employee.job.id} ");
                    }
                    if(employee.getName() != null && !employee.getName().equals("")){
                        WHERE("  NAME LIKE CONCAT ('%',#{employee.name},'%') ");
                    }
                    if(employee.getPhone() != null && !employee.getPhone().equals("")){
                        WHERE(" phone LIKE CONCAT ('%',#{employee.phone},'%') ");
                    }
                    if(employee.getCardId() != null && !employee.getCardId().equals("") ){
                        WHERE(" card_id LIKE CONCAT ('%',#{employee.cardId},'%') ");
                    }
                    if(employee.getSex()!= null && employee.getSex() != 0){
                        WHERE("sex = #{employee.sex}");
                    }
                }
            }
        }.toString();
        if(params.get("pageModel") != null){
            sql += " limit #{pageModel.firstLimitParam} , #{pageModel.pageSize}  ";
        }
        return sql;
    }
    // 动态查询总数量
    public String count(Map<String, Object> params){
        return new SQL(){
            {
                SELECT("count(*)");
                FROM(EMPLOYEETABLE);
                if(params.get("employee") != null){
                    Employee employee = (Employee)params.get("employee");
```

```
                if(employee.getDept() != null && employee.getDept().getId() !=
                    null && employee.getDept().getId() != 0){
                    WHERE(" DEPT_ID = #{employee.dept.id} ");
                }
                if(employee.getJob() != null && employee.getJob().getId() != null
                    && employee.getJob().getId() != 0){
                    WHERE(" JOB_ID = #{employee.job.id} ");
                }
                if(employee.getName() != null && !employee.getName().equals("")){
                    WHERE("  NAME LIKE CONCAT ('%',#{employee.name},'%') ");
                }
                if(employee.getPhone() != null && !employee.getPhone().equals("")){
                    WHERE(" phone LIKE CONCAT ('%',#{employee.phone},'%') ");
                }
                if(employee.getCardId() != null && !employee.getCardId().equals("") ){
                    WHERE(" card_id LIKE CONCAT ('%',#{employee.cardId},'%') ");
                }
                if(employee.getSex()!= null && employee.getSex() != 0){
                    WHERE("sex = #{employee.sex}");
                }
            }
        }
    }.toString();
}
// 动态插入
public String insertEmployee(Employee employee){
    return new SQL(){
        {
            INSERT_INTO(EMPLOYEETABLE);
            if(employee.getName() != null){
                VALUES("name", "#{name}");
            }
            if(employee.getCardId() != null){
                VALUES("card_id", "#{cardId}");
            }
            if(employee.getAddress()!= null){
                VALUES("address", "#{address}");
            }
            if(employee.getPostCode()!= null){
                VALUES("post_code", "#{postCode}");
            }
            if(employee.getTel()!= null){
                VALUES("tel", "#{tel}");
            }
            if(employee.getPhone()!= null){
                VALUES("phone", "#{phone}");
            }
            if(employee.getQqNum()!= null){
                VALUES("qq_num", "#{qqNum}");
            }
            if(employee.getEmail()!= null){
                VALUES("email", "#{email}");
            }
            if(employee.getSex()!= null){
                VALUES("sex", "#{sex}");
            }
            if(employee.getParty()!= null){
                VALUES("party", "#{party}");
            }
            if(employee.getBirthday()!= null){
                VALUES("birthday", "#{birthday}");
            }
            if(employee.getRace()!= null){
                VALUES("race", "#{race}");
            }
```

```
            if(employee.getEducation()!= null){
                VALUES("education", "#{education}");
            }
            if(employee.getSpeciality()!= null){
                VALUES("speciality", "#{speciality}");
            }
            if(employee.getHobby()!= null){
                VALUES("hobby", "#{hobby}");
            }
            if(employee.getRemark()!= null){
                VALUES("remark", "#{remark}");
            }
            if(employee.getCreateDate()!= null){
                VALUES("create_Date", "#{createDate}");
            }
            if(employee.getDept()!= null){
                VALUES("dept_id", "#{dept.id}");
            }
            if(employee.getJob()!= null){
                VALUES("job_id", "#{job.id}");
            }
        }
    }.toString();
}
// 动态更新
public String updateEmployee(Employee employee){
    return new SQL(){
        {
            UPDATE(EMPLOYEETABLE);
            if(employee.getName() != null){
                SET(" name = #{name} ");
            }
            if(employee.getCardId() != null){
                SET(" card_id = #{cardId} ");
            }
            if(employee.getAddress()!= null){
                SET(" address = #{address} ");
            }
            if(employee.getPostCode()!= null){
                SET(" post_code = #{postCode} ");
            }
            if(employee.getTel()!= null){
                SET(" tel = #{tel} ");
            }
            if(employee.getPhone()!= null){
                SET(" phone = #{phone} ");
            }
            if(employee.getQqNum()!= null){
                SET(" qq_num = #{qqNum} ");
            }
            if(employee.getEmail()!= null){
                SET(" email = #{email} ");
            }
            if(employee.getSex()!= null){
                SET(" sex = #{sex} ");
            }
            if(employee.getParty()!= null){
                SET(" party = #{party} ");
            }
            if(employee.getBirthday()!= null){
                SET(" birthday = #{birthday} ");
            }
            if(employee.getRace()!= null){
                SET(" race = #{race} ");
            }
```

```
                if(employee.getEducation()!= null){
                    SET(" education = #{education} ");
                }
                if(employee.getSpeciality()!= null){
                    SET(" speciality = #{speciality} ");
                }
                if(employee.getHobby()!= null){
                    SET(" hobby = #{hobby} ");
                }
                if(employee.getRemark()!= null){
                    SET(" remark = #{remark} ");
                }
                if(employee.getCreateDate()!= null){
                    SET(" create_Date = #{createDate} ");
                }
                if(employee.getDept()!= null){
                    SET(" dept_id = #{dept.id} ");
                }
                if(employee.getJob()!= null){
                    SET(" job_id = #{job.id} ");
                }
                WHERE(" id = #{id} ");
            }
        }.toString();
    }
}
```

下面是 NoticeDao 接口的源代码。

程序清单：codes/13/hrmapp/src/org/fkit/hrm/dao/NoticeDao.java

```
import java.util.List;
import java.util.Map;
import org.apache.ibatis.annotations.*;
import org.apache.ibatis.mapping.FetchType;
import org.fkit.hrm.dao.provider.NoticeDynaSqlProvider;
import org.fkit.hrm.domain.Notice;
import static org.fkit.hrm.util.common.HrmConstants.NOTICETABLE;
public interface NoticeDao {
    // 动态查询
    @SelectProvider(type=NoticeDynaSqlProvider.class,method="selectWhitParam")
    @Results({
        @Result(id=true,column="id",property="id"),
        @Result(column="CREATE_DATE",property="createDate",javaType=
          java.util.Date.class),
        @Result(column="USER_ID",property="user",
            one=@One(select="org.fkit.hrm.dao.UserDao.selectById",
        fetchType=FetchType.EAGER))
    })
    List<Notice> selectByPage(Map<String, Object> params);
    @SelectProvider(type=NoticeDynaSqlProvider.class,method="count")
    Integer count(Map<String, Object> params);
    @Select("select * from "+NOTICETABLE+" where ID = #{id}")
    Notice selectById(int id);
    // 根据 id 删除公告
    @Delete(" delete from "+NOTICETABLE+" where id = #{id} ")
    void deleteById(Integer id);
    // 动态插入公告
    @SelectProvider(type=NoticeDynaSqlProvider.class,method="insertNotice")
    void save(Notice notice);
    // 动态修改公告
    @SelectProvider(type=NoticeDynaSqlProvider.class,method="updateNotice")
    void update(Notice notice);
}
```

在 NoticeDao 接口中使用了动态 SQL 提供类 NoticeDynaSqlProvider。

程序清单：codes/13/hrmapp/src/org/fkit/hrm/dao/provider/NoticeDynaSqlProvider.java

```
import java.util.Map;
import org.apache.ibatis.jdbc.SQL;
import org.fkit.hrm.domain.Notice;
import static org.fkit.hrm.util.common.HrmConstants.NOTICETABLE;
public class NoticeDynaSqlProvider {
    // 分页动态查询
    public String selectWhitParam(Map<String, Object> params){
        String sql =  new SQL(){
            {
                SELECT("*");
                FROM(NOTICETABLE);
                if(params.get("notice") != null){
                    Notice notice = (Notice)params.get("notice");
                    if(notice.getTitle() != null && !notice.getTitle().equals("")){
                        WHERE("  title LIKE CONCAT ('%',#{notice.title},'%') ");
                    }
                    if(notice.getContent() != null && !notice.getContent().equals("")){
                        WHERE("  content LIKE CONCAT ('%',#{notice.content},'%') ");
                    }
                }
            }
        }.toString();

        if(params.get("pageModel") != null){
            sql += " limit #{pageModel.firstLimitParam} , #{pageModel.pageSize}  ";
        }
        return sql;
    }
    // 动态查询总数量
    public String count(Map<String, Object> params){
        return new SQL(){
            {
                SELECT("count(*)");
                FROM(NOTICETABLE);
                if(params.get("notice") != null){
                    Notice notice = (Notice)params.get("notice");
                    if(notice.getTitle() != null && !notice.getTitle().equals("")){
                        WHERE("  title LIKE CONCAT ('%',#{notice.title},'%') ");
                    }
                    if(notice.getContent() != null && !notice.getContent().equals("")){
                        WHERE("  content LIKE CONCAT ('%',#{notice.content},'%') ");
                    }
                }
            }
        }.toString();
    }
    // 动态插入
    public String insertNotice(Notice notice){
        return new SQL(){
            {
                INSERT_INTO(NOTICETABLE);
                if(notice.getTitle() != null && !notice.getTitle().equals("")){
                    VALUES("title", "#{title}");
                }
                if(notice.getContent() != null && !notice.getContent().equals("")){
                    VALUES("content", "#{content}");
                }
                if(notice.getUser() != null && notice.getUser().getId() != null){
```

```
                VALUES("user_id", "#{user.id}");
            }
        }
    }.toString();
}
// 动态更新
public String updateNotice(Notice notice){
    return new SQL(){
        {
            UPDATE(NOTICETABLE);
            if(notice.getTitle() != null && !notice.getTitle().equals("")){
                SET(" title = #{title} ");
            }
            if(notice.getContent() != null && !notice.getContent().equals("")){
                SET(" content = #{content} ");
            }
            if(notice.getUser() != null && notice.getUser().getId() != null){
                SET(" user_id = #{user.id} ");
            }
            WHERE(" id = #{id} ");
        }
    }.toString();
}
}
```

下面是 DocumentDao 接口的源代码。

程序清单：codes/13/hrmapp/src/org/fkit/hrm/dao/DocumentDao.java

```
import java.util.List;
import java.util.Map;
import org.apache.ibatis.annotations.*;
import org.apache.ibatis.mapping.FetchType;
import org.fkit.hrm.dao.provider.DocumentDynaSqlProvider;
import org.fkit.hrm.domain.Document;
import static org.fkit.hrm.util.common.HrmConstants.DOCUMENTTABLE;
public interface DocumentDao {
    // 动态查询
    @SelectProvider(type=DocumentDynaSqlProvider.class,method="selectWhitParam")
    @Results({
        @Result(id=true,column="id",property="id"),
        @Result(column="CREATE_DATE",property="createDate",javaType=java.util.
          Date.class),
        @Result(column="USER_ID",property="user",
           one=@One(select="org.fkit.hrm.dao.UserDao.selectById",
        fetchType=FetchType.EAGER))
    })
    List<Document> selectByPage(Map<String, Object> params);
    @SelectProvider(type=DocumentDynaSqlProvider.class,method="count")
    Integer count(Map<String, Object> params);
    // 动态插入文档
    @SelectProvider(type=DocumentDynaSqlProvider.class,method="insertDocument")
    void save(Document document);
    @Select("select * from "+DOCUMENTTABLE+" where ID = #{id}")
    Document selectById(int id);
    // 根据 id 删除文档
    @Delete(" delete from "+DOCUMENTTABLE+" where id = #{id} ")
    void deleteById(Integer id);
    // 动态修改文档
    @SelectProvider(type=DocumentDynaSqlProvider.class,method="updateDocument")
    void update(Document document);
}
```

在 DocumentDao 接口中使用了动态 SQL 提供类 DocumentDynaSqlProvider。

程序清单：codes/13/hrmapp/src/org/fkit/hrm/dao/provider/DocumentDynaSqlProvider.java

```
import java.util.Map;
import org.apache.ibatis.jdbc.SQL;
import org.fkit.hrm.domain.Document;
import static org.fkit.hrm.util.common.HrmConstants.DOCUMENTTABLE;
public class DocumentDynaSqlProvider {
    // 分页动态查询
    public String selectWhitParam(Map<String, Object> params){
        String sql =  new SQL(){
            {
                SELECT("*");
                FROM(DOCUMENTTABLE);
                if(params.get("document") != null){
                    Document document = (Document) params.get("document");
                    if(document.getTitle() != null && !document.getTitle().equals("")){
                        WHERE("  title LIKE CONCAT ('%',#{document.title},'%') ");
                    }
                }
            }
        }.toString();
        if(params.get("pageModel") != null){
            sql += " limit #{pageModel.firstLimitParam} , #{pageModel.pageSize}  ";
        }
        return sql;
    }
    // 动态查询总数量
    public String count(Map<String, Object> params){
        return new SQL(){
            {
                SELECT("count(*)");
                FROM(DOCUMENTTABLE);
                if(params.get("document") != null){
                    Document document = (Document) params.get("document");
                    if(document.getTitle() != null && !document.getTitle().equals("")){
                        WHERE("  title LIKE CONCAT ('%',#{document.title},'%') ");
                    }
                }
            }
        }.toString();
    }
    // 动态插入
    public String insertDocument(Document document){
        return new SQL(){
            {
                INSERT_INTO(DOCUMENTTABLE);
                if(document.getTitle() != null && !document.getTitle().equals("")){
                    VALUES("title", "#{title}");
                }
                if(document.getFileName() != null && !document.getFileName().equals("")){
                    VALUES("filename", "#{fileName}");
                }
                if(document.getRemark() != null && !document.getRemark().equals("")){
                    VALUES("remark", "#{remark}");
                }
                if(document.getUser() != null && document.getUser().getId() != null){
                    VALUES("user_id", "#{user.id}");
                }
            }
        }.toString();
    }
    // 动态更新
```

```
    public String updateDocument(Document document){
        return new SQL(){
            {
                UPDATE(DOCUMENTTABLE);
                if(document.getTitle() != null && !document.getTitle().equals("")){
                    SET(" title = #{title} ");
                }
                if(document.getFileName() != null && !document.getFileName().equals("")){
                    SET(" filename = #{fileName} ");
                }
                if(document.getRemark() != null && !document.getRemark().equals("")){
                    SET("remark = #{remark}");
                }
                if(document.getUser() != null && document.getUser().getId() != null){
                    SET("user_id = #{user.id}");
                }
                WHERE(" id = #{id} ");
            }
        }.toString();
    }
}
```

13.3.3 部署 DAO 层

通过前面的介绍不难发现，MyBatis 的持久化 DAO 接口只需要通过 SqlSession 的 getMapper 方法获得对应的接口实例，即可以调用接口的方法完成数据库操作。而在 Spring 和 MyBatis 整合的应用中，由 Spring 容器负责生成并管理 DAO 组件。

MyBatis 社区自己开发了一个 Mybatis-Spring 中间件用来满足 MyBatis 用户整合 Spring 的需求。通过 SqlSessionFactoryBean 类，可以将 SqlSessionFactory 纳入 IoC 容器内。

在使用 Spring 管理 DAO 组件之前，必须要为其提供对应的数据源，本应用使用 C3P0 数据源。

程序清单：codes/13/hrmapp/src/db.properties

```
dataSource.driverClass=com.mysql.jdbc.Driver
dataSource.jdbcUrl=jdbc:mysql://127.0.0.1:3306/hrm_db
dataSource.user=root
dataSource.password=root
dataSource.maxPoolSize=20
dataSource.maxIdleTime = 1000
dataSource.minPoolSize=6
dataSource.initialPoolSize=5
```

配置了所需的数据源之后，程序就可以在此数据源基础上配置 SqlSessionFactory 对象。配置 SqlSessionFactory 的代码如下：

程序清单：codes/13/hrmapp/WebConten/WEB-INF/applicationContext.xml

```
<?xml version="1.0" encoding="UTF-8"?>
<beans xmlns="http://www.springframework.org/schema/beans"
   xmlns:mybatis="http://mybatis.org/schema/mybatis-spring"
   xmlns:xsi="http://www.w3.org/2001/XMLSchema-instance"
   xmlns:p="http://www.springframework.org/schema/p"
   xmlns:context="http://www.springframework.org/schema/context"
   xmlns:mvc="http://www.springframework.org/schema/mvc"
   xmlns:tx="http://www.springframework.org/schema/tx"
   xsi:schemaLocation="http://www.springframework.org/schema/beans
                  http://www.springframework.org/schema/beans/spring-beans.xsd
                  http://www.springframework.org/schema/context
                  http://www.springframework.org/schema/context/spring-context.xsd
```

```
                    http://www.springframework.org/schema/mvc
                    http://www.springframework.org/schema/mvc/spring-mvc.xsd
                    http://www.springframework.org/schema/tx
                    http://www.springframework.org/schema/tx/spring-tx.xsd
                    http://mybatis.org/schema/mybatis-spring
                    http://mybatis.org/schema/mybatis-spring.xsd ">

     <!-- mybatis:scan 会扫描 org.fkit.dao 包里的所有接口当作 Spring 的 bean 配置，之后可以
      进行依赖注入-->
    <mybatis:scan base-package="org.fkit.hrm.dao"/>

     <!-- 扫描 org.fkit 包下面的 java 文件,若有 Spring 的相关注解的类,则把这些类注册为 Spring
      的 bean -->
    <context:component-scan base-package="org.fkit.hrm"/>

    <!-- 使用 PropertyOverrideConfigurer 后处理器加载数据源参数 -->
    <context:property-override location="classpath:db.properties"/>

    <!-- 配置 c3p0 数据源 -->
    <bean id="dataSource" class="com.mchange.v2.c3p0.ComboPooledDataSource"/>

    <!-- 配置 SqlSessionFactory,org.mybatis.spring.SqlSessionFactoryBean 是 Mybatis
     社区开发用于整合 Spring 的 bean -->
    <bean id="sqlSessionFactory" class="org.mybatis.spring.SqlSessionFactoryBean"
       p:dataSource-ref="dataSource"/>

    <!-- JDBC 事务管理器 -->
    <bean id="transactionManager"
    class="org.springframework.jdbc.datasource.DataSourceTransactionManager"
        p:dataSource-ref="dataSource"/>

    <!-- 启用支持 annotation 注解方式事务管理 -->
    <tx:annotation-driven transaction-manager="transactionManager"/>

</beans>
```

<mybatis:scan base-package="org.fkit.hrm.dao"/>元素会扫描 org.fkit.hrm.dao 包下面的所有接口作为 Spring 的 Bean 配置，之后可以进行依赖注入。

13.4 实现 Service 持久层

本系统只使用了一个业务逻辑组件：HrmService。该组件作为门面封装 6 个 DAO 组件，系统使用这个业务逻辑组件将这些 DAO 对象封装在一起。

13.4.1 业务逻辑组件的设计

业务逻辑组件是 DAO 组件的门面，所以也可以理解为业务逻辑组件需要依赖于 DAO 组件。DAO 组件与 HrmService（业务逻辑组件）直接的关系如图 13.3 所示。

在 HrmService 接口中定义了大量的业务方法，这些方法的实现依赖于 DAO 组件。由于每个业务方法要涉及多个 DAO 操作，DAO 操作是单条数据记录的操作，而业务逻辑方法的访问，则需要涉及多个 DAO 操作，因此每个业务逻辑方法可能需要涉及多条记录的访问。

业务逻辑组件面向 DAO 接口编程，可以让业务逻辑组件从 DAO 组件的实现中分离。因此业务逻辑组件只关心业务逻辑的实现，无须关心数据访问逻辑的实现。

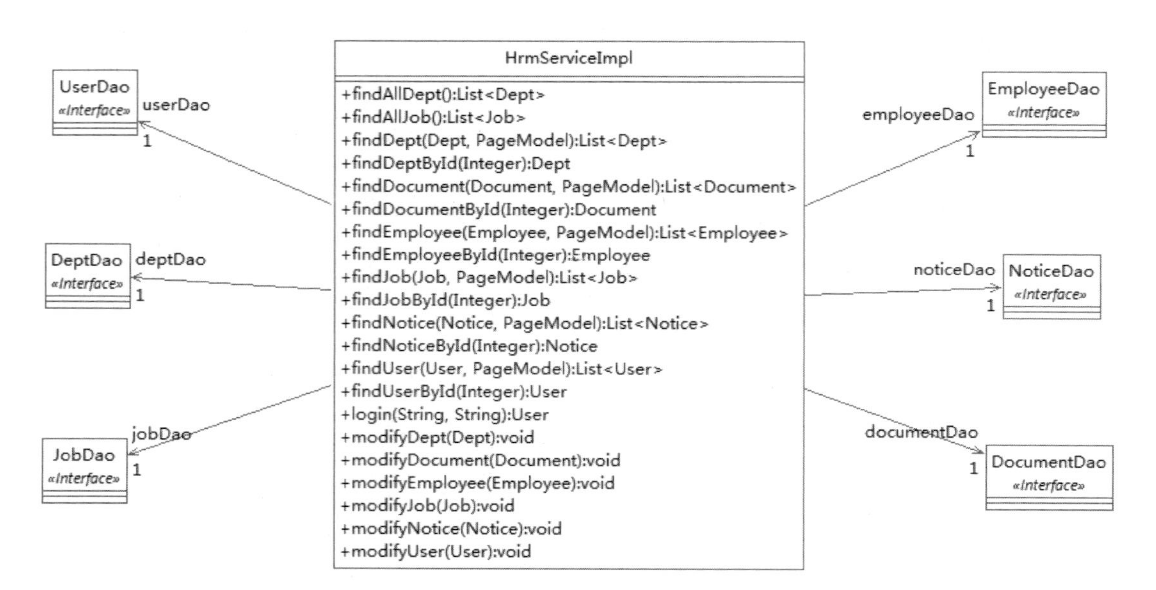

图 13.3　HrmService 与 DAO 组件接口的类关系图

13.4.2　实现业务逻辑组件

业务逻辑组件负责实现系统所需的业务方法，系统有多少个业务方法，业务逻辑组件就提供多少个对应方法，业务逻辑方法完全由业务逻辑组件负责实现。

业务逻辑组件只负责业务逻辑上的变化，而持久层上的变化则交给 DAO 层负责，因此业务逻辑组件必须依赖于 DAO 组件。

为了简化分页功能，设计了一个分页的 JSP 标签，只需要在页面使用分页标签，就可以完成所有页面的分页功能。下面是分页标签的源代码。

程序清单：codes/13/hrmapp/src/org/fkit/hrm/util/tag/PageModel.java

```
import org.fkit.hrm.util.common.HrmConstants;
/**
 *  分页实体
 */
public class PageModel {
    /** 分页总数据条数  */
    private int recordCount;
    /** 当前页面 */
    private int pageIndex ;
    /** 每页分多少条数据   */
    private int pageSize = HrmConstants.PAGE_DEFAULT_SIZE = 4;
    /** 总页数  */
    private int totalSize;
    public int getRecordCount() {
        this.recordCount = this.recordCount <= 0 ? 0:this.recordCount;
        return recordCount;
    }
    public void setRecordCount(int recordCount) {
        this.recordCount = recordCount;
    }
    public int getPageIndex() {
        this.pageIndex = this.pageIndex <= 0?1:this.pageIndex;
        /** 判断当前页面是否超过了总页数:如果超过了默认将最后一页作为当前页  */
        this.pageIndex = this.pageIndex>=this.getTotalSize()?this.getTotalSize():
this.pageIndex;
```

```
        return pageIndex;
    }
    public void setPageIndex(int pageIndex) {
        this.pageIndex = pageIndex;
    }
    public int getPageSize() {
        this.pageSize = this.pageSize <= HrmConstants.PAGE_DEFAULT_
         SIZE?HrmConstants.PAGE_DEFAULT_SIZE:this.pageSize;
        return pageSize;
    }
    public void setPageSize(int pageSize) {
        this.pageSize = pageSize;
    }
    public int getTotalSize() {
        if(this.getRecordCount() <=0){
            totalSize = 0 ;
        }else{
            totalSize = (this.getRecordCount() -1)/this.getPageSize() + 1;
        }
        return totalSize;
    }
    public int getFirstLimitParam(){
        return (this.getPageIndex()-1)*this.getPageSize() ;
    }
}
```

程序清单：codes/13/hrmapp/src/org/fkit/hrm/util/tag/PagerTag.java

```
import java.io.IOException;
import javax.servlet.jsp.JspException;
import javax.servlet.jsp.tagext.SimpleTagSupport;
/**
 * 分页标签
 */
public class PagerTag extends SimpleTagSupport {
    /** 定义请求 URL 中的占位符常量 */
    private static final String TAG = "{0}";
    /** 当前页码 */
    private int pageIndex;
    /** 每页显示的数量 */
    private int pageSize;
    /** 总记录条数 */
    private int recordCount;
    /** 请求 URL page.action?pageIndex={0}*/
    private String submitUrl;
    /** 样式 */
    private String style = "sabrosus";
    /** 定义总页数 */
    private int totalPage = 0;
    /**  在页面上引用自定义标签就会触发一个标签处理类    */
    @Override
    public void doTag() throws JspException, IOException {
        /** 定义它拼接最终的结果 */
        StringBuilder res = new StringBuilder();
        /** 定义它拼接中间的页码 */
        StringBuilder str = new StringBuilder();
        /** 判断总记录条数 */
        if (recordCount > 0){   //1499 / 15  = 100
            /** 需要显示分页标签，计算出总页数 需要分多少页 */
            totalPage = (this.recordCount - 1) / this.pageSize + 1;
            /** 判断上一页或下一页需不需要加 a 标签 */
            if (this.pageIndex == 1){ // 首页
                str.append("<span class='disabled'>上一页</span>");
```

```
                /** 计算中间的页码 */
                this.calcPage(str);
                /** 下一页需不需要 a 标签 */
                if (this.pageIndex == totalPage){
                    /** 只有一页 */
                    str.append("<span class='disabled'>下一页</span>");
                }else{
                    String tempUrl = this.submitUrl.replace(TAG, String.valueOf(pageIndex + 1));
                    str.append("<a href='"+ tempUrl +"'>下一页</a>");
                }
            }else if (this.pageIndex == totalPage){ // 尾页
                String tempUrl = this.submitUrl.replace(TAG, String.valueOf(pageIndex - 1));
                str.append("<a href='"+ tempUrl +"'>上一页</a>");
                /** 计算中间的页码 */
                this.calcPage(str);
                str.append("<span class='disabled'>下一页</span>");
            }else{ // 中间
                String tempUrl = this.submitUrl.replace(TAG, String.valueOf(pageIndex - 1));
                str.append("<a href='"+ tempUrl +"'>上一页</a>");
                /** 计算中间的页码 */
                this.calcPage(str);
                tempUrl = this.submitUrl.replace(TAG, String.valueOf(pageIndex + 1));
                str.append("<a href='"+ tempUrl +"'>下一页</a>");
            }
            /** 拼接其他的信息 */
            res.append("<table width='100%' align='center' style='font-size:13px;'
class='"+ style +"'>");
            res.append("<tr><td   style='COLOR:   #0061de;   MARGIN-RIGHT:   3px;
PADDING-TOP: 2px; TEXT-DECORATION: none'>" + str.toString());
            res.append("  跳 转 到   <input  style='text-align:
center;BORDER-RIGHT: #aaaadd 1px solid; PADDING-RIGHT: 5px; BORDER-TOP: #aaaadd 1px
solid; PADDING-LEFT: 5px; PADDING-BOTTOM: 2px; MARGIN: 2px; BORDER-LEFT: #aaaadd 1px
solid;  COLOR:  #000099;  PADDING-TOP:  2px;  BORDER-BOTTOM:  #aaaadd  1px  solid;
TEXT-DECORATION: none' type='text' size='2' id='pager_jump_page_size'/>");
            res.append(" <input      type='button'      style='text-align:
center;BORDER-RIGHT: #dedfde 1px solid; PADDING-RIGHT: 6px; BACKGROUND-POSITION: 50%
bottom;  BORDER-TOP:  #dedfde  1px  solid;  PADDING-LEFT:  6px;  PADDING-BOTTOM:  2px;
BORDER-LEFT: #dedfde 1px solid; COLOR: #0061de; MARGIN-RIGHT: 3px; PADDING-TOP: 2px;
BORDER-BOTTOM:  #dedfde  1px  solid;  TEXT-DECORATION:  none'  value=' 确 定 '
id='pager_jump_btn'/>");
            res.append("</td></tr>");
            res.append("<tr  align='center'><td  style='font-size:13px;'><tr><td
style='COLOR: #0061de; MARGIN-RIGHT: 3px; PADDING-TOP: 2px; TEXT-DECORATION: none'>");
            /** 开始条数 */
            int startNum = (this.pageIndex - 1) * this.pageSize + 1;
            /** 结束条数 */
            int endNum = (this.pageIndex == this.totalPage) ? this.recordCount :
this.pageIndex * this.pageSize;
            res.append("总共<font color='red'>"+ this.recordCount +"</font>条记录,
当前显示"+ startNum +"-"+ endNum +"条记录。");
            res.append("</td></tr>");
            res.append("</table>");
            res.append("<script type='text/javascript'>");
            res.append("    document.getElementById('pager_jump_btn').onclick  =
function(){");
            res.append("     var page_size = document.getElementById('pager_jump_
page_size').value;");
            res.append("     if (!/^[1-9]\\d*$/.test(page_size) || page_size < 1 ||
page_size > "+ this.totalPage +"){");
            res.append("        alert('请输入[1-"+ this.totalPage +"]之间的页码！');");
            res.append("     }else{");
            res.append("       var submit_url = '" + this.submitUrl + "';");
```

```
            res.append("        window.location = submit_url.replace('"+ TAG +"',
             page_size);");
            res.append("      }");
            res.append("}");
            res.append("</script>");
        }else{
            res.append("<table align='center' style='font-size:13px;'><tr>
             <td style='COLOR: #0061de; MARGIN-RIGHT: 3px; PADDING-TOP: 2px;
              TEXT-DECORATION: none'>总共<font color='red'>0</font>条记录，当前显示 0-0
             条记录。</td></tr></table>");
        }
        this.getJspContext().getOut().print(res.toString());
    }
    /** 计算中间页码的方法 */
    private void calcPage(StringBuilder str) {
        /** 判断总页数 */
        if (this.totalPage <= 11){
            /** 一次性显示全部的页码 */
            for (int i = 1; i <= this.totalPage; i++){
                if (this.pageIndex == i){
                    /** 当前页码 */
                    str.append("<span class='current'>"+ i +"</span>");
                }else{
                    String tempUrl = this.submitUrl.replace(TAG, String.valueOf(i));
                    str.append("<a href='"+ tempUrl +"'>"+ i +"</a>");
                }
            }
        }else{
            /** 靠近首页 */
            if (this.pageIndex <= 8){
                for (int i = 1; i <= 10; i++){
                    if (this.pageIndex == i){
                        /** 当前页码 */
                        str.append("<span class='current'>"+ i +"</span>");
                    }else{
                        String tempUrl = this.submitUrl.replace(TAG, String.valueOf(i));
                        str.append("<a href='"+ tempUrl +"'>"+ i +"</a>");
                    }
                }
                str.append("...");
                String tempUrl = this.submitUrl.replace(TAG, String.valueOf
                 (this. totalPage));
                str.append("<a href='"+ tempUrl +"'>"+ this.totalPage +"</a>");
            }
            /** 靠近尾页 */
            else if (this.pageIndex + 8 >= this.totalPage){
                String tempUrl = this.submitUrl.replace(TAG, String.valueOf(1));
                str.append("<a href='"+ tempUrl +"'>1</a>");
                str.append("...");
                for (int i = this.totalPage - 10; i <= this.totalPage; i++){
                    if (this.pageIndex == i){
                        /** 当前页码 */
                        str.append("<span class='current'>"+ i +"</span>");
                    }else{
                        tempUrl = this.submitUrl.replace(TAG, String.valueOf(i));
                        str.append("<a href='"+ tempUrl +"'>"+ i +"</a>");
                    }
                }
            }
            /** 在中间 */
            else{
                String tempUrl = this.submitUrl.replace(TAG, String.valueOf(1));
                str.append("<a href='"+ tempUrl +"'>1</a>");
```

```
                str.append("...");
                for (int i = this.pageIndex - 4; i <= this.pageIndex + 4; i++){
                    if (this.pageIndex == i){
                        /** 当前页码 */
                        str.append("<span class='current'>"+ i +"</span>");
                    }else{
                        tempUrl = this.submitUrl.replace(TAG, String.valueOf(i));
                        str.append("<a href='"+ tempUrl +"'>"+ i +"</a>");
                    }
                }
                str.append("...");
                tempUrl = this.submitUrl.replace(TAG, String.valueOf(this.totalPage));
                str.append("<a href='"+ tempUrl +"'>"+ this.totalPage +"</a>");
            }
        }
    }
    /** setter 方法 */
    public void setPageIndex(int pageIndex) {
        this.pageIndex = pageIndex;
    }
    public void setPageSize(int pageSize) {
        this.pageSize = pageSize;
    }
    public void setRecordCount(int recordCount) {
        this.recordCount = recordCount;
    }
    public void setSubmitUrl(String submitUrl) {
        this.submitUrl = submitUrl;
    }
    public void setStyle(String style) {
        this.style = style;
    }
}
```

要使用 JSP 的标签还需要在 WEB-INF 下增加一个 tld 标签文件。

程序清单：codes/13/hrmapp/WebContent/WEB-INF/page.tld

```
<?xml version="1.0" encoding="utf-8"?>
<taglib xmlns="http://java.sun.com/xml/ns/javaee"
    xmlns:xsi="http://www.w3.org/2001/XMLSchema-instance"
    xsi:schemaLocation="http://java.sun.com/xml/ns/javaee
                    http://java.sun.com/xml/ns/javaee/web-jsptaglibrary_2_1.xsd"
                    version="2.1">
  <!-- 描述 自定义标签版本的一种描述 -->
  <description>Pager 1.0 core library</description>
  <!-- 显示的名称 导包进行的一个展示 -->
  <display-name>Pager core</display-name>
  <!-- 版本号 -->
  <tlib-version>1.0</tlib-version>
  <!-- 短名 -->
  <short-name>fkjava</short-name>
  <!-- uri : 导包 -->
  <uri>/pager-tags</uri>
  <!-- 定义一个标签 -->
  <tag>
        <!-- 标签名 -->
        <name>pager</name>
        <!-- 标签处理类 -->
        <tag-class>org.fkit.hrm.util.tag.PagerTag</tag-class>
        <!-- 设置标签为空 -->
        <body-content>empty</body-content>

        <!-- 定义标签的属性 -->
```

```
        <attribute>
            <!-- 属性名 表示分页的第几页 -->
            <name>pageIndex</name>
            <!-- 必须的 -->
            <required>true</required>
            <!-- run time expression value 为true 支持EL表达式 -->
            <rtexprvalue>true</rtexprvalue>
        </attribute>
        <!-- 定义标签的属性 -->
        <attribute>
            <!-- 属性名 表示分页标签 ，每页显示多少条数据 -->
            <name>pageSize</name>
            <!-- 必须的 -->
            <required>true</required>
            <!-- run time expression value 为true 支持EL表达式 -->
            <rtexprvalue>true</rtexprvalue>
        </attribute>
        <!-- 定义标签的属性 -->
        <attribute>
            <!-- 属性名  记录分页的总数 -->
            <name>recordCount</name>
            <!-- 必须的 -->
            <required>true</required>
            <!-- run time expression value 为true 支持EL表达式 -->
            <rtexprvalue>true</rtexprvalue>
        </attribute>
        <!-- 定义标签的属性 -->
        <attribute>
            <!-- 属性名 -->
            <name>submitUrl</name>
            <!-- 必须的 -->
            <required>true</required>
            <!-- run time expression value 为true 支持EL表达式 -->
            <rtexprvalue>true</rtexprvalue>
        </attribute>
        <!-- 定义标签的属性 -->
        <attribute>
            <!-- 属性名 -->
            <name>style</name>
            <!-- 必须的 -->
            <required>false</required>
            <!-- run time expression value 为true 支持EL表达式 -->
            <rtexprvalue>true</rtexprvalue>
        </attribute>
    </tag>
</taglib>
```

下面是 HrmService 接口的源代码。

程序清单：codes/13/hrmapp/src/org/fkit/hrm/service/HrmService.java

```
import java.util.List;
import org.fkit.hrm.domain.Dept;
import org.fkit.hrm.domain.Document;
import org.fkit.hrm.domain.Employee;
import org.fkit.hrm.domain.Job;
import org.fkit.hrm.domain.Notice;
import org.fkit.hrm.domain.User;
import org.fkit.hrm.util.tag.PageModel;
public interface HrmService {
    /**
     * 用户登录
     * @param  loginname
```

```
     * @param  password
     * @return User 对象
     * */
    User login(String loginname,String password);
    /**
     * 根据 id 查询用户
     * @param id
     * @return 用户对象
     * */
    User findUserById(Integer id);
    /**
     * 获得所有用户
     * @return User 对象的 List 集合
     * */
    List<User> findUser(User user,PageModel pageModel);
    /**
     * 根据 id 删除用户
     * @param id
     * */
    void removeUserById(Integer id);
    /**
     * 修改用户
     * @param User 用户对象
     * */
    void modifyUser(User user);
    /**
     * 添加用户
     * @param User 用户对象
     * */
    void addUser(User user);
    /**
     * 获取所有员工
     * @param employee 查询条件
     * @param pageModel 分页对象
     * @return Dept 对象的 List 集合
     * */
    List<Employee> findEmployee(Employee employee,PageModel pageModel);

    /**
     * 根据 id 删除员工
     * @param id
     * */
    void removeEmployeeById(Integer id);
    /**
     * 根据 id 查询员工
     * @param id
     * @return 员工对象
     * */
    Employee findEmployeeById(Integer id);
    /**
     * 添加员工
     * @param employee 员工对象
     * */
    void addEmployee(Employee employee);
    /**
     * 修改员工
     * @param employee 员工对象
     * */
    void modifyEmployee(Employee employee);
    /**
     * 获取所有部门，分页查询
     * @return Dept 对象的 List 集合
```

```
 * */
List<Dept> findDept(Dept dept,PageModel pageModel);
/**
 * 获取所有部门
 * @return Dept 对象的 List 集合
 * */
List<Dept> findAllDept();
/**
 * 根据 id 删除部门
 * @param id
 * */
public void removeDeptById(Integer id);
/**
 * 添加部门
 * @param dept 部门对象
 * */
void addDept(Dept dept);
/**
 * 根据 id 查询部门
 * @param id
 * @return 部门对象
 * */
Dept findDeptById(Integer id);
/**
 * 修改部门
 * @param dept 部门对象
 * */
void modifyDept(Dept dept);
/**
 * 获取所有职位
 * @return Job 对象的 List 集合
 * */
List<Job> findAllJob();
/**
 * 获取所有职位，分页查询
 * @return Job 对象的 List 集合
 * */
List<Job> findJob(Job job,PageModel pageModel);
/**
 * 根据 id 删除职位
 * @param id
 * */
public void removeJobById(Integer id);
/**
 * 添加职位
 * @param Job 部门对象
 * */
void addJob(Job job);
/**
 * 根据 id 查询职位
 * @param id
 * @return 职位对象
 * */
Job findJobById(Integer id);
/**
 * 修改职位
 * @param dept 部门对象
 * */
void modifyJob(Job job);
/**
 * 获取所有公告
 * @return Notice 对象的 List 集合
```

```
     * */
    List<Notice> findNotice(Notice notice,PageModel pageModel);
    /**
     * 根据 id 查询公告
     * @param id
     * @return 公告对象
     * */
    Notice findNoticeById(Integer id);
    /**
     * 根据 id 删除公告
     * @param id
     * */
    public void removeNoticeById(Integer id);
    /**
     * 添加公告
     * @param Notice 公告对象
     * */
    void addNotice(Notice notice);
    /**
     * 修改公告
     * @param Notice 公告对象
     * */
    void modifyNotice(Notice notice);
    /**
     * 获取所有文档
     * @return Document 对象的 List 集合
     * */
    List<Document> findDocument(Document document,PageModel pageModel);
    /**
     * 添加文档
     * @param Document 文件对象
     * */
    void addDocument(Document document);
    /**
     * 根据 id 查询文档
     * @param id
     * @return 文档对象
     * */
    Document findDocumentById(Integer id);
    /**
     * 根据 id 删除文档
     * @param id
     * */
    public void removeDocumentById(Integer id);
    /**
     * 修改文档
     * @param Document 公告对象
     * */
    void modifyDocument(Document document);
}
```

HrmService 接口中是本系统所有业务逻辑方法的定义，下面是这些业务逻辑方法的实现。

程序清单：codes/13/hrmapp/src/org/fkit/hrm/service/impl/HrmServiceImpl.java

```
import java.util.HashMap;
import java.util.List;
import java.util.Map;
import org.fkit.hrm.dao.*;
import org.fkit.hrm.domain.*;
import org.fkit.hrm.service.HrmService;
import org.fkit.hrm.util.tag.PageModel;
import org.springframework.beans.factory.annotation.Autowired;
```

```
import org.springframework.stereotype.Service;
/**
 *人事管理系统服务层接口实现类
 */
@Transactional(propagation=Propagation.REQUIRED,isolation=Isolation.DEFAULT)
@Service("hrmService")
public class HrmServiceImpl implements HrmService{
    /**
     * 自动注入持久层 Dao 对象
     * */
    @Autowired
    private UserDao userDao;
    @Autowired
    private DeptDao deptDao;
    @Autowired
    private EmployeeDao employeeDao;
    @Autowired
    private JobDao jobDao;
    @Autowired
    private NoticeDao noticeDao;
    @Autowired
    private DocumentDao documentDao;
    /****************用户服务接口实现************************************/
    /**
     * HrmServiceImpl 接口 login 方法实现
     *  @see { HrmService }
     * */
    @Transactional(readOnly=true)
    @Override
    public User login(String loginname, String password) {
        System.out.println("HrmServiceImpl login -- >>");
        return userDao.selectByLoginnameAndPassword(loginname, password);
    }
    /**
     * HrmServiceImpl 接口 findUser 方法实现
     * @see { HrmService }
     * */
    @Transactional(readOnly=true)
    @Override
    public List<User> findUser(User user,PageModel pageModel) {
        /** 当前需要分页的总数据条数  */
        Map<String,Object> params = new HashMap<>();
        params.put("user", user);
        int recordCount = userDao.count(params);
        System.out.println("recordCount -->> " + recordCount);
        pageModel.setRecordCount(recordCount);
        if(recordCount > 0){
            /** 开始分页查询数据：查询第几页的数据 */
            params.put("pageModel", pageModel);
        }
        List<User> users = userDao.selectByPage(params);
        return users;
    }
    /**
     * HrmServiceImpl 接口 findUserById 方法实现
     * @see { HrmService }
     * */
    @Transactional(readOnly=true)
    @Override
    public User findUserById(Integer id) {
        return userDao.selectById(id);
    }
    /**
     * HrmServiceImpl 接口 removeUserById 方法实现
```

```
     * @see { HrmService }
     * */
    @Override
    public void removeUserById(Integer id) {
        userDao.deleteById(id);
    }
    /**
     * HrmServiceImpl 接口 addUser 方法实现
     * @see { HrmService }
     * */
    @Override
    public void modifyUser(User user) {
        userDao.update(user);
    }
    /**
     * HrmServiceImpl 接口 modifyUser 方法实现
     * @see { HrmService }
     * */
    @Override
    public void addUser(User user) {
        userDao.save(user);
    }
    /*****************部门服务接口实现*************************************/
    @Override
    @Transactional(readOnly=true)
    public List<Dept> findAllDept() {
        return deptDao.selectAllDept();
    }
    /**
     * HrmServiceImpl 接口 findDept 方法实现
     * @see { HrmService }
     * */
    @Transactional(readOnly=true)
    @Override
    public List<Dept> findDept(Dept dept,PageModel pageModel) {
        /** 当前需要分页的总数据条数  */
        Map<String,Object> params = new HashMap<>();
        params.put("dept", dept);
        int recordCount = deptDao.count(params);
        System.out.println("recordCount -->> " + recordCount);
        pageModel.setRecordCount(recordCount);
        if(recordCount > 0){
            /** 开始分页查询数据：查询第几页的数据 */
            params.put("pageModel", pageModel);
        }
        List<Dept> depts = deptDao.selectByPage(params);
        return depts;
    }
    /**
     * HrmServiceImpl 接口 removeUserById 方法实现
     * @see { HrmService }
     * */
    @Override
    public void removeDeptById(Integer id) {
        deptDao.deleteById(id);
    }
    /**
     * HrmServiceImpl 接口 addDept 方法实现
     * @see { HrmService }
     * */
    @Override
    public void addDept(Dept dept) {
        deptDao.save(dept);
    }
```

```
    /**
     * HrmServiceImpl 接口 findDeptById 方法实现
     * @see { HrmService }
     * */
    @Transactional(readOnly=true)
    @Override
    public Dept findDeptById(Integer id) {
        return deptDao.selectById(id);
    }
    /**
     * HrmServiceImpl 接口 modifyDept 方法实现
     * @see { HrmService }
     * */
    @Override
    public void modifyDept(Dept dept) {
        deptDao.update(dept);
    }
    /*****************员工服务接口实现*************************************/
    /**
     * HrmService 接口 findEmployee 方法实现
     * @see { HrmService }
     * */
    @Transactional(readOnly=true)
    @Override
    public List<Employee> findEmployee(Employee employee,PageModel pageModel) {
        /** 当前需要分页的总数据条数  */
        Map<String,Object> params = new HashMap<>();
        params.put("employee", employee);
        int recordCount = employeeDao.count(params);
        System.out.println("recordCount -->> " + recordCount);
        pageModel.setRecordCount(recordCount);

        if(recordCount > 0){
            /** 开始分页查询数据：查询第几页的数据 */
            params.put("pageModel", pageModel);
        }
        List<Employee> employees = employeeDao.selectByPage(params);
        return employees;
    }
    /**
     * HrmService 接口 removeEmployeeById 方法实现
     * @see { HrmService }
     * */
    @Override
    public void removeEmployeeById(Integer id) {
        employeeDao.deleteById(id);
    }
    /**
     * HrmService 接口 findEmployeeById 方法实现
     * @see { HrmService }
     * */
    @Transactional(readOnly=true)
    @Override
    public Employee findEmployeeById(Integer id) {
        return employeeDao.selectById(id);
    }
    /**
     * HrmService 接口 addEmployee 方法实现
     * @see { HrmService }
     * */
    @Override
    public void addEmployee(Employee employee) {
        employeeDao.save(employee);
    }
```

```
/**
 * HrmService 接口 modifyEmployee 方法实现
 * @see { HrmService }
 * */
@Override
public void modifyEmployee(Employee employee) {
    employeeDao.update(employee);
}
/*****************职位接口实现*************************************/
/**
 * HrmService 接口 findAllJob 方法实现
 * @see { HrmService }
 * */
@Transactional(readOnly=true)
@Override
public List<Job> findAllJob() {
    return jobDao.selectAllJob();
}
/**
 * HrmService 接口 findJob 方法实现
 * @see { HrmService }
 * */
@Transactional(readOnly=true)
@Override
public List<Job> findJob(Job job, PageModel pageModel) {
    /** 当前需要分页的总数据条数  */
    Map<String,Object> params = new HashMap<>();
    params.put("job", job);
    int recordCount = jobDao.count(params);
    System.out.println("recordCount -->> " + recordCount);
    pageModel.setRecordCount(recordCount);
    if(recordCount > 0){
        /** 开始分页查询数据：查询第几页的数据 */
        params.put("pageModel", pageModel);
    }
    List<Job> jobs = jobDao.selectByPage(params);
    return jobs;
}
/**
 * HrmService 接口 removeJobById 方法实现
 * @see { HrmService }
 * */
@Override
public void removeJobById(Integer id) {
    jobDao.deleteById(id);
}
/**
 * HrmService 接口 addJob 方法实现
 * @see { HrmService }
 * */
@Override
public void addJob(Job job) {
    jobDao.save(job);
}
/**
 * HrmService 接口 findJobById 方法实现
 * @see { HrmService }
 * */
@Transactional(readOnly=true)
@Override
public Job findJobById(Integer id) {
    return jobDao.selectById(id);
}
/**
```

```
     * HrmService 接口 modifyJob 方法实现
     * @see { HrmService }
     * */
    @Override
    public void modifyJob(Job job) {
        jobDao.update(job);
    }

    /*****************公告接口实现*************************************/
    @Transactional(readOnly=true)
    @Override
    public List<Notice> findNotice(Notice notice, PageModel pageModel) {
        /** 当前需要分页的总数据条数  */
        Map<String,Object> params = new HashMap<>();
        params.put("notice", notice);
        int recordCount = noticeDao.count(params);
        System.out.println("recordCount -->> " + recordCount);
        pageModel.setRecordCount(recordCount);
        if(recordCount > 0){
            /** 开始分页查询数据：查询第几页的数据 */
            params.put("pageModel", pageModel);
        }
        List<Notice> notices = noticeDao.selectByPage(params);
        return notices;
    }
    /**
     * HrmService 接口 findNoticeById 方法实现
     * @see { HrmService }
     * */
    @Transactional(readOnly=true)
    @Override
    public Notice findNoticeById(Integer id) {
        return noticeDao.selectById(id);
    }
    /**
     * HrmService 接口 removeNoticeById 方法实现
     * @see { HrmService }
     * */
    @Override
    public void removeNoticeById(Integer id) {
        noticeDao.deleteById(id);
    }
    /**
     * HrmService 接口 addNotice 方法实现
     * @see { HrmService }
     * */
    @Override
    public void addNotice(Notice notice) {
        noticeDao.save(notice);
    }
    /**
     * HrmService 接口 modifyNotice 方法实现
     * @see { HrmService }
     * */
    @Override
    public void modifyNotice(Notice notice) {
        noticeDao.update(notice);
    }
    /*****************文件接口实现*************************************/
    /**
     * HrmService 接口 findDocument 方法实现
     * @see { HrmService }
     * */
```

```
    @Transactional(readOnly=true)
    @Override
    public List<Document> findDocument(Document document, PageModel pageModel) {
        /** 当前需要分页的总数据条数  */
        Map<String,Object> params = new HashMap<>();
        params.put("document", document);
        int recordCount = documentDao.count(params);
        System.out.println("recordCount -->> " + recordCount);
        pageModel.setRecordCount(recordCount);
        if(recordCount > 0){
            /** 开始分页查询数据：查询第几页的数据 */
            params.put("pageModel", pageModel);
        }
        List<Document> documents = documentDao.selectByPage(params);
        return documents;
    }
    /**
     * HrmService 接口 addDocument 方法实现
     * @see { HrmService }
     * */
    @Override
    public void addDocument(Document document) {
        documentDao.save(document);
    }
    /**
     * HrmService 接口 removeDocumentById 方法实现
     * @see { HrmService }
     * */
    @Override
    public void removeDocumentById(Integer id) {
        documentDao.deleteById(id);
    }
    /**
     * HrmService 接口 modifyDocument 方法实现
     * @see { HrmService }
     * */
    @Override
    public void modifyDocument(Document document) {
        documentDao.update(document);
    }
    /**
     * HrmService 接口 findDocumentById 方法实现
     * @see { HrmService }
     * */
    @Transactional(readOnly=true)
    @Override
    public Document findDocumentById(Integer id) {
        return documentDao.selectById(id);
    }
}
```

在 HrmServiceImpl 类中实现了服务接口 HrmService 中定义的所有业务逻辑方法，并且在 HrmServiceImpl 类上使用了两个注解：

- @Transactional(propagation=Propagation.REQUIRED,isolation=Isolation.DEFAULT)。表示该类需要 Spring 加入事务，Propagation.REQUIRED 属性指有事务就处于当前事务中，没有事务就创建一个事务；isolation=Isolation.DEFAULT 属性表示使用事务数据库的默认隔离级别。
- @Service("hrmService")。将该类配置成一个 Spring 的 Bean，标识符是 hrmService。
- @Autowired。在 HrmServiceImpl 类中业务方法的实现依赖于 DAO 组件，在配置文件中使用的注解<mybatis:scan base-package="org.fkit.hrm.dao"/>会将 org.fkit.hrm.dao 文件夹下的所有 MyBatis 文件配置成 Spring 的 Bean，Bean 的 id 就是文件的类的名称。@Autowired 注解默认使用 byType 自动装配将 6 个持久层的 DAO 注入给 HrmServiceImpl 类对应依赖的 DAO 组件。

13.4.3 事务管理

与所有的Java EE应用类似，本系统的事务管理负责管理业务逻辑组件里的业务逻辑方法，只有对业务逻辑方法添加事务管理才有实际意义，对于单个 DAO 方法（基本的 CRUD 方法）增加事务管理是没有太大实际意义的。

下面是本应用中事务管理的配置代码：

程序清单：codes/13/hrmapp/WebConten/WEB-INF/applicationContext.xml

```
<!-- JDBC 事务管理器 -->
<bean id="transactionManager"
   class="org.springframework.jdbc.datasource.DataSourceTransactionManager"
   p:dataSource-ref="dataSource"/>
<!-- 启用支持 annotation 注解方式事务管理 -->
<tx:annotation-driven transaction-manager="transactionManager"/>
```

借助于 Spring Scheam 所提供的 tx 命名空间的帮助，系统可以非常方便地为业务逻辑组件配置事务管理。其中，tx 命名空间下的<tx:annotation-driven.../>元素用于支持事务注解，transaction-manager 属性用于指定使用哪个事务管理器。之后就可以在 Java 类中使用@Transactional 注解给 Spring 的 Bean 添加事务管理了。

13.4.4 部署业务逻辑组件

在 applicationContext.xml 中增加<context:component-scan.../>扫描 Spring 的相关注解类，就可以通过@Controller、@Service 等 Spring 提供的注解将 Java 类配置成 Spring 的 Bean，<context:component-scan.../>元素的 base-package 属性用于指定扫描哪个包（包含子包）下面的 Java 类。

程序清单：codes/13/hrmapp/WebConten/WEB-INF/applicationContext.xml

```
<!-- 扫描 org.fkit 包下面的 java 文件，若有 Spring 的相关注解的类，则把这些类注册为 Spring 的
  bean -->
<context:component-scan base-package="org.fkit.hrm"/>
```

13.5 实现 Web 层

前面部分已经实现了本应用的所有中间层，系统的所有业务逻辑组件也都部署在 Spring 容器中了。接下来应该为应用实现 Web 层了。通常而言，系统的控制器和 JSP 在一起设计。因为当 JSP 页面发出请求后，该请求被控制器接收，然后控制器负责调用业务逻辑组件来处理请求。从这个意义上来说，控制器是 JSP 页面和业务逻辑组件之间的纽带。

13.5.1 控制器的处理顺序

当控制器接收到用户请求后，控制器并不会处理用户请求，只是对用户的请求参数进行解析处理，然后调用业务逻辑方法来处理用户请求；当请求被处理完成后，控制器负责将处理结果通过 JSP 页面呈现给用户。

对于使用 Spring MVC 的应用而言，控制器实际上由两个部分组成：系统的核心控制器 DispatcherServlet 和业务控制器 Controller。

程序清单：codes/13/hrmapp/WebConten/WEB-INF/web.xml

```
<?xml version="1.0" encoding="UTF-8"?>
<web-app xmlns:xsi="http://www.w3.org/2001/XMLSchema-instance"
   xmlns="http://xmlns.jcp.org/xml/ns/javaee"
   xsi:schemaLocation="http://xmlns.jcp.org/xml/ns/javaee
   http://xmlns.jcp.org/xml/ns/javaee/web-app_3_1.xsd"
   id="WebApp_ID" version="3.1">
   <!-- 配置 Spring 核心监听器，默认会以 /WEB-INF/applicationContext.xml 作为配置文件 -->
   <listener>
      <listener-class>org.springframework.web.context.ContextLoaderListener
       </listener-class>
   </listener>
   <!-- contextConfigLocation 参数用来指定 Spring 的配置文件 -->
   <context-param>
      <param-name>contextConfigLocation</param-name>
      <param-value>/WEB-INF/applicationContext*.xml</param-value>
   </context-param>
   <!-- 定义 Spring MVC 的前端控制器 -->
  <servlet>
   <servlet-name>springmvc</servlet-name>
   <servlet-class>
      org.springframework.web.servlet.DispatcherServlet
   </servlet-class>
   <init-param>
     <param-name>contextConfigLocation</param-name>
     <param-value>/WEB-INF/springmvc-config.xml</param-value>
   </init-param>
   <load-on-startup>1</load-on-startup>
  </servlet>
  <!-- 让 Spring MVC 的前端控制器拦截所有请求 -->
  <servlet-mapping>
   <servlet-name>springmvc</servlet-name>
   <url-pattern>/</url-pattern>
  </servlet-mapping>
  <!-- 编码过滤器 -->
  <filter>
      <filter-name>characterEncodingFilter</filter-name>
      <filter-class>org.springframework.web.filter.CharacterEncodingFilter
</filter-class>
      <init-param>
          <param-name>encoding</param-name>
          <param-value>UTF-8</param-value>
      </init-param>
 </filter>
   <filter-mapping>
      <filter-name>characterEncodingFilter</filter-name>
      <url-pattern>/*</url-pattern>
   </filter-mapping>
   <!-- jsp 的配置 -->
  <jsp-config>
   <jsp-property-group>
       <!-- 配置拦截所有的 jsp 页面  -->
     <url-pattern>*.jsp</url-pattern>
      <!-- 可以使用 el 表达式  -->
     <el-ignored>false</el-ignored>
     <!-- 不能在页面使用 java 脚本 -->
     <scripting-invalid>true</scripting-invalid>
     <!-- 给所有的 jsp 页面导入要依赖的库，tablib.jsp 就是一个全局的标签库文件  -->
     <include-prelude>/WEB-INF/jsp/taglib.jsp</include-prelude>
```

```
    </jsp-property-group>
  </jsp-config>
  <error-page>
    <error-code>404</error-code>
    <location>/404.html</location>
  </error-page>
  <welcome-file-list>
    <welcome-file>index.jsp</welcome-file>
  </welcome-file-list>
</web-app>
```

程序清单：codes/13/hrmapp/WebConten/WEB-INF/springmvc-config.xml

```
<?xml version="1.0" encoding="UTF-8"?>
<beans xmlns="http://www.springframework.org/schema/beans"
    xmlns:xsi="http://www.w3.org/2001/XMLSchema-instance"
    xmlns:mvc="http://www.springframework.org/schema/mvc"
    xmlns:context="http://www.springframework.org/schema/context"
    xsi:schemaLocation="
        http://www.springframework.org/schema/beans
        http://www.springframework.org/schema/beans/spring-beans.xsd
        http://www.springframework.org/schema/mvc
        http://www.springframework.org/schema/mvc/spring-mvc.xsd
        http://www.springframework.org/schema/context
        http://www.springframework.org/schema/context/spring-context.xsd">
    <!-- 自动扫描该包，SpringMVC 会将包下用了@controller 注解的类注册为 Spring 的
     controller -->
    <context:component-scan base-package="org.fkit.hrm.controller"/>
    <!-- 设置默认配置方案 -->
    <mvc:annotation-driven/>
    <!-- 使用默认的 Servlet 来响应静态文件 -->
    <mvc:default-servlet-handler/>
    <!-- 定义 Spring MVC 的拦截器 -->
    <mvc:interceptors>
        <mvc:interceptor>
            <!-- 拦截所有请求 -->
            <mvc:mapping path="/*"/>
            <!-- 自定义判断用户权限的拦截类 -->
             <bean class=" org.fkit.hrm.interceptor.AuthorizedInterceptor "/>
        </mvc:interceptor>
    </mvc:interceptors>
    <!-- 视图解析器  -->
     <bean id="viewResolver"
          class="org.springframework.web.servlet.view.InternalResourceViewResolver"
          p:prefix="/WEB-INF/jsp/" p:suffix=".jsp"/>

<!-- 文件上传下载  -->
<bean id="multipartResolver"
        class="org.springframework.web.multipart.commons.CommonsMultipartResolver">
        <!-- 上传文件大小上限，单位为字节（10MB） -->
        <property name="maxUploadSize">
            <value>10485760</value>
        </property>
        <!-- 请求的编码格式，必须和 jSP 的 pageEncoding 属性一致，以便正确读取表单的内容，默认
         为 ISO-8859-1 -->
        <property name="defaultEncoding">
            <value>UTF-8</value>
        </property>
    </bean>
</beans>
```

在 springmvc-config.xml 文件中配置了一个拦截器，用于判断用户是否登录，如果其没有登录，则用户不能访问网站，跳回登录页面重新登录。

程序清单：codes/13/hrmapp/src/org/fkit/hrm/interceptor/AuthorizedInterceptor.java

```
import javax.servlet.http.HttpServletRequest;
import javax.servlet.http.HttpServletResponse;
import org.fkit.hrm.domain.User;
import org.fkit.hrm.util.common.HrmConstants;
import org.springframework.web.servlet.HandlerInterceptor;
import org.springframework.web.servlet.ModelAndView;
/**
 * 判断用户权限的 Spring MVC 的拦截器
 */
public class AuthorizedInterceptor  implements HandlerInterceptor {
    /** 定义不需要拦截的请求 */
    private static final String[] IGNORE_URI = {"/loginForm", "/login","/404.html"};
     /**
     * 该方法需要 preHandle 方法的返回值为 true 时才会执行。
     * 该方法将在整个请求完成之后执行，主要作用是用于清理资源。
     */
    @Override
    public void afterCompletion(HttpServletRequest request,
            HttpServletResponse response, Object handler, Exception exception)
            throws Exception {
    }
     /**
     * 这个方法在 preHandle 方法返回值为 true 的时候才会执行。
     * 执行时间是在处理器处理之后，也就是在 Controller 的方法调用之后执行。
     */
    @Override
    public void postHandle(HttpServletRequest request, HttpServletResponse response,
            Object handler, ModelAndView mv) throws Exception {
    }
     /**
     * preHandle 方法是进行处理器拦截用的，该方法将在 Controller 处理之前进行调用，
     * 当 preHandle 的返回值为 false 的时候整个请求就结束了。
     * 如果 preHandle 的返回值为 true，则会继续执行 postHandle 和 afterCompletion。
     */
    @Override
    public boolean preHandle(HttpServletRequest request, HttpServletResponse response,
            Object handler) throws Exception {
        /** 默认用户没有登录 */
        boolean flag = false;
        /** 获得请求的 ServletPath */
        String servletPath = request.getServletPath();
        /**  判断请求是否需要拦截 */
        for (String s : IGNORE_URI) {
            if (servletPath.contains(s)) {
                flag = true;
                break;
            }
        }
        /** 拦截请求 */
        if (!flag){
            /** 1.获取 session 中的用户  */
            User user = (User) request.getSession().getAttribute(HrmConstants.USER_
             SESSION);
            /** 2.判断用户是否已经登录 */
            if(user == null){
                 /** 如果用户没有登录，跳转到登录页面 */
                request.setAttribute("message", "请先登录再访问网站!");
                request.getRequestDispatcher(HrmConstants.LOGIN).forward(request,
                 response);
                return flag;
            }else{
```

```
                flag = true;
            }
        }
        return flag;
    }
}
```

下面通过用例来介绍控制层的实现。

13.5.2　用户管理

本系统的所有 JSP 页面都放在 WEB-INF/jsp 目录下，登录页面是 loginForm.jsp。用户提交登录请求后，用户输入的登录名、密码被提交到 UserController 的 login 方法，UserController 将会根据请求参数决定呈现哪个视图资源。

当 UserController 处理登录请求后，程序可以返回两个逻辑视图，其中，loginForm 是登录失败后的逻辑视图。当用户登录成功后，转入 main 页面。

处理用户的 UserController 代码如下：

程序清单：codes/13/hrmapp/src/org/fkit/hrm/controller/UserController.java

```
import java.util.List;
import javax.servlet.http.HttpSession;
import org.fkit.hrm.domain.User;
import org.fkit.hrm.service.HrmService;
import org.fkit.hrm.util.common.HrmConstants;
import org.fkit.hrm.util.tag.PageModel;
import org.springframework.beans.factory.annotation.Autowired;
import org.springframework.beans.factory.annotation.Qualifier;
import org.springframework.stereotype.Controller;
import org.springframework.ui.Model;
import org.springframework.web.bind.annotation.ModelAttribute;
import org.springframework.web.bind.annotation.RequestMapping;
import org.springframework.web.bind.annotation.RequestParam;
import org.springframework.web.servlet.ModelAndView;
/**
 * 处理用户请求控制器
 * */
@Controller
public class UserController {
    /**
     * 自动注入 UserService
     * */
    @Autowired
    @Qualifier("hrmService")
    private HrmService hrmService;
    /**
     * 处理登录请求
     * @param String loginname  登录名
     * @param String password 密码
     * @return 跳转的视图
     * */
    @RequestMapping(value="/login")
     public ModelAndView login(@RequestParam("loginname") String loginname,
            @RequestParam("password") String password,
            HttpSession session,
            ModelAndView mv){
        // 调用业务逻辑组件判断用户是否可以登录
        User user = hrmService.login(loginname, password);
        if(user != null){
            // 将用户保存到 HttpSession 当中
            session.setAttribute(HrmConstants.USER_SESSION, user);
```

```
            // 客户端跳转到 main 页面
            mv.setViewName("redirect:/main");
        }else{
            // 设置登录失败提示信息
            mv.addObject("message", "登录名或密码错误!请重新输入");
            // 服务器内部跳转到登录页面
            mv.setViewName("forward:/loginForm");
        }
        return mv;
    }
    /**
     * 处理查询请求
     * @param pageIndex 请求的是第几页
     * @param employee 模糊查询参数
     * @param Model model
     * */
    @RequestMapping(value="/user/selectUser")
     public String selectUser(Integer pageIndex,
             @ModelAttribute User user,
             Model model){
        System.out.println("user = " + user);
        PageModel pageModel = new PageModel();
        if(pageIndex != null){
            pageModel.setPageIndex(pageIndex);
        }
        /** 查询用户信息     */
        List<User> users = hrmService.findUser(user, pageModel);
        model.addAttribute("users", users);
        model.addAttribute("pageModel", pageModel);
        return "user/user";
    }
    /**
     * 处理删除用户请求
     * @param String ids 需要删除的 id 字符串
     * @param ModelAndView mv
     * */
    @RequestMapping(value="/user/removeUser")
     public ModelAndView removeUser(String ids,ModelAndView mv){
        // 分解 id 字符串
        String[] idArray = ids.split(",");
        for(String id : idArray){
            // 根据 id 删除员工
            hrmService.removeUserById(Integer.parseInt(id));
        }
        // 设置客户端跳转到查询请求
        mv.setViewName("redirect:/user/selectUser");
        // 返回 ModelAndView
        return mv;
    }
    /**
     * 处理修改用户请求
     * @param String flag 标记， 1 表示跳转到修改页面，2 表示执行修改操作
     * @param User user  要修改用户的对象
     * @param ModelAndView mv
     * */
    @RequestMapping(value="/user/updateUser")
     public ModelAndView updateUser(
             String flag,
             @ModelAttribute User user,
             ModelAndView mv){
        if(flag.equals("1")){
            // 根据 id 查询用户
```

```
            User target = hrmService.findUserById(user.getId());
            // 设置 Model 数据
            mv.addObject("user", target);
            // 返回修改员工页面
            mv.setViewName("user/showUpdateUser");
        }else{
            // 执行修改操作
            hrmService.modifyUser(user);
            // 设置客户端跳转到查询请求
            mv.setViewName("redirect:/user/selectUser");
        }
        // 返回
        return mv;
    }
    /**
     * 处理添加请求
     * @param String flag 标记， 1 表示跳转到添加页面，2 表示执行添加操作
     * @param User user  要添加用户的对象
     * @param ModelAndView mv
     * */
    @RequestMapping(value="/user/addUser")
    public ModelAndView addUser(
            String flag,
            @ModelAttribute User user,
            ModelAndView mv){
        if(flag.equals("1")){
            // 设置跳转到添加页面
            mv.setViewName("user/showAddUser");
        }else{
            // 执行添加操作
            hrmService.addUser(user);
            // 设置客户端跳转到查询请求
            mv.setViewName("redirect:/user/selectUser");
        }
        // 返回
        return mv;
    }
}
```

部署 hrmapp 这个 Web 应用，在浏览器中输入如下 URL 来测试应用：

```
http://localhost:8080/hrmapp
```

会看到如图 13.4 所示的界面。这表示 Spring MVC 成功跳转到初始登录页面 loginForm.jsp。

图 13.4 登录界面

输入正确的登录名“admin”、密码“123456”，若登录成功，则跳转到如图 13.1 所示的人事管理系统界面。

单击左侧菜单“用户管理”下面的“添加用户”命令，跳转到“添加用户”界面，如图 13.5 所示。

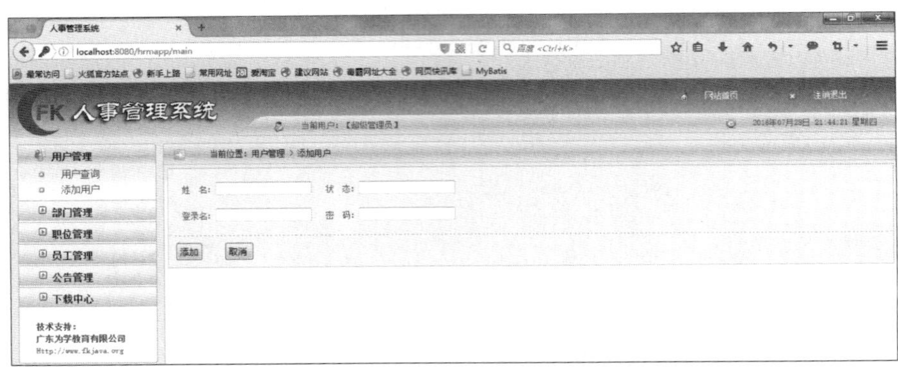

图 13.5　添加用户界面

输入需要添加的用户姓名、状态、登录名和密码，单击“添加”按钮，若添加成功则跳转到如图 13.6 所示的用户查询界面，显示所有用户信息。

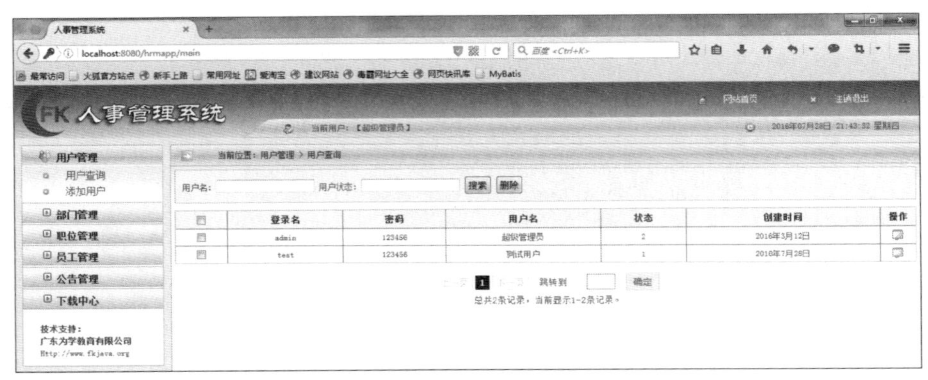

图 13.6　用户查询界面

输入用户名、用户状态，单击“搜索”按钮可以完成模糊查询功能。

选择每一行最后一列的“操作”按钮，可以进入修改页面，对选中的用户进行修改操作。

选择每一行第一列的复选框，单击“删除”按钮，则可以对选中的用户进行删除操作。

13.5.3　部门管理

处理部门的 DeptController 代码如下：

程序清单：codes/13/hrmapp/src/org/fkit/hrm/controller/DeptController.java

```
import java.util.List;
import org.fkit.hrm.domain.Dept;
import org.fkit.hrm.service.HrmService;
import org.fkit.hrm.util.tag.PageModel;
import org.springframework.beans.factory.annotation.Autowired;
import org.springframework.beans.factory.annotation.Qualifier;
import org.springframework.stereotype.Controller;
import org.springframework.ui.Model;
import org.springframework.web.bind.annotation.ModelAttribute;
import org.springframework.web.bind.annotation.RequestMapping;
```

```
import org.springframework.web.servlet.ModelAndView;
@Controller
public class DeptController {
    /**
     * 自动注入 UserService
     * */
    @Autowired
    @Qualifier("hrmService")
    private HrmService hrmService;

    /**
     * 处理/login 请求
     * */
    @RequestMapping(value="/dept/selectDept")
     public String selectDept(Model model,Integer pageIndex,
            @ModelAttribute Dept dept){
        System.out.println("selectDept -->>");
        System.out.println("pageIndex = " + pageIndex);
        System.out.println("dept = " + dept);
        PageModel pageModel = new PageModel();
        System.out.println("getPageIndex = " + pageModel.getPageIndex());
        System.out.println("getPageSize = " + pageModel.getPageSize());
        System.out.println("getRecordCount = " + pageModel.getRecordCount());
        if(pageIndex != null){
            pageModel.setPageIndex(pageIndex);
        }
        /** 查询用户信息     */
        List<Dept> depts = hrmService.findDept(dept, pageModel);
        model.addAttribute("depts", depts);
        model.addAttribute("pageModel", pageModel);
        return "dept/dept";
    }
    /**
     * 处理删除部门请求
     * @param String ids 需要删除的 id 字符串
     * @param ModelAndView mv
     * */
    @RequestMapping(value="/dept/removeDept")
     public ModelAndView removeDept(String ids,ModelAndView mv){
        // 分解 id 字符串
        String[] idArray = ids.split(",");
        for(String id : idArray){
            // 根据 id 删除部门
            hrmService.removeDeptById(Integer.parseInt(id));
        }
        // 设置客户端跳转到查询请求
        mv.setViewName("redirect:/dept/selectDept");
        // 返回 ModelAndView
        return mv;
    }
    /**
     * 处理添加请求
     * @param String flag 标记， 1 表示跳转到添加页面，2 表示执行添加操作
     * @param Dept  dept  要添加的部门对象
     * @param ModelAndView mv
     * */
    @RequestMapping(value="/dept/addDept")
     public ModelAndView addDept(
            String flag,
            @ModelAttribute Dept dept,
            ModelAndView mv){
        if(flag.equals("1")){
            // 设置跳转到添加页面
```

```
            mv.setViewName("dept/showAddDept");
        }else{
            // 执行添加操作
            hrmService.addDept(dept);
            // 设置客户端跳转到查询请求
            mv.setViewName("redirect:/dept/selectDept");
        }
        // 返回
        return mv;
    }
    /**
     * 处理修改部门请求
     * @param String flag 标记， 1 表示跳转到修改页面，2 表示执行修改操作
     * @param Dept dept 要修改部门的对象
     * @param ModelAndView mv
     * */
    @RequestMapping(value="/dept/updateDept")
     public ModelAndView updateDpet(
            String flag,
            @ModelAttribute Dept dept,
            ModelAndView mv){
        if(flag.equals("1")){
            // 根据 id 查询部门
            Dept target = hrmService.findDeptById(dept.getId());
            // 设置 Model 数据
            mv.addObject("dept", target);
            // 设置跳转到修改页面
            mv.setViewName("dept/showUpdateDept");
        }else{
            // 执行修改操作
            hrmService.modifyDept(dept);
            // 设置客户端跳转到查询请求
            mv.setViewName("redirect:/dept/selectDept");
        }
        // 返回
        return mv;
    }
}
```

单击左侧菜单“部门管理”下面的“添加部门”命令，跳转到“添加部门”界面，如图 13.7 所示。

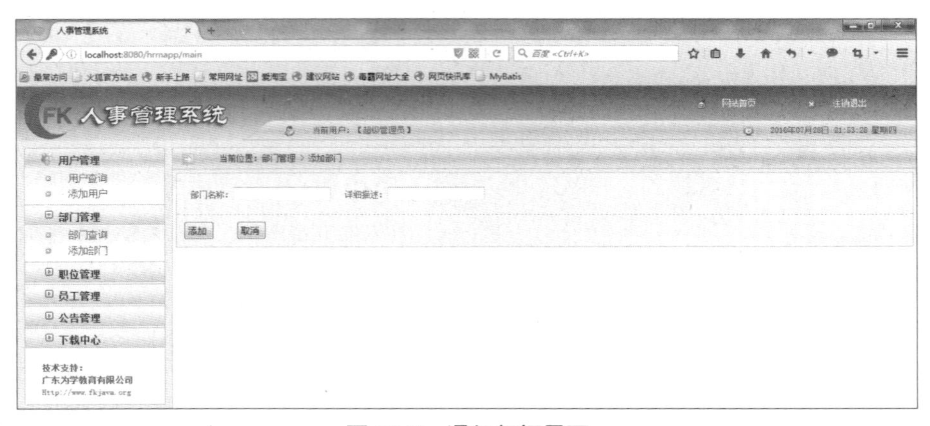

图 13.7　添加部门界面

输入需要添加的部门名称和详细描述，单击“添加”按钮，若添加成功则跳转到如图 13.8 所示的部门查询界面，显示所有部门信息。

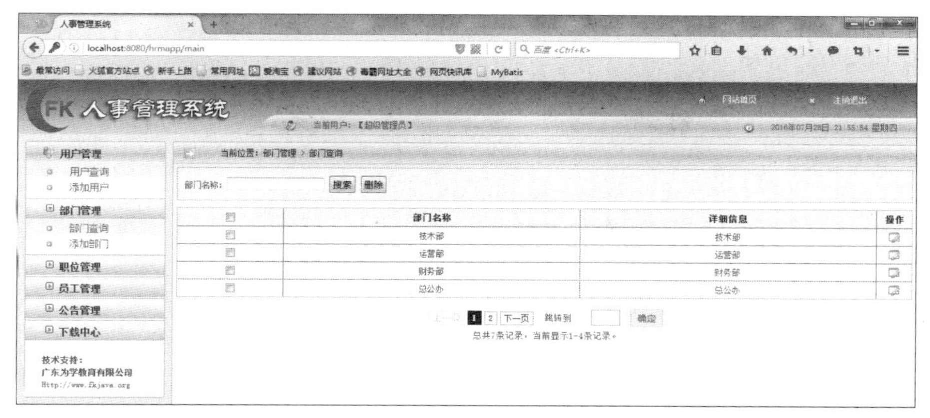

图 13.8 部门查询界面

输入部门名称，单击“搜索”按钮可以完成模糊查询功能。

选择每一行最后一列的“操作”按钮，可以进入修改页面，对选中的部门进行修改操作。

选择每一行第一列的复选框，单击“删除”按钮，则可以对选中的部门进行删除操作。

13.5.4 职位管理

处理职位的 JobController 代码如下：

程序清单：codes/13/hrmapp/src/org/fkit/hrm/controller/JobController.java

```
import java.util.List;
import org.fkit.hrm.domain.Job;
import org.fkit.hrm.service.HrmService;
import org.fkit.hrm.util.tag.PageModel;
import org.springframework.beans.factory.annotation.Autowired;
import org.springframework.beans.factory.annotation.Qualifier;
import org.springframework.stereotype.Controller;
import org.springframework.ui.Model;
import org.springframework.web.bind.annotation.ModelAttribute;
import org.springframework.web.bind.annotation.RequestMapping;
import org.springframework.web.servlet.ModelAndView;
@Controller
public class JobController {
    /**
     * 自动注入 UserService
     * */
    @Autowired
    @Qualifier("hrmService")
    private HrmService hrmService;
    /**
     * 处理/login 请求
     * */
    @RequestMapping(value="/job/selectJob")
     public String selectJob(Model model,Integer pageIndex,
            @ModelAttribute Job job){
       System.out.println("selectJob -->> " + job);
       PageModel pageModel = new PageModel();
       if(pageIndex != null){
          pageModel.setPageIndex(pageIndex);
       }
       /** 查询用户信息     */
       List<Job> jobs = hrmService.findJob(job, pageModel);
```

```
        model.addAttribute("jobs", jobs);
        model.addAttribute("pageModel", pageModel);
        return "job/job";
    }
    /**
     * 处理删除职位请求
     * @param String ids 需要删除的 id 字符串
     * @param ModelAndView mv
     * */
    @RequestMapping(value="/job/removeJob")
    public ModelAndView removeJob(String ids,ModelAndView mv){
        // 分解 id 字符串
        String[] idArray = ids.split(",");
        for(String id : idArray){
            // 根据 id 删除职位
            hrmService.removeJobById(Integer.parseInt(id));
        }
        // 设置客户端跳转到查询请求
        mv.setViewName("redirect:/job/selectJob");
        // 返回 ModelAndView
        return mv;
    }
    /**
     * 处理添加请求
     * @param String flag 标记， 1 表示跳转到添加页面，2 表示执行添加操作
     * @param Job  job  要添加的职位对象
     * @param ModelAndView mv
     * */
    @RequestMapping(value="/job/addJob")
    public ModelAndView addJob(
            String flag,
            @ModelAttribute Job job,
            ModelAndView mv){
        if(flag.equals("1")){
            // 设置跳转到添加页面
            mv.setViewName("job/showAddJob");
        }else{
            // 执行添加操作
            hrmService.addJob(job);
            // 设置客户端跳转到查询请求
            mv.setViewName("redirect:/job/selectJob");
        }
        // 返回
        return mv;
    }
    /**
     * 处理修改职位请求
     * @param String flag 标记， 1 表示跳转到修改页面，2 表示执行修改操作
     * @param Job job 要修改部门的对象
     * @param ModelAndView mv
     * */
    @RequestMapping(value="/job/updateJob")
    public ModelAndView updateDpet(
            String flag,
            @ModelAttribute Job job,
            ModelAndView mv){
        if(flag.equals("1")){
            // 根据 id 查询部门
            Job target = hrmService.findJobById(job.getId());
```

```
            // 设置 Model 数据
            mv.addObject("job", target);
            // 设置跳转到修改页面
            mv.setViewName("job/showUpdateJob");
        }else{
            // 执行修改操作
            hrmService.modifyJob(job);
            // 设置客户端跳转到查询请求
            mv.setViewName("redirect:/job/selectJob");
        }
        // 返回
        return mv;
    }
}
```

单击左侧菜单“职位管理”下面的“添加职位”命令，跳转到添加职位界面，如图 13.9 所示。

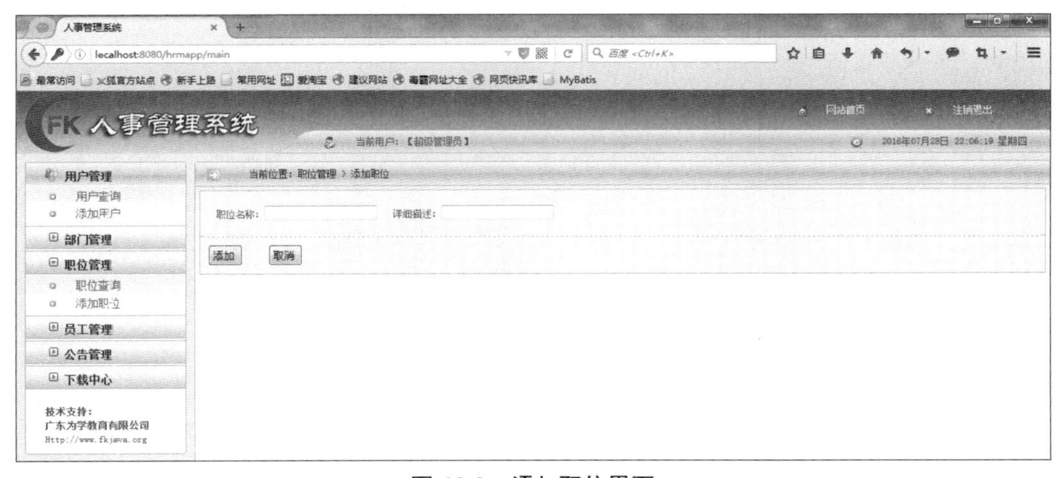

图 13.9　添加职位界面

输入需要添加的职位名称和详细描述，单击“添加”按钮，若添加成功则跳转到如图 13.10 所示的职位查询界面，显示所有职位信息。

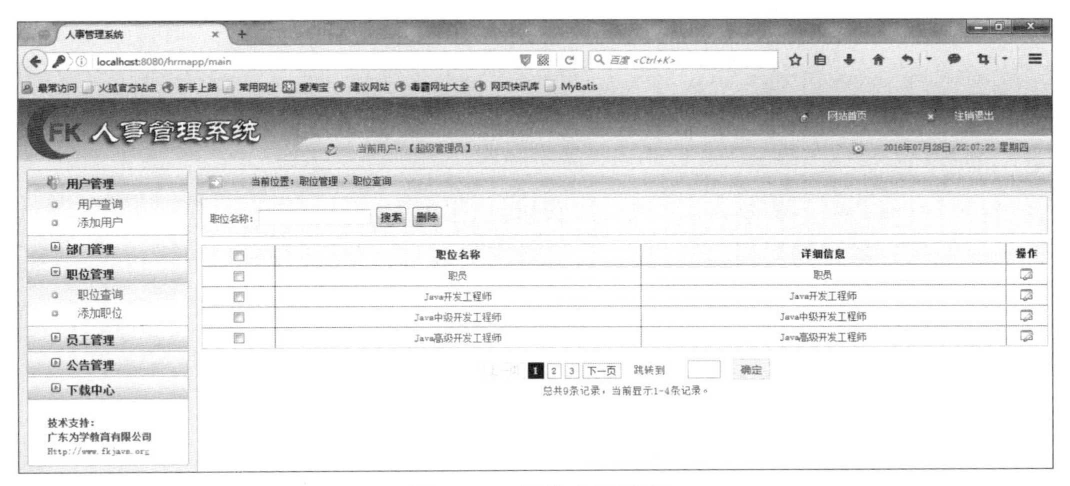

图 13.10　职位查询界面

输入职位名称，单击“搜索”按钮可以完成模糊查询功能。

选择每一行最后一列的“操作”按钮，可以进入修改页面，对选中的职位进行修改操作。

选择每一行第一列的复选框，单击“删除”按钮，则可以对选中的职位进行删除操作。

13.5.5 员工管理

处理员工的 EmployeeController 代码如下：

程序清单：codes/13/hrmapp/src/org/fkit/hrm/controller/EmployeeController.java

```java
import java.util.List;
import org.fkit.hrm.domain.Dept;
import org.fkit.hrm.domain.Employee;
import org.fkit.hrm.domain.Job;
import org.fkit.hrm.service.HrmService;
import org.fkit.hrm.util.tag.PageModel;
import org.springframework.beans.factory.annotation.Autowired;
import org.springframework.beans.factory.annotation.Qualifier;
import org.springframework.stereotype.Controller;
import org.springframework.ui.Model;
import org.springframework.web.bind.annotation.ModelAttribute;
import org.springframework.web.bind.annotation.RequestMapping;
import org.springframework.web.servlet.ModelAndView;
@Controller
public class EmployeeController {
    /**
     * 自动注入 hrmService
     * */
    @Autowired
    @Qualifier("hrmService")
    private HrmService hrmService;
    /**
     * 处理查询请求
     * @param pageIndex 请求的是第几页
     * @param String job_id 职位编号
     * @param String dept_id 部门编号
     * @param employee 模糊查询参数
     * @param Model model
     * */
    @RequestMapping(value="/employee/selectEmployee")
     public String selectEmployee(Integer pageIndex,
            Integer job_id,Integer dept_id,
            @ModelAttribute Employee employee,
            Model model){
        // 模糊查询时判断是否有关联对象传递，如果有，创建并封装关联对象
        this.genericAssociation(job_id, dept_id, employee);
        // 创建分页对象
        PageModel pageModel = new PageModel();
        // 如果参数 pageIndex 不为 null，则设置 pageIndex，即显示第几页
        if(pageIndex != null){
            pageModel.setPageIndex(pageIndex);
        }
        // 查询职位信息，用于模糊查询
        List<Job> jobs = hrmService.findAllJob();
        // 查询部门信息 ，用于模糊查询
        List<Dept> depts = hrmService.findAllDept();
        // 查询员工信息
        List<Employee> employees = hrmService.findEmployee(employee,pageModel);
        // 设置 Model 数据
        model.addAttribute("employees", employees);
```

```
        model.addAttribute("jobs", jobs);
        model.addAttribute("depts", depts);
        model.addAttribute("pageModel", pageModel);
        // 返回员工页面
        return "employee/employee";
    }
    /**
     * 处理添加员工请求
     * @param String flag 标记， 1 表示跳转到添加页面，2 表示执行添加操作
     * @param String job_id 职位编号
     * @param String dept_id 部门编号
     * @param Employee employee 接收添加参数
     * @param ModelAndView mv
     * */
    @RequestMapping(value="/employee/addEmployee")
     public ModelAndView addEmployee(
             String flag,
             Integer job_id,Integer dept_id,
             @ModelAttribute Employee employee,
             ModelAndView mv){
        if(flag.equals("1")){
            // 查询职位信息
            List<Job> jobs = hrmService.findAllJob();
            // 查询部门信息
            List<Dept> depts = hrmService.findAllDept();
            // 设置 Model 数据
            mv.addObject("jobs", jobs);
            mv.addObject("depts", depts);
            // 返回添加员工页面
            mv.setViewName("employee/showAddEmployee");
        }else{
            // 判断是否有关联对象传递，如果有，创建关联对象
            this.genericAssociation(job_id, dept_id, employee);
            // 添加操作
            hrmService.addEmployee(employee);
            // 设置客户端跳转到查询请求
            mv.setViewName("redirect:/employee/selectEmployee");
        }
        // 返回
        return mv;
    }
    /**
     * 处理删除员工请求
     * @param String ids 需要删除的 id 字符串
     * @param ModelAndView mv
     * */
    @RequestMapping(value="/employee/removeEmployee")
     public ModelAndView removeEmployee(String ids,ModelAndView mv){
        // 分解 id 字符串
        String[] idArray = ids.split(",");
        for(String id : idArray){
            // 根据 id 删除员工
            hrmService.removeEmployeeById(Integer.parseInt(id));
        }
        // 设置客户端跳转到查询请求
        // mv.setView(new RedirectView("/hrmapp/employee/selectEmployee"));
        // mv.setViewName("forward:/employee/selectEmployee");
        mv.setViewName("redirect:/employee/selectEmployee");
        // 返回 ModelAndView
        return mv;
```

```
    }
    /**
     * 处理修改员工请求
     * @param String flag 标记，1 表示跳转到修改页面，2 表示执行修改操作
     * @param String job_id 职位编号
     * @param String dept_id 部门编号
     * @param Employee employee  要修改员工的对象
     * @param ModelAndView mv
     * */
    @RequestMapping(value="/employee/updateEmployee")
     public ModelAndView updateEmployee(
             String flag,
             Integer job_id,Integer dept_id,
             @ModelAttribute Employee employee,
             ModelAndView mv){
        if(flag.equals("1")){
            // 根据 id 查询员工
            Employee target = hrmService.findEmployeeById(employee.getId());
            // 需要查询职位信息
            List<Job> jobs = hrmService.findAllJob();
            // 需要查询部门信息
            List<Dept> depts = hrmService.findAllDept();
            // 设置 Model 数据
            mv.addObject("jobs", jobs);
            mv.addObject("depts", depts);
            mv.addObject("employee", target);
            // 返回修改员工页面
            mv.setViewName("employee/showUpdateEmployee");
        }else{
            // 创建并封装关联对象
            this.genericAssociation(job_id, dept_id, employee);
            System.out.println("updateEmployee -->> " + employee);
            // 执行修改操作
            hrmService.modifyEmployee(employee);
            // 设置客户端跳转到查询请求
            mv.setViewName("redirect:/employee/selectEmployee");
        }
        // 返回
        return mv;
    }
    /**
     * 由于部门和职位在 Employee 中是对象关联映射，
     * 所以不能直接接受参数，需要创建 Job 对象和 Dept 对象
     * */
    private void genericAssociation(Integer job_id,
            Integer dept_id,Employee employee){
        if(job_id != null){
            Job job = new Job();
            job.setId(job_id);
            employee.setJob(job);
        }
        if(dept_id != null){
            Dept dept = new Dept();
            dept.setId(dept_id);
            employee.setDept(dept);
        }
    }
}
```

单击左侧菜单“员工管理”下面的“添加员工”命令，跳转到添加员工界面，如图 13.11 所示。

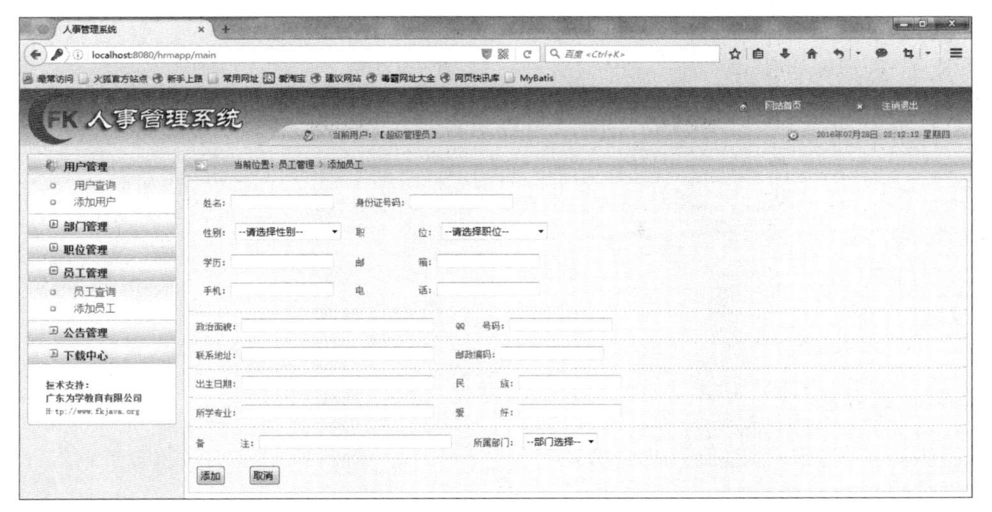

图 13.11 添加员工界面

输入需要添加的员工信息，单击“添加”按钮，若添加成功则跳转到如图 13.12 所示的员工查询界面，显示所有员工信息。

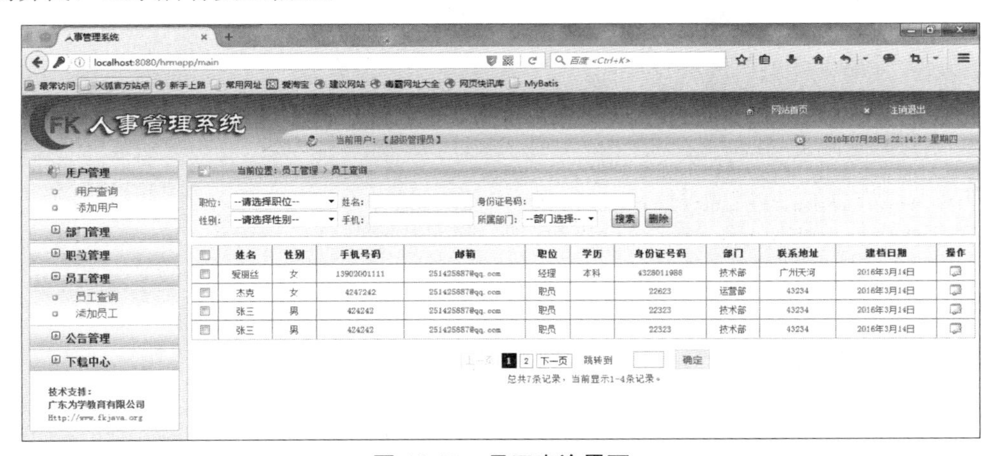

图 13.12 员工查询界面

选中职位、性别、所属部门，输入姓名、身份证号码、手机等，单击“搜索”按钮可以完成模糊查询功能。

选择每一行最后一列的“操作”按钮，可以进入修改页面，对选中的员工进行修改操作。

选择每一行第一列的复选框，单击“删除”按钮，则可以对选中的员工进行删除操作。

13.5.6 公告管理

处理公告的 NoticeController 代码如下：

程序清单：codes/13/hrmapp/src/org/fkit/hrm/controller/NoticeController.java

```
import java.util.List;
import javax.servlet.http.HttpSession;
import org.fkit.hrm.domain.Notice;
import org.fkit.hrm.domain.User;
import org.fkit.hrm.service.HrmService;
import org.fkit.hrm.util.common.HrmConstants;
```

```
import org.fkit.hrm.util.tag.PageModel;
import org.springframework.beans.factory.annotation.Autowired;
import org.springframework.beans.factory.annotation.Qualifier;
import org.springframework.stereotype.Controller;
import org.springframework.ui.Model;
import org.springframework.web.bind.annotation.ModelAttribute;
import org.springframework.web.bind.annotation.RequestMapping;
import org.springframework.web.servlet.ModelAndView;
@Controller
public class NoticeController {
    /**
     * 自动注入 UserService
     * */
    @Autowired
    @Qualifier("hrmService")
    private HrmService hrmService;
    /**
     * 处理/login 请求
     * */
    @RequestMapping(value="/notice/selectNotice")
     public String selectNotice(Model model,Integer pageIndex,
            @ModelAttribute Notice notice){
        PageModel pageModel = new PageModel();
        if(pageIndex != null){
            pageModel.setPageIndex(pageIndex);
        }
        /** 查询用户信息     */
        List<Notice> notices = hrmService.findNotice(notice, pageModel);
        model.addAttribute("notices", notices);
        model.addAttribute("pageModel", pageModel);
        return "notice/notice";
    }
    /**
     * 处理添加请求
     * @param Integer id  要显示的公告 id
     * @param Model model
     * */
    @RequestMapping(value="/notice/previewNotice")
     public String previewNotice(
            Integer id,Model model){
        Notice notice = hrmService.findNoticeById(id);
        model.addAttribute("notice", notice);
        // 返回
        return "notice/previewNotice";
    }
    /**
     * 处理删除公告请求
     * @param String ids 需要删除的 id 字符串
     * @param ModelAndView mv
     * */
    @RequestMapping(value="/notice/removeNotice")
     public ModelAndView removeNotice(String ids,ModelAndView mv){
        // 分解 id 字符串
        String[] idArray = ids.split(",");
        for(String id : idArray){
            // 根据 id 删除公告
            hrmService.removeNoticeById(Integer.parseInt(id));
```

```
        }
        // 设置客户端跳转到查询请求
        mv.setViewName("redirect:/notice/selectNotice");
        // 返回 ModelAndView
        return mv;
    }
    /**
     * 处理添加请求
     * @param String flag 标记， 1 表示跳转到添加页面，2 表示执行添加操作
     * @param Notice notice  要添加的公告对象
     * @param ModelAndView mv
     * */
    @RequestMapping(value="/notice/addNotice")
     public ModelAndView addNotice(
             String flag,
             @ModelAttribute Notice notice,
             ModelAndView mv,
             HttpSession session){
        if(flag.equals("1")){
            mv.setViewName("notice/showAddNotice");
        }else{
            User user = (User) session.getAttribute(HrmConstants.USER_SESSION);
            notice.setUser(user);
            hrmService.addNotice(notice);
            mv.setViewName("redirect:/notice/selectNotice");
        }
        // 返回
        return mv;
    }
    /**
     * 处理添加请求
     * @param String flag 标记， 1 表示跳转到修改页面，2 表示执行修改操作
     * @param Notice notice  要添加的公告对象
     * @param ModelAndView mv
     * */
    @RequestMapping(value="/notice/updateNotice")
     public ModelAndView updateNotice(
             String flag,
             @ModelAttribute Notice notice,
             ModelAndView mv,
             HttpSession session){
        if(flag.equals("1")){
            Notice target = hrmService.findNoticeById(notice.getId());
            mv.addObject("notice",target);
            mv.setViewName("notice/showUpdateNotice");
        }else{
            hrmService.modifyNotice(notice);
            mv.setViewName("redirect:/notice/selectNotice");
        }
        // 返回
        return mv;
    }
}
```

单击左侧菜单“公告管理”下面的“添加公告”命令，跳转到添加公告界面，如图 13.13 所示。

图 13.13　添加公告界面

输入需要添加的公告标题和公告内容，单击“添加”按钮，若添加成功则跳转到如图 13.14 所示的公告查询界面，显示所有公告信息。

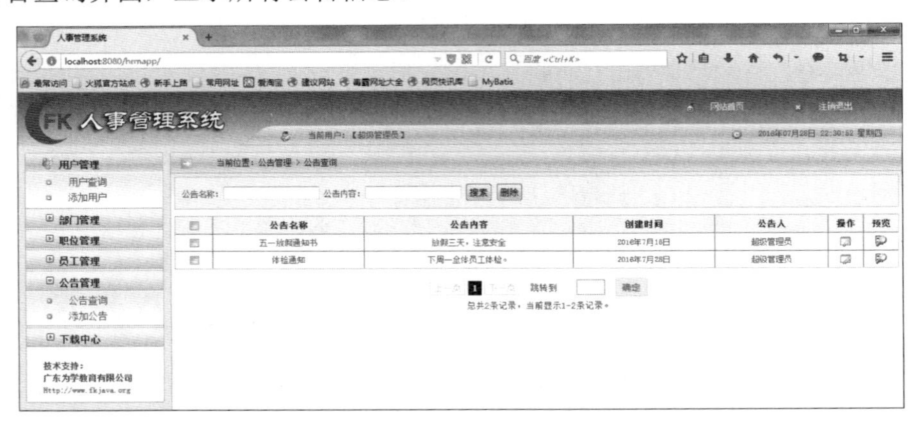

图 13.14　公告查询界面

选择每一行公告最后一列的“预览”按钮，则可以预览公告内容。如图 13.15 所示。

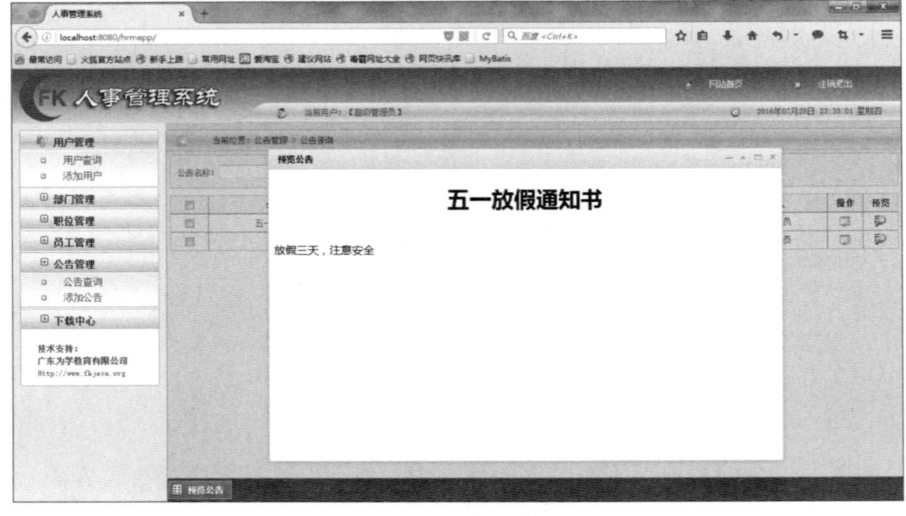

图 13.15　公告预览界面

输入公告名称和公告内容，单击“搜索”按钮可以完成模糊查询功能。

选择每一行的“操作”按钮，可以进入修改页面，对选中的公告进行修改操作。

选择每一行第一列的复选框，单击“删除”按钮，则可以对选中的公告进行删除操作。

13.5.7　下载中心

处理下载中心的 DocumentController 代码如下：

程序清单：codes/13/hrmapp/src/org/fkit/hrm/controller/DocumentController.java

```
import java.io.File;
import java.util.List;
import javax.servlet.http.HttpSession;
import org.apache.commons.io.FileUtils;
import org.fkit.hrm.domain.Document;
import org.fkit.hrm.domain.User;
import org.fkit.hrm.service.HrmService;
import org.fkit.hrm.util.common.HrmConstants;
import org.fkit.hrm.util.tag.PageModel;
import org.springframework.beans.factory.annotation.Autowired;
import org.springframework.beans.factory.annotation.Qualifier;
import org.springframework.http.HttpHeaders;
import org.springframework.http.HttpStatus;
import org.springframework.http.MediaType;
import org.springframework.http.ResponseEntity;
import org.springframework.stereotype.Controller;
import org.springframework.ui.Model;
import org.springframework.web.bind.annotation.ModelAttribute;
import org.springframework.web.bind.annotation.RequestMapping;
import org.springframework.web.servlet.ModelAndView;
@Controller
public class DocumentController {
    /**
     * 自动注入 UserService
     * */
    @Autowired
    @Qualifier("hrmService")
    private HrmService hrmService;
    /**
     * 处理/login 请求
     * */
    @RequestMapping(value="/document/selectDocument")
     public String selectDocument(
            Model model,Integer pageIndex,
            @ModelAttribute  Document document){
       PageModel pageModel = new PageModel();
       if(pageIndex != null){
          pageModel.setPageIndex(pageIndex);
       }
       /** 查询用户信息     */
       List<Document> documents = hrmService.findDocument(document, pageModel);
       model.addAttribute("documents", documents);
       model.addAttribute("pageModel", pageModel);
       return "document/document";

    }
    /**
     * 处理添加请求
```

```
 * @param String flag 标记， 1 表示跳转到上传页面，2 表示执行上传操作
 * @param Notice notice  要添加的公告对象
 * @param ModelAndView mv
 * */
@RequestMapping(value="/document/addDocument")
 public ModelAndView addDocument(
         String flag,
         @ModelAttribute Document document,
         ModelAndView mv,
         HttpSession session)throws Exception{
    if(flag.equals("1")){
        mv.setViewName("document/showAddDocument");
    }else{
        // 上传文件路径
        String path = session.getServletContext().getRealPath(
                "/upload/");
        // 上传文件名
        String fileName = document.getFile().getOriginalFilename();
        // 将上传文件保存到一个目标文件当中
        document.getFile().transferTo(new File(path+File.separator+ fileName));

        // 插入数据库
        // 设置 fileName
        document.setFileName(fileName);
        // 设置关联的 User 对象
        User user = (User) session.getAttribute(HrmConstants.USER_SESSION);
        document.setUser(user);
        // 插入数据库
        hrmService.addDocument(document);
        // 返回
        mv.setViewName("document/selectDocument");
    }
    // 返回
    return mv;
}
/**
 * 处理删除文档请求
 * @param String ids 需要删除的 id 字符串
 * @param ModelAndView mv
 * */
@RequestMapping(value="/document/removeDocument")
 public ModelAndView removeDocument(String ids,ModelAndView mv){
    // 分解 id 字符串
    String[] idArray = ids.split(",");
    for(String id : idArray){
        // 根据 id 删除文档
        hrmService.removeDocumentById(Integer.parseInt(id));
    }
    // 设置客户端跳转到查询请求
    mv.setViewName("redirect:/document/selectDocument");
    // 返回 ModelAndView
    return mv;
}
/**
 * 处理修改文档请求
 * @param String flag 标记， 1 表示跳转到修改页面，2 表示执行修改操作
 * @param Document document 要修改文档的对象
 * @param ModelAndView mv
```

```
     * */
    @RequestMapping(value="/document/updateDocument")
     public ModelAndView updateDocument(
            String flag,
            @ModelAttribute Document document,
            ModelAndView mv){
        if(flag.equals("1")){
            // 根据 id 查询文档
            Document target = hrmService.findDocumentById(document.getId());
            // 设置 Model 数据
            mv.addObject("document", target);
            // 设置跳转到修改页面
            mv.setViewName("document/showUpdateDocument");
        }else{
            // 执行修改操作
            hrmService.modifyDocument(document);
            // 设置客户端跳转到查询请求
            mv.setViewName("redirect:/document/selectDocument");
        }
        // 返回
        return mv;
    }
    /**
     * 处理文档下载请求
     * @param String flag 标记， 1 表示跳转到修改页面，2 表示执行修改操作
     * @param Document document 要修改文档的对象
     * @param ModelAndView mv
     * */
    @RequestMapping(value="/document/downLoad")
     public ResponseEntity<byte[]>  downLoad(
            Integer id,
            HttpSession session) throws Exception{
        // 根据 id 查询文档
        Document target = hrmService.findDocumentById(id);
        String fileName = target.getFileName();
        // 上传文件路径
        String path = session.getServletContext().getRealPath(
             "/upload/");
        // 获得要下载文件的 File 对象
        File file = new File(path+File.separator+ fileName);
        // 创建 springframework 的 HttpHeaders 对象
        HttpHeaders headers = new HttpHeaders();
        // 下载显示的文件名，解决中文名称乱码问题
        String downloadFielName = new String(fileName.getBytes("UTF-8"),"iso-8859-1");
        // 通知浏览器以 attachment （下载方式）打开图片
        headers.setContentDispositionFormData("attachment", downloadFielName);
        // application/octet-stream ： 二进制流数据（最常见的文件下载）。
        headers.setContentType(MediaType.APPLICATION_OCTET_STREAM);
        // 201 HttpStatus.CREATED
        return new ResponseEntity<byte[]>(FileUtils.readFileToByteArray(file),
             headers, HttpStatus.CREATED);
    }
}
```

单击左侧菜单“下载中心”下面的“上传文档”命令，跳转到上传文档界面，如图 13.16 所示。

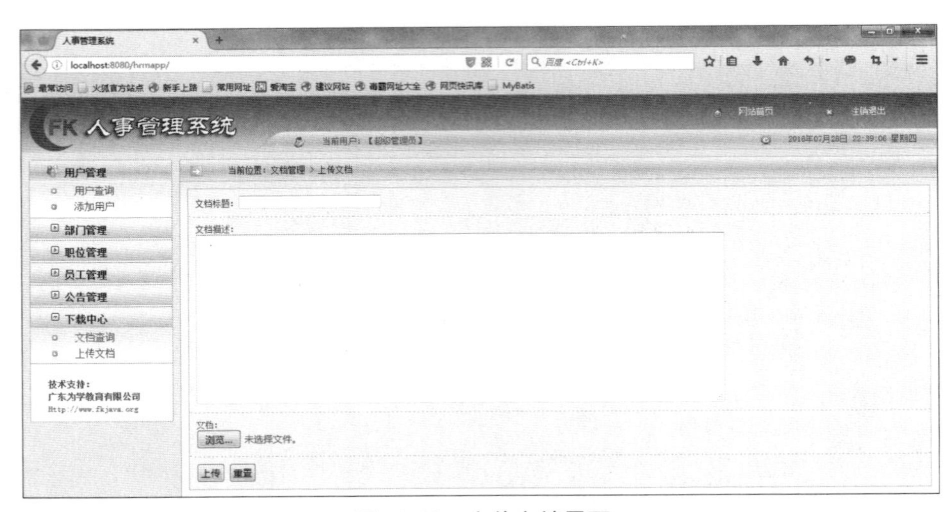

图 13.16　上传文档界面

输入需要上传的文档标题和文档概述，选择要上传的文档，单击“上传”按钮，若上传成功则跳转到如图 13.17 所示的文档查询界面，显示所有文档信息。

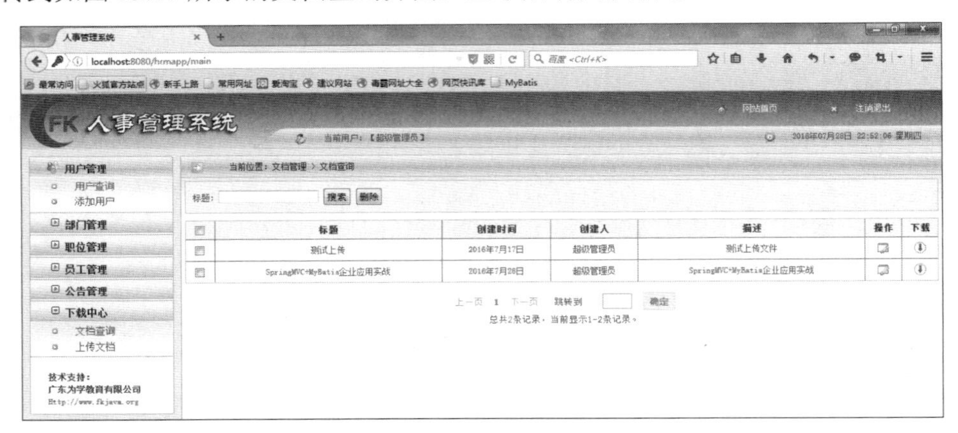

图 13.17　文档查询界面

选择每一行最后的“下载”按钮，会弹出下载页面，选择下载文档保存的路径，即可以下载文档，如图 13.18 所示。

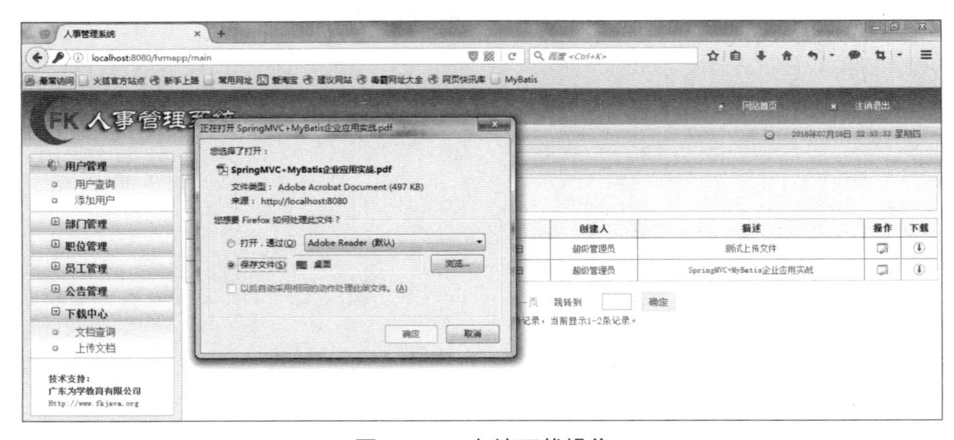

图 13.18　文档下载操作

输入文档标题，单击“搜索”按钮可以完成模糊查询功能。

选择每一行的“操作”按钮，可以进入修改页面，对选中的文档进行修改操作。

选择每一行第一列的复选框，单击“删除”按钮，则可以对选中的文档进行删除操作。

13.6 本章小结

本章详细介绍了一个完整的 Java EE 项目：人事管理系统，在此基础上可以扩展出企业的 HRM 系统、OA 系统等。因为企业平台本身的复杂性，所以本项目涉及的表达到 6 个，而且各个模块的业务逻辑也比较复杂，这些对初学者可能有一定难度，但只要读者先认真阅读本书前面章节所介绍的知识，并结合本章的讲解，再配合资源文件中的案例代码，则一定可以掌握本章所介绍的内容。

本章介绍的 Java EE 应用综合使用了前面介绍的两个框架：Spring MVC + MyBatis 3，因此本章内容既是对前面知识点的回顾和复习，也是将理论知识应用到实际开发的典范。一旦读者掌握了本章案例的开发方法，就会对实际Java EE企业应用的开发产生一种豁然开朗的感觉。

APPENDIX

A

附录 A EL 表达式和 JSTL 标签库

本附录要点

JSTL 标签库由标签库和 EL 表达式两个部分组成。EL 表达式在 JSTL1.0 规范中被引入，它配合 JSTL 标签库可以简化 JSP 页面的开发。

在 JSP 页面编程中，使用标签库替代传统的 Java 片段语言来实现页面的显示逻辑是非常流行的做法。虽然 SUN 公司提供了自定义标签功能，但是由于自定义标签很容易造成重复定义和非标准的代码，因此，SUN 公司推出了 JSP 标准标签库（Java Server Page Standard Tag Library，JSTL），用来解决创建的页面操作问题，包括输入输出、条件测试、遍历集合等。

A.1 EL 表达式语法

JSP 2.0 最重要的特性之一就是表达式语言 EL（Expression Language），由于受到 ECMAScript 和 XPath 表达式语言的启发，EL 也被设计成可以轻松地编写免脚本的 JSP 页面。使用 EL 表达式，JSP 页面可以不再使用任何的 JSP 声明、脚本和表达式，就可以轻松地访问应用程序的数据，这使得 JSP 页面写起来更加简单。

EL 在最早的时候必须配合 JSP 标准标签库（JSTL）才能使用，在 JSP 2.0 及更高版本中，JSP 容器已经能够独立地解释 EL 表达式，但在许多 Java Web 应用程序中，EL 还是会和 JSTL 配合使用。

EL 借鉴了 JavaScript 脚本语言的多类型转换无关的特点，使用 EL 从 Scope 中得到的参数可以自动转换类型，因此其对于类型的限制更加宽松。Web 服务器对于 request 请求参数通常会以 String 类型来发送，在得到它时使用的 Java 语言脚本应该是 request.getParameter(“XXX”)，这样的话，对于实际应用还必须进行强制类型转换。而 EL 就将用户从这种类型转换的烦琐工作中解放出来，允许用户之间使用 EL 表达式取值，而不用关心它是什么类型。

EL 表达式的主要作用：

- **获取数据**。EL 表达式主要用于替换 JSP 页面中的脚本表达式，以从各种类型的 Web 作用域中检索 Java 对象，获取数据，包括访问 JavaBean 的属性、List 集合、Map 集合、数组等。
- **执行运算**。利用 EL 表达式可以在 JSP 页面中执行一些基本的关系运算、逻辑运算和算术运算，以在 JSP 页面中完成一些简单的逻辑操作。
- **获取 Web 开发常用对**。EL 表达式定义了一些隐式对象，利用这些隐式对象，开发人员可以很轻松地获取对 Web 常用对象的引用，从而获取这些对象中的数据。

A.1.1 ${expression}

EL 表达式以${开头，并以}结束。EL 表达式的结构如下：

```
${expression }
```

例如，表达式 x+y，可以写成：

```
${x+y}
```

它也常用来连接两个表达式。对于一系列的表达式，它们的取值将是从左到右进行，计算结果的类型为 String，并且连接在一起。

${表示是一个 EL 表达式的开头，如果需要的只是文本${，则需要在它前面加一个转义符，如\${。

A.1.2 []和.运算符

EL 表达式可以返回任意类型的值。如果 EL 表达式的结果是一个带有属性的对象，则可以利用[]或者.运算符来访问其属性，即通过 getter 方法获取其属性。[]和.运算符类似，[]是比较规范的形式，.运算符则比较简便快捷。

访问对象的属性，可以使用以下任意一种形式：

```
${object["propertyName"]}
${object.propertyName}
```

如果 propertyName 不是有效的 Java 变量名，则只能使用[]运算符。

例如，访问隐式对象标题中的 HTTP 的 host：

```
${header["host"]}
${header.host}
```

但是，要想访问 accept-language 标题，只能使用[]运算符，因为 accept-language 不是一个合法的 Java 变量名。如果用.运算符访问它，将会导致异常。

如果对象的属性返回的也是一个对象，并且也带有属性，此时既可以使用[]，也可以使用.运算符来访问第二个对象的属性。

例如，隐式对象 pageContext 是表示当前 JSP 的 PageContext 对象。它有一个 request 属性，表示 HttpServletRequest。HttpServletRequest 带有 servletPath 属性。下列几个表达式的结果相同，都可以得到 pageContext 中 HttpServletRequest 的 servletPath 属性值：

```
${pageContext["request"]["servletPath"]}
${pageContext.request["servletPath"]}
${pageContext.request.servletPath}
${pageContext["request"].servletPath}
```

要访问 HttpSession，可以使用以下语法：

```
${pageContext.session}
```

要获得 HttpSession 的 id 标识符，可以使用下面形式：

```
${pageContext.session.id}
```

A.1.3　取值规则

EL 表达式的取值是从左到右进行的。对于 expr-a[expr-b]形式的表达式，其 EL 表达式的取值方法如下：

（1）先计算 expr-a 得到 value-a。

（2）如果 value-a 为 null，则返回 null。

（3）然后计算 expr-b 得到 value-b。

（4）如果 value-b 为 null，则返回 null。

（5）如果 value-a 为 java.uitl.List，则要进行以下处理：

a. 强制 value-b 为 int，如果不是，则抛出异常。

b. 如果 value-a.get(value-b)抛出 IndexOutBoundsException，则返回 null。

c. 返回 value-a.get(value-b)的值。

（6）如果 value-a 为 java.util.Map，则会查看 value-b 是否为 Map 中的一个 key。如果是，则返回 value-a.get(value-b)；如果不是，则返回 null。

（7）如果 value-a 不是一个集合，那么，value-a 必须是一个 JavaBean。在这种情况下，必须强制 value-b 为 String 类型。如果 value-b 是 value-a 的一个可读属性，则会调用该属性的 getter 方法返回值。如果 getter 方法抛出异常，则该表达式就是无效的。

A.1.4　访问 JavaBean

EL 最常用来访问 JavaBean 的属性，利用[]和.运算符都可以访问 JavaBean 的属性，示例

代码如下：

```
${beanName["propertyName"]}
${beanName.propertyName}
```

如果 JavaBean 的属性是一个带属性的对象，那么同样也可以利用[]或.运算符来访问第二个对象的属性。例如有一个 employee 对象，employee 对象有一个 dept 对象，dept 对象有一个 name 属性，则示例代码如下：

```
${employee.dept.name}
```

如果 JavaBean 的属性是一个集合，则可以使用 5.1.3 节介绍的访问规则。

A.2 EL 表达式的运算符

A.2.1 算术运算符

算术运算符有 5 种：

- 加法（+）
- 减法（-）
- 乘法（*）
- 除法（/）
- 取余/取模（%和 mod）

运算符是按优先级顺序排列的：

、/、%、mod 优先级相同，+、-优先级相同。、/、%、mod 优先级高于+、-。

A.2.2 逻辑运算符

逻辑运算符有 3 种：

- 和（&&和 and）
- 或（||和 or）
- 非（! 和 not）

A.2.3 关系运算符

关系运算符有如下几种：

- 等于（==和 eq）
- 不等于（!=和 ne）
- 大于（>和 gt）
- 大于或等于（>=和 ge）
- 小于（<和 lt）
- 小于或等于（<=和 le）

关系运算符返回的是 boolean 值，例如，表达式${1==2}返回 false，${2<3}返回 true。

EL 关系运算符的语法如下：

```
${statement?A:B}
```

如果 statement 的计算结果为 true，那么该表达式的输出结果是 A，否则是 B。

例如，下面 EL 表达式可以判断用户是否登录。

```
${sessionScope.user.username==null?"您已经登录":"您还没有登录"}
```

A.2.4 empty 运算符

empty 运算符用来检查作用域中是否有该变量存在，然后再判断该变量是否为 null 或者空字符串。示例代码如下：

```
${empty username}
```

（1）empty 首选判断 JSP 的作用域中是否有 username 变量，如果没有，则表达式返回 true。

（2）empty 判断 username 变量是否为 null 或者一个长度为 0 的空字符串，如果是，则表达式返回 true；如果 username 是一个空的集合或者数组，则表达式也返回 true。否则表达式将返回 false。

示例：EL 表达式运算符的使用

程序清单：codes/05/ELTest/WebContent/operator.jsp

```
<h2>测试 EL 运算符</h2>
<hr>
<!-- 算术表达式 -->
1+1: ${1+1 }<br> <!-- 2 -->
10-5:${10-5 }<br><!-- 5 -->
2*2:${2*2 }<br><!-- 4 -->
10/5:${10/5 }<br><!-- 2 -->
10 取模 3:${10%3 }<br><!-- 1 -->
<!-- 关系表达式 -->
<hr>
1 = 1 : ${1==1}<br><!-- true -->
1 != 1 : ${1!=1}<br><!-- false -->
1 > 1 : ${1>1}<br><!-- false -->
1 < 1 : ${1<1}<br><!-- false -->
1 >= 1 : ${1>=1}<br><!-- true -->
1 <= 1 : ${1<=1}<br><!-- true -->
<!-- 逻辑表达式 -->
<hr>
true && true : ${true && true }<br><!-- true -->
true && false : ${true && false }<br><!-- false -->
true || true : ${true || true }<br><!-- true -->
true || false : ${true || false }<br><!-- true -->
<!-- empty 运算符判断作用域中是否有该变量存在,然后再判断该变量是否为 null 或者空字符串-->
<%
    String abc = "a";
    pageContext.setAttribute("abc", abc);
    pageContext.setAttribute("user", null);
    pageContext.setAttribute("password", "");
%>
abc 是否为空: ${empty abc }<br><!-- false -->
user 是否为空: ${empty user }<br><!-- true -->
password 是否为空: ${empty password }<br><!-- true -->
clazz 是否为空: ${empty clazz }<br><!-- true -->
```

operator.jsp 页面可以测试 EL 表达式的运算符，读者可以自行运行它观察结果。

A.3 EL 表达式的隐式对象

在 JSP 页面中，可以利用 JSP 脚本访问 JSP 的对象。在 EL 中，提供了 11 个隐式对象用来访问 Web 开发中常用的 JSP 对象并获取数据。表 A.1 列出了 EL 表达式的隐式对象。

表 A.1　EL 表达式的隐式对象

隐式对象	描述
pageContext	JSP 页面的上下文对象。它可以访问 JSP 隐式对象，如请求、响应、会话、输出、servletContext 等
param	一个包含所有请求参数，并用参数名作为 key 的 Map。每个 key 的值就是指定名称的第一个参数值，表达式${param.name}相当于 request.getParameter (name)
paramValues	一个包含所有请求参数，并用参数名作为 key 的 Map。每个 key 的值是一个字符串数组，其中包含了指定参数名称的所有参数值。它与 param 隐式对象非常类似，但它检索一个字符串数组而不是单个值。表达式 ${paramvalues. name} 相当于 request.getParamterValues(name)
header	包含请求标题，并用标题名作为 key 的 Map。每个 key 的值就是指定标题名称的第一个标题值。表达式${header.name}相当于 request.getHeader(name)
headerValues	包含请求标题，并用标题名作为 key 的 Map。每个 key 的值就是一个字符串数组，其中包含了指定标题名称的所有参数值。它与 head 隐式对象非常类似，但它检索一个字符串数组而不是单个值。表达式 ${headerValues. name} 相当于 request.getHeaderValues(name)
cookie	包含了当前请求对象中所有 Cookie 对象的 Map。Cookie 名称就是 key 名称，每个 key 都映射到一个 Cookie 对象。表达式 ${cookie. name .value} 返回带有特定名称的第一个 Cookie 值
initParam	包含所有环境初始化参数，并用参数名作为 key 的 Map。表达式 ${ initParam. name} 相当于 ServletContext. getInitparameter (name)
pageScope	这是一个 Map，其中包含了页面域内的所有属性（Attribute）。属性名称就是 Map 的 key。表达式 ${pageScope.name} 相当于 pageContext.getAttribute("name")
requestScope	这是一个 Map，其中包含了请求域内的所有属性（Attribute）。属性名称就是 Map 的 key。表达式 ${ requestScope.name} 相当于 request.getAttribute("name")
sessionScope	这是一个 Map，其中包含了会话域内的所有属性（Attribute）。属性名称就是 Map 的 key。表达式 ${ sessionScope.name} 相当于 session.getAttribute("name")
applicationScope	这是一个 Map，其中包含了应用程序上下文域内的所有属性（Attribute）。属性名称就是 Map 的 key。表达式 ${ applicationScope.name} 相当于 application.getAttribute("name")

A.3.1　pageContext

pageContext 对象表示当前 JSP 页面的 javax.servlet.jsp.PageContext，它可以访问所有其他 JSP 的对象。表 A.2 列出了 pageContext 可以访问的 JSP 对象。

表 A.2　pageContext 可访问的 JSP 对象

对象	JSP 类型
request	javax.servlet.http.HttpServletRequest
response	javax.servlet.http.HttpServletResponse
session	javax.servlet.http.HttpSession
application	javax.servlet.ServletContext
config	javax.servlet.ServletConfig
out	javax.servlet.jsp.JspWriter
page	javax.servlet.jsp.HttpJspPage
exception	java.lang.Throwable

EL 的 pageContext 对象虽然表示当前 JSP 页面的 javax.servlet.jsp.PageContext，但是实际上 pageContext 对象只能访问 javax.servlet.jsp.PageContext 对象的 getter 方法。这也不难理解，EL 表达式的设计初衷只是在页面获取并显示 JSP 对象的数据，而不是设置和存储数据。

通过表 A.2 可以发现，EL 的 pageContext 对象可以访问到 JSP 的 9 个内置对象，这是因为，EL 表达式只是提供了 pageContext 隐含对象，并没有提供其他内置对象的隐含对象，其他内置对象可以通过 pageContext 获取。

例如，可以通过下面的表达式获取当前 JSP 页面的请求对象：

```
${pageContext.request}
${pageContext["request"]}
```

可以通过以下任意表达式来获取请求的 method：

```
${pageContext.request.method}
${pageContext.request["method"]}
${pageContext["request"].method}
${pageContext["request"]["method"]}
```

提示

读者一定觉得奇怪，既然通过 pageContext 可以获取 request 对象，那么为什么会提供 param、paramValues 等隐式对象来获取请求参数呢？原因是 request.getParameter (name)和 request.getParamterValues(name)之类的方法并不是标准的 getter 方法，所以提供了 param 等隐式对象来获取请求参数。

A.3.2 param

隐式对象 param 用于获取请求参数值。param 表示包含所有请求参数的 Map。例如，要获取前台页面传递的 loginname 参数，可以使用以下表达式：

```
${param.loginname}
${parma["loginname"]}
```

A.3.3 paramValues

隐式对象 paramValues 用于获取一个请求参数的多个值。paramValues 表示包含所有请求参数的 Map，每个 key 的值是一个字符串数组，其中包含了指定参数名称的所有值。即使该参数只有一个值，它也仍然返回一个带有一个元素的数组。例如，要获取前台页面传递的 selectOptions 参数的第一个和第二个值，则可以使用以下表达式：

```
${param. selectOptions[0]}
${param. selectOptions[1]}
```

A.3.4 header

隐式对象 header 用于获取请求标题。header 表示包含所有请求标题的 Map，其中 header 名称作为 key。例如，要获取 accept-language 这个 hearder 值，可以使用以下表达式：

```
${ header ["accept-language"]}
```

如果 header 名称是一个有效的 Java 变量名，如 connection，则也可以使用.运算符：

```
${header.connection}
```

A.3.5 headerValues

隐式对象 headerValues 用于获取请求标题。headerValues 表示包含所有请求标题的 Map，其中 header 名称作为 key，与 header 不同的是，headerValues 返回的是一个字符串数组。例如，要获取 accept-language 的第一个值，可以使用以下表达式：

```
${ headerValues ["accept-language"][0]}
```

A.3.6 cookie

隐式对象 cookie 用于获取 Cookie 对象。隐式对象 cookie 表示当前请求中所有的 Cookie 的值。例如，获取名为 JSESSIONID 的 Cookie 对象，可以使用以下表达式：

```
${cookie. JSESSIONID}
```

获取 JSESSIONID 的 Cookie 的值，可以使用以下表达式：

```
${ cookie.JSESSIONID.value}
```

A.3.7 initParam

隐式对象 initParam 用于获取 Web 应用程序上下文参数的值。例如，获取名为 username 的上下文参数值，可以使用以下表达式：

```
${initParam. username}
${ initParam["username"] }
```

A.3.8 pageScope、requestScope、sessionScope 和 applicationScope

在 JSP 中，经常通过 JSP 的 page、request、session 和 application 4 个作用域对象获取存储在其中的属性值。而这 4 个对象用来取值的 getAttribute("name")方法并不是一个标准的 getter 方法，其无法通过 EL 的对象获取存储的属性值。所以 EL 表达式提供了 4 个 scope 隐式对象，用于获取 4 个作用域中所存储的属性值，对应的 4 个作用域中的取值方法为 getAttribute("name")。

4 个作用域的范围从小到大依次为：pageScope、requestScope、sessionScope 和 applicationScope。假设 4 个作用域中都存在有名称为 username 的属性，则可以使用以下表达式：

```
${pageScope.username}
${requestScope.username}
${sessionScope.username}
${applicationScope.username}
```

EL 表达式也支持默认访问，即省略 scope 对象，直接访问 username 变量，默认范围为从小到大取值。

```
${username}
```

即首先取 pageScope 中的 username，如果为 null，则取 requestScope 中的 username，依此类推。

示例：EL 表达式隐式对象的使用

程序清单：codes/05/ELTest/WebContent/pageContext.jsp

```
<h2>测试 pageContext 隐式对象</h2>
请求方式：${pageContext.request.method}<br>
请求方式：${pageContext["request"]["method"]}<br>
请求上下文路径：${pageContext.request.contextPath}<br>
请求上下文路径：${pageContext["request"]["contextPath"]}<br>
请求 URL：${pageContext.request.requestURL}<br>
请求 URL：${pageContext["request"]["requestURL"]}<br>
```

在浏览器中输入如下 URL 来测试应用：

```
http://localhost:8080/ELTest/pageContext.jsp
```

测试结果如图 A.1 所示。

图 A.1　测试 pageContext 隐式对象

程序清单：codes/05/ELTest/WebContent/registerForm.jsp

```
<h2>注册页面</h2>
<form action="register.jsp" method="post">
    <table>
        <tr>
            <td>登录名：</td>
            <td><input type="text" name="loginname"></td>
        </tr>
        <tr>
            <td>选择课程：</td>
            <td>
                <input type="checkbox" name="courses" value="JAVAEE">JAVAEE 
                <input type="checkbox" name="courses" value="Spring">Spring 
                <input type="checkbox" name="courses" value="Mybatis">Mybatis 
            </td>
        </tr>
        <tr>
            <td><input type="submit" value="提交"></td>
        </tr>
    </table>
</form>
```

程序清单：codes/05/ELTest/WebContent/registerForm.jsp

```
<h2>测试 EL 隐式对象</h2>
param 隐式对象获取单个参数：${param.loginname}<br>
paramValues 隐式对象获取多个参数：${paramValues.courses[0]}-${paramValues.courses[1]}
-${paramValues.courses[2]}<br>
header 隐式对象获取标题：${header.connection}<br>
header 隐式对象获取标题：${header["accept-language"]}<br>
headerValues 隐式对象获取标题：${headerValues["accept-language"][0]}<br>
cookie 隐式对象获取 JSESSIONID 对象：${cookie.JSESSIONID}<br>
cookie 隐式对象获取 JSESSIONID 对象的名称：${cookie.JSESSIONID.name}<br>
cookie 隐式对象获取 JSESSIONID 对象的值：${cookie.JSESSIONID.value}<br>
initParam 隐式对象获取：${initParam.username}<br>
```

在 registerForm.jsp 中使用 initParam 隐式对象获取 username 的值，需要在 web.xml 文件中设置一个初始参数。

程序清单：codes/05/ELTest/WebContent/WEB-INF/web.xml

```
<context-param>
    <param-name>username</param-name>
    <param-value>fkit</param-value>
</context-param>
```

在浏览器中输入如下 URL 来测试应用：

```
http://localhost:8080/ELTest/registerForm.jsp
```

测试结果如图 A.2 所示。

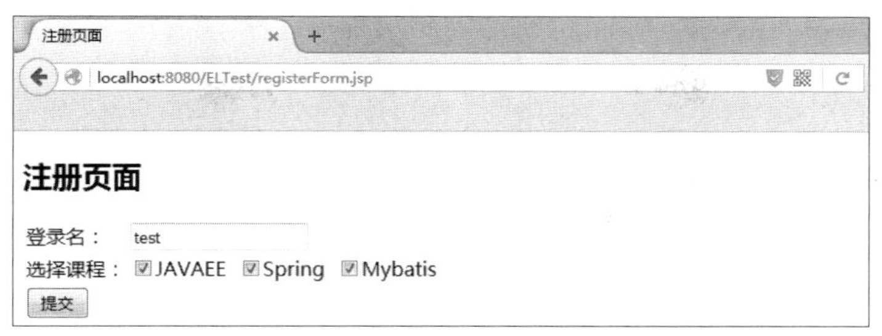

图 A.2　测试隐式对象

单击“提交”按钮，页面跳转到 register.jsp，如图 A.3 所示。

图 A.3　测试隐式对象

示例：EL 表达式 4 个 Scope 的使用

程序清单：codes/05/ELTest/WebContent/scope.jsp

```
<%
    // 往 jsp 的 4 个 scope 中分别设置了一个名称为 key 的值
    pageContext.setAttribute("key", "pageContext"); // 页面
    request.setAttribute("key", "request");  // 请求
    session.setAttribute("key", "session");  // 会话
    application.setAttribute("key", "application"); // 上下文
%>
pageScope 隐式对象取值：${pageScope.key}<br>
requestScope 隐式对象取值：${requestScope.key}<br>
sessionScope 隐式对象取值：${sessionScope.key}<br>
applicationScope 隐式对象取值：${applicationScope.key}<br>
scope 默认取值：${key}<br>
<h2>测试 scope 访问 JavaBean</h2>
<jsp:useBean id="user" class="org.fkit.domain.User" scope="page"/>
用户名：${pageScope.user.username }<br>
部门名称：${pageScope.user.dept.name }<br>
```

在 scope.jsp 中测试 scope 访问 JavaBean，需要在后台增加 User 和 Dept 两个 Java 类。

程序清单：codes/05/ELTest/src/org/fkit/domain/Dept

```
public class Dept {
    private String name;
    public Dept() {
        super();
    }
```

```
    public Dept(String name) {
        super();
        this.name = name;
    }
    public String getName() {
        return name;
    }
    public void setName(String name) {
        this.name = name;
    }
}
```

程序清单：codes/05/ELTest/src/org/fkit/domain/User

```
public class User {
    private String username;
    private Dept dept;
    public User() {
        super();
        this.username="周杰伦";
        dept = new Dept("开发部");
    }
    public String getUsername() {
        return username;
    }
    public void setUsername(String username) {
        this.username = username;
    }
    public Dept getDept() {
        return dept;
    }
    public void setDept(Dept dept) {
        this.dept = dept;
    }
}
```

在浏览器中输入如下 URL 来测试应用：

```
http://localhost:8080/ELTest/scope.jsp
```

测试结果如图 A.4 所示。

图 A.4　测试 4 个 scope 隐式对象

 A.4　JSTL 标签库

JSTL 是一个不断完善的开放源代码的 JSP 标签库，其由 Apache 的 jakarta 小组维护，在 JSP 2.0 中作为标准支持。

JSTL 目前最新版本是 1.2，读者可以在官网下载：

```
https://jstl.java.net/
```

其中，JSTL API 和 JSTL Implementation 是必须下载的。JSTL API 中包含 javax.servlet.jsp.jstl-api-1.2.1.jar 包，里面包含了 JSTL 规范中定义的类型。JSTL Implementation 中包含 javax.servlet.jsp.jstl-1.2.1 包，里面包含了 JSTL 的实现类。这两个 jar 包都必须复制到 Web 应用的 WEB-INF/lib 目录下。

JSTL 中的标签可以分为 5 大类，如表 A.3 所示。

表 A.3　JSTL 标签库

标签库	作用	URI	前缀
核心	包含 Web 应用的常见工作，如循环、输入输出等	http://java.sun.com/jsp/jstl/core	c
国际化	语言区域、消息、数字和日期格式化	http://java.sun.com/jsp/jstl/fmt	fmt
XML	访问 XML 文件	http://java.sun.com/jsp/jstl/xml	x
数据库	访问数据库	http://java.sun.com/jsp/jstl/sql	sql
函数	集合长度、字符串操作	http://java.sun.com/jsp/jstl/functions	fn

在 JSP 页面中使用 JSTL 标签库，必须通过以下格式使用 taglib 指令：

```
<%@ taglib uri="uri" prefix="prefix" %>
```

例如，要使用核心的 core 标签库，必须在 JSP 页面的开头处引入 tablib：

```
<%@ taglib uri="http://java.sun.com/jsp/jstl/core" prefix="c" %>
```

前缀 prefix 可以是任意的，但是采用惯例能够使团队的其他开发人员更容易读懂代码，建议采用表 A.3 的前缀字符。

A.5　Core 标签库

Core 标签库，又被称为核心标签库，该标签库的工作是对于 JSP 页面一般处理的封装。该标签库中一共有 14 个标签，被分为了 4 类，分别是：

- **多用途核心标签。**<c:out>、<c:set>、<c:remove>、<c:catch>
- **条件控制标签。**<c:if>、<c:choose>、<c:when>、<c:otherwise>
- **循环控制标签。**<c:forEach>、<c:forTokens>
- **URL 相关标签。**<c:import>、<c:url>、<c:redirect>、<c:param>

下面分别介绍各个标签的用途和属性以及简单示例。

A.5.1　<c:out>

<c:out>标签是一个最常用的标签，用于在 JSP 中显示数据。<c:out>标签可指定如表 A.4 所示的属性。

表 A.4　<c:out>标签的属性

属性	描述
value	输出到页面的数据，可以是 EL 表达式和常量
default	当 value 为空时显示的数据
escapeXml	当设置为 true 时会主动更换特殊字符。表示结果中的字符<、>、&等将会转换成 lt、gt、amp 等

下列代码将输出作用域中变量 username 的值：

```
<c:out value="${username }" />
```

下列代码输出 HttpSession 中的 username 的值，如果没有找到，则输出默认值。

```
<c:out value="${sessionScope.username }" default="游客"/>
```

A.5.2 <c:set>

<c:set>标签用于为变量或 JavaBean 中的变量属性赋值。<c:set>标签可指定如表 A.5 所示的属性。

表 A.5 <c:set>标签的属性

属性	描述
value	要设置的值，可以是 EL 表达式和常量
var	要创建的变量名
scope	新创建的变量的作用域，默认为 page
target	要被赋值的 JavaBean 实例的名称
property	JavaBean 实例的属性名称

下列代码将在 page 作用域中新设置一个变量 username，值是 fkit：

```
<c:set var="username" value="fkit" scope="page"/>
```

下列代码将给 page 作用域中 JavaBean 的实例 user 的 username 属性赋值 fkit：

```
<c:set target="${pageScope.user}" property="username" value="fkit"/>
```

A.5.3 <c:remove>

<c:remove>标签用于删除存在于 scope 中的变量。<c:remove>标签可指定如表 A.6 所示的属性。

表 A.6 <c:remove>标签的属性

属性	描述
var	要删除的变量名
scope	新创建的变量的作用域，默认为 page

下列代码将在 page 作用域中的变量 username 删除：

```
<c:remove var="username" scope="page"/>
```

A.5.4 <c:catch>

<c:catch >标签用于在 JSP 页面中捕获异常。它包含一个 var 属性，是一个描述异常的变量。如果没有 var 属性，则仅仅捕获异常而不做任何事情；如果定义了 var 属性，则可以利用 var 所定义的异常变量进行判断从而转发到其他页面或提示异常信息。

下列代码的 JSP 脚本会抛出 ArithmeticException，而该异常会被<c:catch >标签捕获并保存到 myerr 变量：

```
<c:catch var="myerr">
   <% int i = 5/0; %>
</c:catch>
${myerr }
```

A.5.5 <c:if>

<c:if >标签用于对某一个条件进行测试，假如结果为 ture，就处理它的 body content。测试结果可以保持在 var 变量当中，之后可以引用该变量。<c:if>标签可指定如表 A.7 所示的属性。

表 A.7　<c:if>标签的属性

属性	描述
test	需要判断的条件
var	保存判断结果的变量名，该变量的值为 true 或 false
scope	变量的作用域，默认为 page

下列代码判断用户登录，如果请求参数正确则显示“登录成功，欢迎访问网站!”：

```
<c:if test="${param.loginname == 'fkit' && param.password == '123456'}">
    登录成功，欢迎访问网站！
</c:if>
```

A.5.6　<c:choose>、<c:when>和<c:otherwise>

JSTL 没有提供 else 标签，复杂的条件判断使用<c:choose>、<c:when>和<c:otherwise>标签。

<c:choose>标签没有属性，只是作为父标签存在，<c:when>和<c:otherwise>作为其子标签来使用。<c:choose>标签中可以嵌套一个或者多个<c:when>标签，每个<c:when>标签都表示一种可以处理的情况。

<c:when>标签等价于 if 语句，它包含一个 test 属性，该属性表示需要判断的条件。

<c:otherwise>标签没有属性，它等价于 else 语句。

下列代码判断用户登录，如果请求参数正确则显示“登录成功，欢迎访问网站!”，否则显示“登录失败!”：

```
<c:choose>
    <c:when test="${param.loginname == 'fkit' && param.password == '123456'}">
        登录成功，欢迎访问网站！
    </c:when>
    <c:otherwise>
        登录失败！
    </c:otherwise>
</c:choose>
```

A.5.7　<c:forEach>

<c:forEach>标签于循环控制，可以遍历的对象包括 java.util.Collection 和 java.util.Map 的所有实现类以及数组。<c:forEach>标签可指定如表 A.8 所示的属性。

表 A.8　<c:forEach>标签的属性

属性	描述
begin	开始条件，如果指定 items，循环将从 begin 指定的索引处开始。begin 的值必须大于或者等于 0
end	结束条件，如果指定 items，循环将在 end 指定的索引处结束。end 的值必须大于或者等于 0
step	循环的步长，默认为 1
items	遍历的集合对象
var	循环集合中的元素的变量名
varStatus	保存循环状态的变量，类型为 javax.servlet.jsp.core.LoopTagStatus

示例：<c:forEach>标签的使用

程序清单：codes/06/JSTLTest/src/org/fkit/domain/Book

```
public class Book implements Serializable {
    private Integer id;
    private String name;
    private String author;
    public Book() {
```

```
        super();
    }
    public Book(Integer id, String name, String author) {
        super();
        this.id = id;
        this.name = name;
        this.author = author;
    }
    public Integer getId() {
        return id;
    }
    public void setId(Integer id) {
        this.id = id;
    }
    public String getName() {
        return name;
    }
    public void setName(String name) {
        this.name = name;
    }
    public String getAuthor() {
        return author;
    }
    public void setAuthor(String author) {
        this.author = author;
    }
}
```

程序清单：codes/06/JSTLTest/src/org/fkit/controller/TestController

```
@Controller
public class TestController{
    @RequestMapping(value="/forEachTest",method=RequestMethod.GET)
    public String registerForm(Model model) {
        // 页面用于遍历的 List 集合
        List<String> courseList = new ArrayList<String>();
        courseList.add("JAVAEE");
        courseList.add("Mybatis");
        courseList.add("Spring");
        // 页面用于遍历的 List 集合<Book>
        List<Book> bookList = new ArrayList<Book>();
        bookList.add(new Book(1, "疯狂 Java 讲义", "李刚"));
        bookList.add(new Book(2, "疯狂 Android 讲义", "李刚"));
        bookList.add(new Book(3, "疯狂 iOS 讲义", "李刚"));
        bookList.add(new Book(4, "轻量级 Java EE 企业应用实战", "李刚"));
        bookList.add(new Book(5,"Spring MVC 企业应用实战","肖文吉"));
        // 页面用于遍历的 Map 集合<Integer,Book>
        Map<Integer, Book> bookMap = new HashMap<Integer, Book>();
        bookMap.put(1, new Book(1, "疯狂 Java 讲义", "李刚"));
        bookMap.put(2, new Book(2, "疯狂 Android 讲义", "李刚"));
        bookMap.put(3, new Book(3, "疯狂 iOS 讲义", "李刚"));
        bookMap.put(4, new Book(4, "轻量级 Java EE 企业应用实战", "李刚"));
        bookMap.put(5, new Book(5,"Spring MVC 企业应用实战","肖文吉"));
        // 向 model 中添加属性
        model.addAttribute("courseList",courseList);
        model.addAttribute("bookList",bookList);
        model.addAttribute("bookMap",bookMap);
        return "forEachTest";
    }
}
```

程序清单：codes/06/JSTLTest/WebContent/content/forEachTest.jsp

```
<font color="red">普通循环</font><br>
```

```
    <table width="400" border="0" cellspacing="0" cellpadding="0">
        <tr><td>index</td><td>count</td><td>isFirst</td><td>isLast</td></tr>
        <c:forEach begin="1" end="5" step="2" varStatus="vs">
            <tr align="center">
                <td>${vs.index}</td><td>${vs.count}</td>
                <td>${vs.first}</td><td>${vs.last}</td>
            </tr>
        </c:forEach>
    </table>
    <br>
    <font color="red">遍历的 List 集合元素是 String</font><br>
    <table width="400" border="0" cellspacing="0" cellpadding="0">
        <tr>
            <c:forEach items="${requestScope.courseList }" var="course">
                 <td>${course}</td>
            </c:forEach>
        </tr>
    </table>
    <br>
    <font color="red">遍历的 List 集合元素是对象</font><br>
    <table width="400" border="0" cellspacing="0" cellpadding="0">
        <tr><td>编号</td><td>书名</td><td>作者</td></tr>
        <c:forEach items="${requestScope.bookList }" var="book">
            <tr>
                <td>${book.id}</td><td>${book.name}</td><td>${book.author}</td>
            </tr>
        </c:forEach>
    </table>
    <br>
    <font color="red">遍历 Map 集合</font><br>
    <table width="400" border="0" cellspacing="0" cellpadding="0">
        <tr><td>Key</td><td>编号</td><td>书名</td><td>作者</td></tr>
        <c:forEach items="${requestScope.bookMap }" var="entry">
            <tr>
                <td>${entry.key}</td><td>${entry.value.id}</td>
                <td>${entry.value.name}</td><td>${entry.value.author}</td>
            </tr>
        </c:forEach>
    </table>
```

web.xml 文件和 springmvc-config.xml 文件和前面章节中讲述的一致，此处不再赘述。

部署 FormTest 这个 Web 应用，在浏览器中输入如下 URL 来测试应用：

```
http://localhost:8080/JSTLTest
/forEachTest
```

会看到如图 A.5 所示的界面。

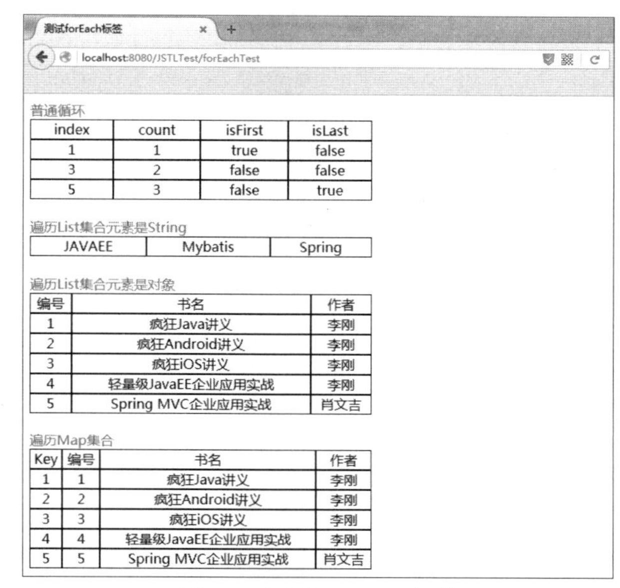

index	count	isFirst	isLast
1	1	true	false
3	2	false	false
5	3	false	true

JAVAEE	Mybatis	Spring

编号	书名	作者
1	疯狂Java讲义	李刚
2	疯狂Android讲义	李刚
3	疯狂iOS讲义	李刚
4	轻量级JavaEE企业应用实战	李刚
5	Spring MVC企业应用实战	肖文吉

Key	编号	书名	作者
1	1	疯狂Java讲义	李刚
2	2	疯狂Android讲义	李刚
3	3	疯狂iOS讲义	李刚
4	4	轻量级JavaEE企业应用实战	李刚
5	5	Spring MVC企业应用实战	肖文吉

图 A.5 测试 forEach 标签

A.5.8 <c:forTokens>

<c:forTokens>标签可以使用某个分隔符分隔指定的字符串。<c:forTokens >标签可指定如表 A.9 所示的属性。

表 A.9 <c:forTokens >标签的属性

属性	描述
begin	开始条件，如果指定 items，循环将从 begin 指定的索引处开始。begin 的值必须大于或者等于 0
end	结束条件，如果指定 items，循环将在 end 指定的索引处结束。end 的值必须大于或者等于 0
step	循环的步长，默认为 1
items	分隔的 EL 表达式或常量字符串
var	循环集合中的元素变量名
varStatus	保存循环状态的变量，类型为 javax.servlet.jsp.core.LoopTagStatus
delims	分隔符
scope	变量的作用域，默认为 page

下列代码使用<c:forTokens >分隔一个字符串：

```
<c:forTokens items="aa,bb,cc,dd" delims="," var="s">
    ${s}<br>
</c:forTokens>
```

A.5.9 <c:import >

<c:import >标签用于包含另一个 JSP 页面到本页面中来。示例代码如下：

```
<c:import url="title.jsp"/>
```

以上代码将 title.jsp 包含到本页面。

A.5.10 <c:url >

<c:url >标签用于组合一个正确的 URL 地址。<c:url >标签可指定如表 A.10 所示的属性。

表 A.10 <c:url >标签的属性

属性	描述
value	页面的 URL 地址
var	存储 URL 的变量名
context	该属性用于得到不同的 context 下的 URL 地址。当使用 context 属性时，必须以“/”开头
scope	变量的作用域，默认为 page

下列代码组合一个当前 Web 应用程序的 URL：

```
<c:url var="myurl" value="index.jsp"></c:url>
<a href="${myurl }">index</a><br><br>
```

下列代码组合一个其他 Web 应用程序的 URL：

```
<c:url var="myurl2" value="/idnex.jsp" context="/ELTest"/>
<a href="${myurl2 }">operator</a><br><br>
```

由于在实际项目开发中，国际化、XML、数据库和函数标签使用比较少，故本书不再赘述，读者可以自行查阅相关文档。

A.6 小结

本附录介绍了 EL 表达式，它是 JSP 2.0 及更高版本中的重要特性之一。它能够更加简便地访问 JavaBean 和 JSP 的常用对象。使用 EL 能够编写更简捷、更高效的 JSP 页面。

还重点介绍了 JSTL 的 Core 标签库，使用它配合 EL 表达式可以简化 JSP 页面的开发。